Methods in Molecular Biology

Series Editor
John M. Walker
School of Life and Medical Sciences
University of Hertfordshire
Hatfield, Hertfordshire, UK

For further volumes:
http://www.springer.com/series/7651

For over 35 years, biological scientists have come to rely on the research protocols and methodologies in the critically acclaimed *Methods in Molecular Biology* series. The series was the first to introduce the step-by-step protocols approach that has become the standard in all biomedical protocol publishing. Each protocol is provided in readily-reproducible step-by-step fashion, opening with an introductory overview, a list of the materials and reagents needed to complete the experiment, and followed by a detailed procedure that is supported with a helpful notes section offering tips and tricks of the trade as well as troubleshooting advice. These hallmark features were introduced by series editor Dr. John Walker and constitute the key ingredient in each and every volume of the *Methods in Molecular Biology* series. Tested and trusted, comprehensive and reliable, all protocols from the series are indexed in PubMed.

Plant Hormones

Methods and Protocols

Edited by

Shahid Mukhtar

Department of Genetics & Biochemistry, Clemson University, Clemson, SC, USA

Humana Press

Editor
Shahid Mukhtar
Department of Genetics &
Biochemistry
Clemson University
Clemson, SC, USA

ISSN 1064-3745 ISSN 1940-6029 (electronic)
Methods in Molecular Biology
ISBN 978-1-0716-5213-8 ISBN 978-1-0716-5214-5 (eBook)
https://doi.org/10.1007/978-1-0716-5214-5

This Humana imprint is published by the registered company Springer Science+Business Media, LLC, part of Springer Nature.
The registered company address is: 1 New York Plaza, New York, NY 10004, U.S.A.

Preface

Plant hormones play an indispensable role in governing various aspects of plant growth, development, reproduction, and responses to environmental cues. Understanding the functions and intricate crosstalk of these phytohormones has been a cornerstone of modern plant science. The book is a comprehensive compilation designed to highlight both established and cutting-edge experimental methods and protocols in plant hormone research. Each chapter offers insights into recent discoveries in phytohormones, ranging from the initial isolation and quantification techniques to cutting-edge molecular methodologies. While targeted to specialized audience, this book can serve as a graduate textbook or standard laboratory protocols.

Clemson, SC, USA *Shahid Mukhtar*

Contents

Contributors

PRIANKA ACHARJYA • *Department of Biology, Texas State University, San Marcos, TX, USA*

EMRE AKSOY • *Department of Biological Sciences, Faculty of Arts and Science, Middle East Technical University, Ankara, Turkey*

KAWTHAR F. ALASHOOR • *The BioActives Lab, Center for Desert Agriculture, King Abdullah University of Science and Technology (KAUST), Thuwal, Kingdom of Saudi Arabia; Centre of Excellence for Sustainable Food Security, King Abdullah University of Science and Technology, Thuwal, Kingdom of Saudi Arabia; The Plant Science Program, Biological and Environmental Science and Engineering Division, King Abdullah University of Science and Technology (KAUST), Thuwal, Kingdom of Saudi Arabia*

SALIM AL-BABILI • *The BioActives Lab, Center for Desert Agriculture, King Abdullah University of Science and Technology (KAUST), Thuwal, Kingdom of Saudi Arabia; Centre of Excellence for Sustainable Food Security, King Abdullah University of Science and Technology, Thuwal, Kingdom of Saudi Arabia; The Plant Science Program, Biological and Environmental Science and Engineering Division, King Abdullah University of Science and Technology (KAUST), Thuwal, Kingdom of Saudi Arabia*

ALFONSO ALBACETE • *Instituto Murciano de Investigación y Desarrollo Agrario y Alimentario: La Alberca, Murcia, Spain*

CRISTIANA T. ARGUESO • *Department of Agricultural Biology, Colorado State University, Fort Collins, CO, USA*

SEFA AYTEN • *Department of Biological Sciences, Faculty of Arts and Science, Middle East Technical University, Ankara, Turkey; Department of Breeding and Genetics, Field Crops Central Research Institute, Ankara, Turkey*

BASTIAAN O. R. BARGMANN • *School of Plant and Environmental Sciences, Virginia Polytechnic Institute and State University, Blacksburg, VA, USA*

ELENA BENCUROVA • *Department of Bioinformatics, University of Wuerzburg, Wuerzburg, Germany*

YOGENDRA BORDIYA • *Biosciences Division, Thermo Fisher Scientific, Austin, TX, USA; Pattern Bioscience, Austin, TX, USA*

CHRISTIAN DANVE M. CASTROVERDE • *Department of Biology, Wilfrid Laurier University, Waterloo, ON, Canada*

GUAN-TING ERICA CHEN • *The BioActives Lab, Center for Desert Agriculture, King Abdullah University of Science and Technology (KAUST), Thuwal, Kingdom of Saudi Arabia; Centre of Excellence for Sustainable Food Security, King Abdullah University of Science and Technology, Thuwal, Kingdom of Saudi Arabia; The Plant Science Program, Biological and Environmental Science and Engineering Division, King Abdullah University of Science and Technology (KAUST), Thuwal, Kingdom of Saudi Arabia*

SAMUEL CHISHOLM • *Department of Biological Systems Engineering, Polytechnic Institute and State University, Blacksburg, VA, USA*

TANZILA KAMAL CHOITY • *Department of Biology, Texas State University, San Marcos, TX, USA*

THOMAS DANDEKAR • *Department of Bioinformatics, University of Wuerzburg, Wuerzburg, Germany*

Carlos de Ollas Valverde • *Departament de Biologia, Bioquimica i Ciències Naturals, Universitat Jaume I (UJI), Castelló de la Plana, Spain*
Nihal Dharmasiri • *Department of Biology, Texas State University, San Marcos, TX, USA*
Sunethra Dharmasiri • *Department of Biology, Texas State University, San Marcos, TX, USA*
Colleen J. Doherty • *Department of Molecular and Structural Biochemistry, North Carolina State University, Raleigh, NC, USA*
Nicole A. Freeman • *Department of Genetics & Biochemistry, Clemson University, Clemson, SC, USA*
Aurelio Gómez-Cadenas • *Departament de Biologia, Bioquimica i Ciències Naturals, Universitat Jaume I (UJI), Castelló de la Plana, Spain*
Adrián González Ortega-Villaizán • *Centro de Biotecnología y Genómica de Plantas, Universidad Politécnica de Madrid (UPM)–Instituto Nacional de Investigación y Tecnología Agraria y Alimentación (INIA/CSIC), Campus de Montegancedo, Pozuelo de Alarcón (Madrid), Spain*
Sydney A. Greer • *Department of Genetics & Biochemistry, Clemson University, Clemson, SC, USA*
Omar Hasannin • *Department of Biological Sciences, Auburn University, Auburn, AL, USA*
T. Blake Horton • *Department of Molecular and Structural Biochemistry, North Carolina State University, Raleigh, NC, USA*
In-Cheol Jang • *Disruptive and Sustainable Technologies for Agricultural Precision, Singapore-MIT Alliance for Research and Technology, Singapore, Singapore; Temasek Life Sciences Laboratory, 1 Research Link, National University of Singapore, Singapore, Singapore; Department of Biological Sciences, National University of Singapore, Singapore, Singapore*
Grace A. Johnston • *Department of Agricultural Biology, Colorado State University, Fort Collins, CO, USA*
Hong-Gu Kang • *Department of Biology, Texas State University, San Marcos, TX, USA*
Dilara Sedef Karagöz • *Department of Molecular Biology and Genetics, Faculty of Science and Letters, Başkent University, Ankara, Türkiye*
Ceyhun Kayihan • *Department of Molecular Biology and Genetics, Faculty of Science and Letters, Başkent University, Ankara, Turkey*
Risheek R. Khanna • *Department of Biological Sciences, Auburn University, Auburn, AL, USA*
Duc Thinh Khong • *Disruptive and Sustainable Technologies for Agricultural Precision, Singapore-MIT Alliance for Research and Technology, Singapore, Singapore*
Feras Lafi • *College of Natural and Health Sciences, Department of Environmental Science and Sustainability, Zayed University, Abu Dhabi, UAE*
Kanjana Laosuntisuk • *Department of Molecular and Structural Biochemistry, North Carolina State University, Raleigh, NC, USA; Department of Biochemistry, Faculty of Medicine, Siriraj Hospital, Mahidol University, Bangkok, Thailand*
Kit Xi Liew • *The BioActives Lab, Division of Biological and Environmental Sciences and Engineering, King Abdullah University of Science and Technology (KAUST), Jeddah, Saudi Arabia*
Qianwei Liu • *Joseph Gottlieb Kölreuter Institute for Plant Sciences (JKIP), Karlsruhe Institute of Technology (KIT), Karlsruhe, Germany*
Susan M. Magdaleno • *Biosciences Division, Thermo Fisher Scientific, Austin, TX, USA*

Silvia Martínez-Fenoll • *Centro de Biotecnología y Genómica de Plantas, Universidad Politécnica de Madrid (UPM)–Instituto Nacional de Investigación y Tecnología Agraria y Alimentación (INIA/CSIC), Campus de Montegancedo, Pozuelo de Alarcón (Madrid), Spain*

Binoop Mohan • *Department of Genetics & Biochemistry, Clemson University, Clemson, SC, USA*

Khalid Muhammad • *Department of Biology, College of Science, UAE University, Al Ain, UAE*

Shahid Mukhtar • *Department of Genetics & Biochemistry, Clemson University, Clemson, SC, USA*

Muhammad Naseem • *College of Natural and Health Sciences, Department of Environmental Science and Sustainability, Zayed University, Abu Dhabi, UAE; Department of Bioinformatics, University of Wuerzburg, Wuerzburg, Germany*

Ikechukwu E. Opara • *Department of Biology, Texas State University, San Marcos, TX, USA*

Eman M. Othman • *Department of Bioinformatics, University of Wuerzburg, Wuerzburg, Germany; Department of Biochemistry, Faculty of Pharmacy, Minia University, Minia, Egypt; Cancer Therapy Research Center (CTRC), Department of Biochemistry-I, Biocenter, University of Wuerzburg, Wuerzburg, Germany*

Ujjal J. Phukan • *Department of Genetics & Biochemistry, Clemson University, Clemson, SC, USA*

Stephan Pollmann • *Centro de Biotecnología y Genómica de Plantas, Universidad Polité cnica de Madrid (UPM)–Instituto Nacional de Investigación y Tecnología Agraria y Alimentación (INIA/CSIC), Campus de Montegancedo, Pozuelo de Alarcón (Madrid), Spain; Departamento de Biotecnología-Biología Vegetal, Escuela Técnica Superior de Ingeniería Agronómica, Alimentaria y de Biosistemas, Universidad Politécnica de Madrid (UPM), Madrid, Spain*

Jathish Ponnu • *Joseph Gottlieb Kölreuter Institute for Plant Sciences (JKIP), Karlsruhe Institute of Technology (KIT), Karlsruhe, Germany*

Thomas K. Porter • *Department of Chemical Engineering, Massachusetts Institute of Technology, Cambridge, MA, USA*

Dinesh Singh Pujara • *Department of Biology, Texas State University, San Marcos, TX, USA; Plant Pathology and Plant-Microbe Biology Section, School of Integrative Plant Science, College of Agriculture and Life Sciences, Cornell University, Ithaca, NY, USA*

Aaron M. Rashotte • *Department of Biological Sciences, Auburn University, Auburn, AL, USA*

Estefanía Rodríguez-Dobreva • *Centro de Biotecnología y Genómica de Plantas, Universidad Politécnica de Madrid (UPM)–Instituto Nacional de Investigación y Tecnología Agraria y Alimentación (INIA/CSIC), Campus de Montegancedo, Pozuelo de Alarcón (Madrid), Spain*

Mohamed Salem • *The BioActives Lab, Division of Biological and Environmental Sciences and Engineering, King Abdullah University of Science and Technology (KAUST), Jeddah, Saudi Arabia*

Benny Jian Rong Sng • *Disruptive and Sustainable Technologies for Agricultural Precision, Singapore-MIT Alliance for Research and Technology, Singapore, Singapore; Temasek Life Sciences Laboratory, 1 Research Link, National University of Singapore, Singapore, Singapore*

MICHAEL S. STRANO • *Disruptive and Sustainable Technologies for Agricultural Precision, Singapore-MIT Alliance for Research and Technology, Singapore, Singapore; Department of Chemical Engineering, Massachusetts Institute of Technology, Cambridge, MA, USA*

ZHENGZHI TAN • *Department of Genetics & Biochemistry, Clemson University, Clemson, SC, USA*

JOSEPH S. TAYLOR • *School of Plant and Environmental Sciences, Virginia Polytechnic Institute and State University, Blacksburg, VA, USA*

HALIS BATUHAN ÜNAL • *Department of Molecular Biology and Genetics, Faculty of Science and Letters, Başkent University, Ankara, Türkiye*

JIAN YOU WANG • *The BioActives Lab, Center for Desert Agriculture, King Abdullah University of Science and Technology (KAUST), Thuwal, Kingdom of Saudi Arabia; Centre of Excellence for Sustainable Food Security, King Abdullah University of Science and Technology, Thuwal, Kingdom of Saudi Arabia; Biotechnology Center of Southern Taiwan, Academia Sinica, Tainan, Taiwan; Agricultural Biotechnology Research Center, Academia Sinica, Taipei, Taiwan*

YIQING WANG • *Department of Genetics & Biochemistry, Clemson University, Clemson, SC, USA*

HESHINI N. WEERAKKODY • *Department of Biology, Texas State University, San Marcos, TX, USA*

KENESHA WILSON • *College of Natural and Health Sciences, Department of Environmental Science and Sustainability, Zayed University, Abu Dhabi, UAE*

R. CLAY WRIGHT • *Department of Biological Systems Engineering, Polytechnic Institute and State University, Blacksburg, VA, USA*

XIU-FANG XIN • *National Key Laboratory of Plant Molecular Genetics, CAS Center for Excellence in Molecular Plant Sciences, Institute of Plant Physiology and Ecology, Chinese Academy of Sciences, Shanghai, China*

LINGYA YAO • *Shanghai Key Laboratory of Plant Molecular Sciences, College of Life Sciences, Shanghai Normal University, Shanghai, China; National Key Laboratory of Plant Molecular Genetics, CAS Center for Excellence in Molecular Plant Sciences, Institute of Plant Physiology and Ecology, Chinese Academy of Sciences, Shanghai, China*

NESLIHAN YAŞAR • *Department of Molecular Biology and Genetics, Faculty of Science and Letters, Başkent University, Ankara, Türkiye*

Chapter 1

Simultaneous Determination of ABA, IAA, SA, and JA by UPLC-ESI-MS/MS

Carlos de Ollas Valverde and Aurelio Gómez-Cadenas

Abstract

This protocol describes a method for the simultaneous extraction and quantification of the main phytohormones involved in plant stress responses and development—abscisic acid (ABA), indole-3-acetic acid (IAA), jasmonic acid (JA), and salicylic acid (SA)—from small quantities of plant tissue. It combines minimal extraction and purification steps with high-throughput analysis by ultra-high pressure liquid chromatography coupled with tandem mass spectrometry (UPLC-ESI-MS/MS). Surrogate standards are used for accurate quantification, making the method suitable for diverse plant matrices and absolute quantification.

Key words Phytohormones, HPLC-ESI-MS/MS, Abscisic acid, Indole-3-acetic acid, Jasmonic acid, Salicylic acid, Plant extracts

1 Introduction

Plant hormones, also known as phytohormones, are a diverse group of small organic molecules that play critical roles in regulating plant development, physiology, and responses to environmental stimuli. Despite being active at extremely low concentrations—typically in the nanomolar to picomolar range—these compounds exert profound effects on growth processes such as cell division, elongation, organogenesis, senescence, and stress adaptation [1]. Their precise quantification is therefore essential for understanding plant biology at both mechanistic and ecological scales.

The simultaneous quantification of multiple plant hormones using a single extraction protocol represents a transformative advancement in plant hormone analysis, offering unprecedented insights into the complex hormonal crosstalk that governs plant physiology. Furthermore, the ability to generate comprehensive hormone profiles from single tissue samples is particularly valuable when working with limited plant material, such as developing seeds

Shahid Mukhtar (ed.), *Plant Hormones: Methods and Protocols*, Methods in Molecular Biology, vol. 3026,
https://doi.org/10.1007/978-1-0716-5214-5_1,

or specific tissues, where sample availability constrains traditional analytical approaches. This methodology enables a more robust statistical analyses and revealing subtle but biologically significant correlations between hormone levels that would otherwise remain hidden [2, 3].

From an analytical perspective, the study of plant hormones presents several challenges. Their low abundance, chemical diversity, and dynamic turnover require sensitive and selective methods for extraction, separation, and detection. Moreover, the structural variety among hormone classes—including acidic (abscisic acid (ABA), jasmonic acid (JA)), basic (cytokinins (CKs)), neutral (gibberellins, auxins), and lipid-derived molecules (brassinosteroids (BRs), strigolactones (SLs))—requires optimized and sometimes class-specific protocols for accurate recovery and quantification [3].

Considering the heterogeneity of phytohormone concentrations and chemical properties, it is difficult to design a single extraction method that efficiently recovers all target hormones while avoiding interference from the complex plant matrix. The approach presented here provides a practical compromise by focusing on a subset of representative phytohormones (ABA, SA, JA, and indole-3-acetic acid (IAAs)) using a solvent system optimized for broad coverage, reproducibility, and ease of implementation [4].

UPLC-MS/MS with multiple reaction monitoring (MRM) is exceptionally well-suited for phytohormone analysis due to several key analytical challenges that phytohormones present and how this technique addresses them.

This chapter provides a comprehensive methodology for the extraction and quantification of plant hormones using UPLC-MS/MS. It highlights critical steps in sample handling, discusses the chemical properties and stability of different hormone groups, and offers practical guidance for achieving reproducible and biologically meaningful results. The protocol described herein is adaptable to a range of plant species and tissues and is designed to be accessible to researchers with varied experience in chromatographic techniques.

2 Materials

2.1 Chemicals and Solvents

1. Acetonitrile (HPLC-grade).
2. Water (HPLC-grade).
3. Acetic acid (glacial).
4. Diethyl ether.
5. Indole-3-acetic acid (IAA) [2 ppm].
6. (±)-2-cis,4-trans-abscisic acid (ABA) [200 ppm].
7. (±) Jasmonic acid (JA) [200 ppm].
8. Salicylic acid (SA) [200 ppm].

9. Labelled standards: [$^2H_6$]IAA, [$^2H_6$]ABA, [$^2H_6$]JA, [^{13}C] SA.
10. Stock of labelled standard mix: For 10 mL of final volume in a volumetric flask the following volumes of the specified stocks are pipetted: 100 μL ABAd_6, SA C^{13} and JA d_6 at 100 ppm, 100 μL IAAd_6 at 10 ppm.

2.2 Equipment

1. Analytical balance (accuracy in the range of 0.1 mg).
2. Pipettes (200 μL, 1000 μL, multipipette).
3. Eppendorf (2 mL).
4. Centrifuge ($\geq$ 5000× *g*).
5. Lyophilizer.
6. Vortex mixer.
7. pH Meter.
8. Nitrogen evaporator or vacuum concentrator.
9. UPLC-ESI-MS/MS system with triple quadrupole.
10. Reversed-phase C18 HPLC column (100 × 2 mm, 5 μm).
11. 0.22 μm cellulose acetate filters.

3 Methods

3.1 Plant Tissue Preparation

1. Immediately freeze harvested plant tissues in liquid nitrogen (*see* **Note 1**).
2. Lyophilize (if required), grind to fine powder (*see* **Notes 2** and **3**).
3. Weigh 0.1 g (fresh) or 0.01 g (dry) of plant tissue in a 2 mL Eppendorf tube with an analytical balance with the necessary resolution (*see* **Note 4**).
4. Add 50 μL internal standard mix (*see* **Note 5**).
5. Add 800 μL of ddH_2O (*see* **Note 6**).
6. Homogenize in a ball mill at a frequency of 17 rev/s for 5 min (*see* **Note 7**).
7. Centrifuge at 10,000× *g* for 10 min to separate the pellet from the liquid phase (*see* **Note 8**).
8. Collect supernatant into a new 2 mL Eppendorf tube and discard the pellet (*see* **Note 9**).
9. Adjust pH to 2.8–3.2 with 30% acetic acid (*see* **Note 10**).

3.2 Liquid-Liquid Extraction

1. (All steps involving diethyl ether must be carried out in a fume hood.) Add equal volumes of diethyl ether (800 μL of diethyl ether to each sample preferably with a multipipette) (*see* **Note 11**).

2. Mix liquid phases with a vortex for 30 s to assure optimal extraction (*see* **Note 12**).
3. Liquid phase separation: Centrifuge at 1000 *g* for 5 min (at 4 °C) to separate the organic phase (diethyl ether, top) from the aqueous phase (water, bottom).
4. Recover the organic phase into a new 2 mL labelled Eppendorf with a micropipette (*see* **Note 13**).
5. Repeat the liquid-liquid extraction (*see* Subheading 2, **items 1–4**). Discard aqueous phase.
6. Pool ether phases from the liquid-liquid extractions (1400–1500 μL) into a single Eppendorf tube.
7. Evaporate to dryness in a vacuum dryer (*see* **Note 14**).

3.3 Resuspension

1. Pipette 50 μL of methanol (HPLC quality).
2. Ultrasonicate the Eppendorf for 5 min at 4 °C to ensure full resuspension.
3. Add 450 μL of ddH_2O to a final volume of 500 μL and vortex the sample.
4. Filter through a 0.22 μm acetate filter (*see* **Note 15**).
5. Pipette 250 μL to a chromatography vial and keep samples refrigerated (4 °C) until injected (*see* **Note 16**).

3.4 Chromatography

1. Prepare 1 L (each) of acetonitrile and water as mobile phases, both modified by adding 0.1% acetic acid.
2. Prepare a standard calibration curve with known amounts of phytohormones to interpolate our samples and obtain values of absolute quantification. To prepare the calibration curve a mix of stocks of each hormone is used and then subsequently diluted in a series of steps to reach the lowest concentration of standard. Then we will spike each dilution with 50 μL of internal standard. The preparation of the curve will follow the indicative volumes of Table 1.
3. Samples are directly injected into a UPLC system (Acquity SDS, Waters Corp., Milford, MA, USA). Separations are carried out on a C18 column (Nucleodur C18, 1.8 μm particle size, 50 × 2.1 mm, Macherey-Nagel, Düren, Germany). Phytohormones are eluted with a gradient of H_2O:ACN (both supplemented with 0.1% acetic acid) that started from 5:95 (v/v) increasing linearly to 45:55 (v/v) at $t = 2$ min and then from $t = 3$ min linearly decreased to 5:95 (v/v) at 6 min. And finally composition shifts to 95:5 (v/v) and kept isocratic conditions of 95:5 during the last min of the run. The solvent flow rate is 0.3 ml/min with working pressures around 200–400 bar.

Table 1
Volumes used to prepare a calibration curve with serial dilutions from a stock mix of phytohormones and ddH_2O

Standard	Volume to pipette	Volume to dilute
S8	250 μL stock mix phytohormones	960 μL ddH_2O
S7	625 μL S8	375 μL ddH_2O
S6	500 μL S7	500 μL ddH_2O
S5	500 μL S6	500 μL ddH_2O
S4	500 μL S5	500 μL ddH_2O
S3	500 μL S4	500 μL ddH_2O
S2	500 μL S3	500 μL ddH_2O
S1	500 μL S2	500 μL ddH_2O
S0	500 μL ddH_2O	–

4. The UPLC-MS uses an orthogonal Z-spray electrospray interface (Micromass, Manchester, UK), the effluents from the UPLC are introduced into a triple-quadrupole mass spectrometer (Quattro LC, Micromass). Drying gas, as well as nebulizing gas, is nitrogen generated from pressurized air in an NG-7 nitrogen generator (Aquilo, Etten-Leur, The Netherlands). The nebulizer gas flow is set to 80 L/h and the desolvation gas flow to 800–900 L/h. For operation in the tandem MS (MS/MS) mode, the collision gas is 99.995% pure argon with a pressure of 2×10^{-3} mbar in the collision cell. The desolvation gas temperature is set at 350 °C, the source temperature at 120 °C, and the capillary voltage at −3 kV.
5. The mass spectrometer is operated in multiple reaction monitoring (MRM) mode to allow the quantification of multiple compounds including phytohormones and surrogates. The cone voltage and collision energies depended on the compound under investigation and are summarized in Table 2. Masslynx NT version 4.0 (Micromass) software is used to process the chromatograms.

4 Notes

1. Leaves, roots or other material can be stored at −20 °C for short periods (days to weeks) but to assure stability of the hormonal contents long range storage is preferable at −80 °C.
2. Hormones can be extracted from frozen tissue or alternatively from dry frozen/lyophilized material. Each alternative offers

Table 2
List of analytes with the ionization mode, cone voltage (V) used in the detection and collision energy used in the MS/MS detection with the original ion detected, nominal mass and chromatography retention time under the specified conditions

Type of analyte	Compound name	Compound	Ionization mode	Cone voltage (V)	Collision energy (eV)	Precursor ion	Quantification trace	Retention time
Phytohormone	Salicylic acid	SA	–	25	15	137	137 > 93	4.24
Surrogate	Salicylic acid C13	[^{13}C]SA	–	25	15	143	143 > 99	4.24
Phytohormone	Jasmonic acid	Ja	–	40	15	209	209 > 59	4.67
Surrogate	Jasmonic acid d6	[$_2$H^6]ja	–	40	15	215	215 > 59	4.67
Phytohormone	Abscisic acid	ABA	–	25	12	263	263 > 153	4.39
Surrogate	Abscisic acid d6	[$_2$H^6]ABA	–	25	12	269	269 > 159	4.39
Phytohormone	Indole-3-acetic acid	IAA	+	20	10	176	176 > 130	4.24
Surrogate	Indole-3-acetic acid d6	[$_2$H^6]IAA	+	20	10	181	181 > 134	4.24

advantages and has its particular drawbacks. Working with frozen material implies the necessity of keeping the frozen material at the proper temperature and avoid accidental thawing and the use of liquid nitrogen and frozen tools to weight plant material. On the other hand, the use of lyophilized material may allow the storage of samples at room temperature or at −20 °C as long as we keep the samples in a dry environment. If possible, even when lyophilized, it is desirable to keep the samples frozen. Proper grinding of the material can be accomplished with several strategies, mortar and pestle is preferable with lignified roots or challenging material like frozen fruits or embryos. Leaves and not lignified roots can be grinded to fine powder in a ball mill with glass or ceramic beads.

3. If material is limiting (hormones from a single Arabidopsis leaf), it is convenient to weight the intact leaf and then add the beads to grind the material, then without removing the beads perform the following steps of the extraction until solid liquid separation into a new Eppendorf tube.
4. As stated in the previous step, depending on the type of tissue and the species the amount of material can vary. It would be desirable to test the whole protocol range in case we are using a plant species of type of tissue for the first time and then adapt the protocol to maximize the performance of the extraction. Usually 100 mg of frozen material (leaves, roots, stem or fruit material) is enough to quantify the main hormones (ABA, JA, SA, IAA). When dealing with lyophilized material the amount of tissue is 20–10 mg being 10 mg the lower end. With the current sensitivity of the state of the art equipment the amount of material could be further reduced but the difficulty of reliably weighing homogeneous material increases. One mg of error weighing 10 mg is a 10% but decreasing the amount of material implies the use of a scale with accuracy in the range of 0.1 mg and prepare homogeneous material ground to a fine powder to minimize variations in hormonal gradients within a sampled tissue.
5. The internal standard mix can be prepared with methanol (analytical grade) and kept at −20 °C in a labelled flask. It is recommended to prepare a 5 ml working stock, but depending on the number of samples analyzed in a batch the internal standard working solution must cover a volume necessary to prepare all samples and the calibration curve with a proper margin for pipetting errors. It is important to avoid the use of different standard mix in the same batch/experiment even if the recipe and the working solutions are the same.
6. In the original method of [4], 5 mL of ddH_2O were used to extract hormones in a 15 mL conical tube. However in the

iterations and improvements of the original protocol the amount of liquid phase extraction has been reduced to 800 μL of distilled water in a 2 mL Eppendorf tube.

7. There are several strategies to carry out the solid-liquid extraction. The extraction is performed with a ball mill to enhance the extraction from the matrix to the liquid phase. Another option is to use an ultrasound bath (10 min at 5 °C).
8. Centrifugation duration is not critical, a good separation can be achieved in less time if the goal is to process a higher number of samples per day, 10 min allows for a separation while using that time to prepare new Eppendorf tubes and label them with a permanent marker.
9. It is faster to use 200 μL pipettes as it is faster, although you sacrifice some accuracy in pipetting small amounts of plant material is not critical in this step of the protocol. By reducing the volume of the liquid phase to 800 μL it is possible to perform the extraction in 2 mL Eppendorf tubes. By doing this it is possible to use an Eppendorf tube centrifuge at 10,000× *g* for 5 min to get a good separation.
10. It is possible to use concentrated acetic acid or a dilution. In the original protocol pH was measured in each sample to adjust pH. However, this step added a great amount of time to the protocol with little improvement in the overall accuracy. It is preferable to measure the pH in one or two samples to account for the amount of acids used to adjust the pH within range, and then systematically add the same amount to the remaining samples.
11. Diethyl ether does not wet the tip, so it is preferable to pipette before a couple of times to avoid spilling. This step must be carried out in the fume hood to avoid inhalation of diethyl ether due to its high volatility.
12. This can be done with a vortex sample by sample or simply by moving the rack with a proper cover to mix all samples at the same time.
13. 200 μL pipettes are preferable here due to a higher accuracy in order to avoid contamination with the aqueous phase.
14. Diethyl ether is highly volatile, so evaporation to dryness can be relatively fast. We use a Jouan RD1022 speed vac purchased from Thermo Scientific and the 1600 μL pooled from the liquid-liquid extraction can be dried in roughly 2 h. Given the high volatility of the diethyl ether if a speed vac is not available it would be feasible to dry the sample with a flow of N_2.
15. This step is crucial to avoid particles in suspension, filtering assures the exclusion of particles that can obstruct the HPLC

or the chromatography column in this step is likely to lose 50–100 μL in dead volume, getting a final volume of 400–450 μL.

16. Only 20 μL are injected into the HPLC equipment but to ensure a good injection we pipette 200 μL in the chromatography vial which leaves 200–250 μL of resuspension that can be stored at 4 °C for a few days, −20 °C for few weeks or several months or at −80 °C for months or years in case we need to reanalyze the samples.

References

1. Sosnowski J, Truba M, Vasileva V (2023) The impact of auxin and cytokinin on the growth and development of selected crops. Agriculture (Switzerland) 13(3). https://doi.org/10.3390/agriculture13030724
2. Xin P, Guo Q, Li B, Cheng S, Yan J, Chu J (2020) A tailored high-efficiency sample pretreatment method for simultaneous quantification of 10 classes of known endogenous phytohormones. Plant Commun 1(3). https://doi.org/10.1016/j.xplc.2020.100047
3. Šimura J, Antoniadi I, Široká J, Tarkowská D, Strnad M, Ljung K, Novák O (2018) Plant hormonomics: multiple phytohormone profiling by targeted metabolomics. Plant Physiol 177(2):476–489. https://doi.org/10.1104/pp.18.00293
4. Durgbanshi A, Arbona V, Pozo O, Miersch O, Sancho JV, Gómez-Cadenas A (2005) Simultaneous determination of multiple phytohormones in plant extracts by liquid chromatography-electrospray tandem mass spectrometry. J Agric Food Chem 53(22): 8437–8442. https://doi.org/10.1021/jf050884b

Chapter 2

Simultaneous Quantification of Salicylic Acid, Pipecolic Acid, and *N*-Hydroxypipecolic Acid in Plants Using High-Performance Liquid Chromatography Coupled with Tandem Mass Spectrometry

Lingya Yao, Xiu-Fang Xin, and Christian Danve M. Castroverde

Abstract

Plant immune activation in response to disease-causing pathogens entails proper spatiotemporal regulation of plant hormones and defense-activating metabolites. In *Arabidopsis thaliana* and numerous plant species, disease resistance to biotrophic and hemibiotrophic pathogens is underpinned by accumulation of these important immune signals, which include salicylic acid (SA), pipecolic acid (Pip), and *N*-hydroxypipecolic acid (NHP). SA, Pip, and NHP are essential for activating plant defense against various pathogens in local infected tissues, as well as priming distal, uninfected plant tissues in a phenomenon called systemic acquired resistance (SAR). Therefore, accurate quantification of these defense molecules is required to fully investigate the detailed regulation and function of plant immune networks. Here we present a highly sensitive and simple procedure to simultaneously quantify these immune-activating metabolites using state-of-the-art, high-performance liquid chromatography coupled with tandem mass spectrometry (HPLC-MS). A detailed pipeline is presented, encompassing proper plant growth, pathogen infection, metabolite extraction, metabolite quantification, and data analyses. Overall, our current protocol enables accurate measurements of these molecules at low, physiologically relevant concentrations in plant tissues.

Key words Salicylic acid, Pipecolic acid, *N*-hydroxypipecolic acid, Plant immunity, Plant hormone, Metabolite quantification, HPLC, Mass spectrometry, Amino acid, Systemic acquired resistance

1 Introduction

Plant disease resistance to infectious pathogens is governed by constitutive and inducible defense responses [1]. Constitutive defenses consist of physical and chemical barriers that prevent successful invasion and niche establishment of pests and pathogens [2]. In contrast, inducible defenses are activated in response to ongoing pathogen invasion via detection of broadly conserved molecular signatures via pattern-triggered immunity (PTI) and perception of pathogen-specific virulence proteins via effector-

Shahid Mukhtar (ed.), *Plant Hormones: Methods and Protocols*, Methods in Molecular Biology, vol. 3026,
https://doi.org/10.1007/978-1-0716-5214-5_2,

triggered immunity (ETI) [3–5]. PTI and ETI are mutually potentiated to confer robust disease resistance against most virulent pathogens [6–8].

Local immune activation by PTI and ETI eventually results in immune priming of distal, systemic tissues, leading to enhanced disease resistance to future secondary pathogen infections [9, 10]. This systemic acquired resistance (SAR) phenomenon induces the biosynthesis of central immune signals in both the pathogen-infected sites and distal tissues [10–13]. Two branches of SAR are mediated by the central signals salicylic acid (SA), pipecolic acid, and *N*-hydroxypipecolic acid (NHP) [13–16]. SA is perceived by the NPR1 receptor to activate local immune transcriptional reprogramming [17–19], as well as travel to systemic tissues via the apoplast in a cuticle-dependent manner [20]. On the other hand, Pip and its SAR-active form NHP activate the mutual amplification of nitric oxide and reactive oxygen species in local tissues [21, 22] before presumably travelling through the vascular system to the distal plant tissues [23–26]. SA and NHP can eventually be converted into storage and other forms through diverse modifications, such as glycosylation [27–30], methylation [31–33], and/or hydroxylation [34, 35].

Methods to quantify SA and derivatives in plants have ranged from highly sensitive HPLC and gas chromatography (GC) methods coupled with mass spectrometry [36–39], as well as non-MS methods like commercially available ELISA kits and an innovative *Acinetobacter* SA biosensor strain [40, 41]. In contrast, methods to measure Pip and NHP have exclusively relied on technical advances in HPLC-MS and GC-MS [23, 25, 42]. In this article, we present a detailed pipeline to simultaneously quantify SA, SAG, Pip, and NHP in *Arabidopsis thaliana* plants activating immunity against the bacterial pathogen *Pseudomonas syringae* (*see* Fig. 1). Our easy-to-follow protocol encompasses proper growth and care of plants, plant immune elicitation, sample preparation and extraction, metabolite quantification, and complete data analyses. Our method is successful in measuring biologically relevant concentrations within the nanogram-to-microgram (ng-to-μg) range per gram (g) of plant tissue, enabling accurate, precise, and simultaneous measurement of plant metabolite dynamics during immune elicitation.

2 Materials

2.1 Growth of Arabidopsis thaliana Plants

1. *Arabidopsis thaliana* (Columbia ecotype, Col-0) seeds.
2. 1.5 mL centrifuge tubes.
3. 5% sodium hypochlorite solution: sodium hypochlorite:water (5:100, vol/vol).
4. Sterile water.

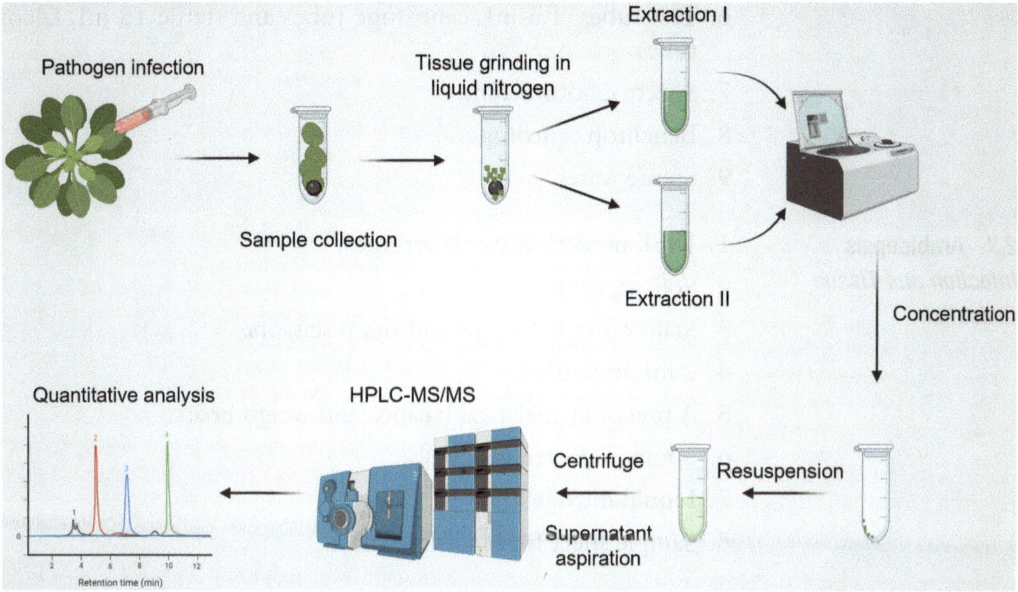

Fig. 1 The schematic representation of the workflow to quantify plant immunity-activating metabolites or SAR signals by HPLC-MS/MS in this protocol

5. 4 °C cold room or refrigerator.
6. Culture soil: "Pindstrup" substrate:vermiculite:perlite (3:9:1, wet/wet/wet).
7. Small pots and trays.
8. Plant growth chamber that can control light, temperature, and humidity.

2.2 Bacterial Inoculum Preparation

1. Bacterial strain: *Pseudomonas syringae* pv. *tomato* (*Pst*) DC3000.
2. Rifampicin stock solution: 50 mg rifampicin powder dissolved in 1 mL dimethylsulfoxide (DMSO), stored at −20 °C.
3. Modified Luria-Bertani (LM) liquid medium with rifampicin: Water with 1% (wt/vol) tryptone, 0.6% (wt/vol) yeast extract, 0.15% (wt/vol) KH_2PO_4, 0.06% (wt/vol) NaCl, 0.04% (wt/vol) $MgSO_4 \cdot 7H_2O$, pH = 7.0, autoclave and allow to cool to room temperature and add 0.1% (vol/vol) rifampicin stock solution.
4. Modified Luria-Bertani (LM) solid medium with rifampicin: Prepare as above with the addition of 1.5% (wet/vol) agar. After autoclaving, cool to approximately 37 °C, then add 0.1% (vol/vol) rifampicin stock solution and pour into disposable sterile petri dishes. Allow medium to solidify before use.
5. Bacteriological incubator set at 28 °C with the test tube rack.

6. Test tubes: 1.5 mL centrifuge tubes and sterile 15 mL falcon tubes.
7. Spectrophotometer.
8. Benchtop centrifuge.
9. Sterile water.

2.3 Arabidopsis Infection and Tissue Collection

1. 1 mL needleless sterile syringes.
2. Soft paper towels.
3. Stainless steel forceps and sharp scissors.
4. 2 mL cryotubes.
5. A precision analytical balance and weigh boats.
6. 5 mm diameter steel balls.
7. Liquid nitrogen.
8. Sample sheet for recording mass.

2.4 Extraction of Plant Metabolites

1. Methanol (MeOH, HPLC grade).
2. Deionized water (H_2O, HPLC grade).
3. Extraction Buffer (EB): methanol:water (4:1, vol/vol) with 0.01% (wt/vol) butylated hydroxytoluene (BHT).
4. 30% MeOH solution: methanol:water (3:7, vol/vol).
5. Liquid nitrogen.
6. Tissue grinder.
7. Fume hood.
8. 50 mL Falcon tubes and 1.5 mL centrifuge tubes.
9. The ice placed in the ice box.
10. Tube rotator in 4 °C cold room.
11. 4 °C refrigerated centrifuge.
12. Pipettes and tips.
13. Vortex.
14. Vacuum concentrator system.
15. Autosample HPLC vials and vial caps.

2.5 Quantification of Plant Metabolites by HPLC–MS/MS

1. Methanol (MeOH, HPLC grade).
2. 200 μM Salicylic Acid (SA, HPLC grade) in MeOH.
3. 200 μM Salicylic acid 2-O-β-D-glucoside (SAG, HPLC grade) in MeOH.
4. 200 μM *N*-Hydroxypipecolic acid (NHP, HPLC grade) in MeOH.
5. 200 μM Pipecolic acid (Pip, HPLC grade) in MeOH.

6. Deionized water (H_2O, HPLC grade).
7. Formic acid (HPLC grade).
8. 1.5 mL centrifuge tubes.
9. Autosample HPLC vials and vial caps.
10. 250 μL inserts.
11. HPLC vial rack.
12. Vortex.
13. SCIEX QTRAP 5500 LC-MS/MS system equipped with an electrospray ionization (ESI) source (AB SCIEX) or a triple quadrupole MS from another manufacturer.
14. A XSelect HSS T3 column (2.5 μm, 3.0 × 100 mm, Waters).
15. The Analyst 1.6.3 software (AB SCIEX) and the MultiQuant 3.0.2 software (AB SCIEX).

3 Methods

3.1 Growth of Arabidopsis thaliana Plants

1. Surface-sterilize *Arabidopsis thaliana* seeds with 5% sodium hypochlorite solution for 5 min in a 1.5 mL centrifuge tube, then wash thoroughly with sterile water.
2. Place seeds in 4 °C cold room for 24–48 h.
3. Put sterile culture soil into small pots and soak soil thoroughly with water in the tray.
4. Sow the seeds on the soil, then transfer pots into the growth chamber with a temperature of 22 °C, a 12-hour light/12-hour dark photoperiod with a light intensity of 100 $\mu E \cdot m^{-2} \cdot s^{-1}$ and a relative humidity of 60%.
5. Grow plants to 3.5–4 weeks for the following experiments.

3.2 Bacterial Inoculum Preparation

1. Grow *Pst* DC3000 bacteria in the LM solid medium (containing 50 mg/L rifampicin) at 28 °C for 48 h.
2. Pick a single colony and inoculate it in the sterile 15 mL falcon tube containing the LM liquid medium (containing 50 mg/L rifampicin) in a 28 °C bacteriological incubator and grow to an optical density (OD) of 0.8–1.0 (mainly about 11 h).
3. Collect bacterial fluid by centrifugation at 2500 *g* for 5 min.
4. Wash and resuspend the bacterial pellet with sterile water (*see* **Note 1**), adjust the suspension to OD of 0.2 using sterile water. Then dilute the suspension 100-fold to achieve a final concentration of 1×10^6 cfu/mL (OD = 0.002).

3.3 Pathogen Infection in *Arabidopsis* and Tissue Collection

1. Infiltrate *Pst* DC3000 bacteria solution into the abaxial surface of *Arabidopsis thaliana* leaves fully using a 1 mL needleless sterile syringe (*see* **Note 2**).
2. Dry *Arabidopsis* leaves with soft paper towels.
3. Label the infiltrated leaves with a marker pen (*see* **Note 3**).
4. Put plants back into the growth chamber.
5. After the specific time (*see* **Note 4**), use sharp scissors to remove 2–4 fully expanded *Arabidopsis* leaves and place them into the weigh boat on balance (*see* **Note 5**).
6. Record the mass as fresh weight (FW) and transfer leaves to labeled cryotube containing a 5 mm diameter steel ball.
7. Snap freeze the cryotube in liquid nitrogen as quickly as possible after capping (*see* **Note 6**).
8. Store frozen plant tissues in −80 °C or process them as described below.

3.4 Extraction of Plant Metabolites

3.4.1 Solution Preparation

1. Add 0.005 g of butylated hydroxytoluene to 10 mL of MeOH in a 50 mL Falcon tube and vortex to mix.
2. In the fume hood, add 30 mL MeOH into the above 50 mL Falcon tube and vortex to mix.
3. Add another 10 mL H_2O into the above 50 mL Falcon tube and vortex to mix. This is the Extraction Buffer (EB, *see* **Note 7**).

3.4.2 Plant Metabolite Extraction

1. Take out the materials from the −80 °C refrigerator and place samples in the pre-frozen grinder adaptors (be sure they are balanced) and beat at 50 r/s for 30 s. Freeze in the liquid nitrogen again the adaptors with samples inside and replace in tissue grinder and beat at 50 r/s for 30 s again. Confirm all tissues are completely ground (*see* **Note 8**).
2. Add an equal volume of cold EB to each sample and vortex immediately to suspend tissue. Around 0.6 mL for each 200 mg tissue works well (*see* **Note 9**).
3. Place the sample in a tube rotator at 35 rpm in a 4 °C cold room for 1.5 hours.
4. Place the sample into a refrigerated centrifuge at 4 °C and centrifuge at 12,000 *g* for 10 min.
5. Take the supernatant into a new 1.5 mL centrifuge tube and temporary storage on ice.
6. Add another 0.6 mL of EB solution to resuspend the precipitation.
7. The sample is again placed on the rotator at 4 °C for another 2.5 h to extract completely.

8. Place the sample into a refrigerated centrifuge at 4 °C and centrifuge at 12,000 *g* for 10 min.
9. Aspirate as much of the supernatant as possible and mix it with the supernatant from **step 5** (*see* **Note 10**).
10. Open the lid of this 1.5 mL centrifuge tube containing all the supernatant and concentrate it in the vacuum concentrator system (be sure they are balanced) at 1500 *r* for 120 min or more at room temperature until the supernatant is concentrated into the powder (*see* **Note 11**).
11. When all samples are concentrated into powder, store the powder in a − 20 °C refrigerator, or perform the following operations (*see* **Note 12**).
12. Add 200 μL 30% MeOH solution into the above tube, pipetting up and down to resuspend the powder. Record this volume as Vol, which is essential for detecting the hormone amount.
13. Put the sample into a 4 °C refrigerated centrifuge and centrifuge it at 12,000 *g* for 10 min.
14. Take 190 μL of the supernatant into a new 1.5 mL centrifuge tube and put it back into a 4 °C refrigerated centrifuge again and centrifuge it at 12,000 *g* for 10 min.
15. Take 180 μL of the supernatant and put it into the labeled HPLC autosampler vial containing 250 μL insert (*see* **Note 13**).

3.5 Quantification of Plant Metabolites by HPLC-MS/MS

3.5.1 Preparation and Serial Dilution of Mixed Standards (See **Note 14**)

1. Prepare mixed standard solution with four standards:
   (a) Label one 1.5 mL tube as "std1," add 700 μL H_2O and 230 μL MeOH, vortexing to mix.
   (b) Add 10 μL 200 μM Pip, 50 μL 200 μM NHP, 5 μL 200 μM SA, and 5 μL 200 μM SAG to the "std1" tube, vortexing to mix after each addition. The "std1" contains 2 μM Pip, 10 μM NHP, 1 μM SA, and 1 μM SAG (*see* **Note 15**).
2. Label one 1.5 mL tube as "std2," add 800 μL 30% MeOH and 200 μL "std1," vortexing to mix.
3. Label one 1.5 mL tube as "std3," add 500 μL 30% MeOH and 500 μL "std2," vortexing to mix.
4. Label one 1.5 mL tube as "std4," add 800 μL 30% MeOH and 200 μL "std3," vortexing to mix.
5. Label one 1.5 mL tube as "std5," add 500 μL 30% MeOH and 500 μL "std4," vortexing to mix.
6. Label one 1.5 mL tube as "std6," add 500 μL 30% MeOH and 500 μL "std5," vortexing to mix.
7. Transfer 180 μL aliquots to labeled HPLC autosampler vials containing 250 μL inserts (*see* **Note 16**).

Table 1
Gradient HPLC for reversed-phase separations

Analytes	Time (min)	Flow rate (mL/min)	A (%)	B (%)
SA, SAG	0–2	0.4	99	1
	2–13	0.4	Decrease linearly to 1	Increase linearly to 99
	13–15	0.4	1	99
	15–15.1	0.4	Increase linearly to 99	Decrease linearly to 1
	15.1–18	0.4	99	1
Pip, NHP	0–2	0.4	99	1
	2–5	0.4	Decrease linearly to 70	Increase linearly to 30
	5–6	0.4	Decrease linearly to 5	Increase linearly to 95
	6–7	0.4	5	95
	7–7.1	0.4	Increase linearly to 99	Decrease linearly to 1
	7.1–10	0.4	99	1

3.5.2 LC-MS/MS Performance and Data Analysis

1. Collect the sample vials (**step 15** of Subheading 3.4.2), standards vials (**step** 7 of Subheading 3.5.1), blank vial (30% MeOH solution) and put them onto the autosampler tray of the SCIEX QTRAP 5500 LC-MS/MS facility.
2. The binary solvent system uses water with 0.1% (vol/vol) formic acid (A) and methanol (B) as mobile phases. Separations are performed by altering the percentage of B as Table 1 (*see* **Note 17**).
3. SA, SAG, Pip, and NHP are separated with a XSelect HSS T3 column (2.5 μm, 3.0 × 100 mm, Waters).
4. The sample injection volume is 1 μL.
5. Set the oven (holding HPLC column) temperature to 40 °C.
6. Choose the negative ion mode (ESI-) to detect SA and SAG, and the optimized electrospray ionization (ESI) operating parameters are: IS, −4500 V; CUR, 35 psi; TEM 400 °C; GS1, 50 psi; GS2, 50 psi. Choose the positive ion mode (ESI +) to detect Pip and NHP, and the optimized ESI operating parameters for positive mode are: IS, 4500 V; CUR, 35 psi; TEM 500 °C; GS1, 55 psi; GS2, 55 psi (*see* **Note 18**).
7. Use the multiple reaction monitoring (MRM) mode to detect all hormones/metabolites, the MS/MS spectra of hormone standards are shown in Fig. 2 and the specific MRM parameters for each analyte are provided in Table 2.
8. Use Analyst 1.6.3 software (AB SCIEX) to perform instrument control and data acquisition according to the software manual (*see* **Note 19**). The MRM chromatograms of SA, SAG, Pip, and NHP all clearly have individual peaks in fresh *Arabidopsis* leaf crude extracts (*see* Fig. 3).
9. Use MultiQuant 3.0.2 software (AB SCIEX) to perform data processing. According to the Quantitative Wizard, the concentrations of hormones can be quantified by calculating the area of each individual peak and comparing this to standard curves.

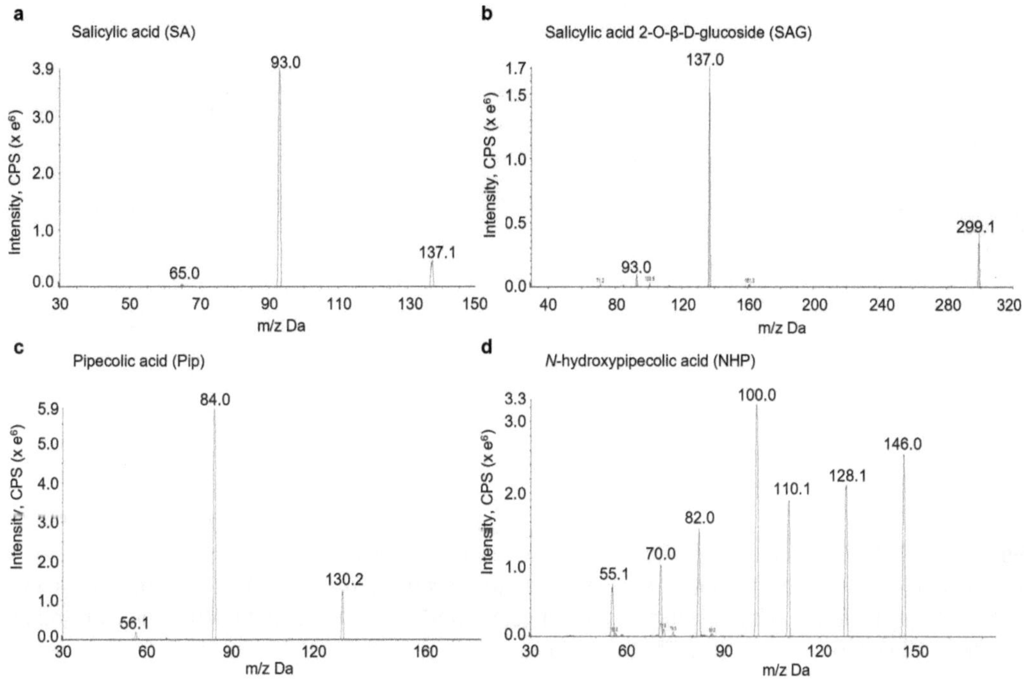

Fig. 2 The MS/MS spectra of hormone/metabolite standards. (**a**) MS/MS spectra of salicylic acid (SA); (**b**) MS/MS spectra of salicylic acid 2-O-β-D-glucoside (SAG); (**c**) MS/MS spectra of pipecolic acid (Pip); (**d**) MS/MS spectra of N-hydroxypipecolic acid (NHP)

Table 2
Selected multiple reaction monitoring parameters for protonated or deprotonated plant hormones ($[M + H]^+$ or $[M - H]^-$)

Analytes	SM	Q1	Q3	DP	CE	RT[a]
SA	−	137.0	93.0	−80	−22	10.00
		137.0	65.0	−80	−44	10.00
SAG	−	299.0	137.0	−80	−17	7.30
		299.0	93.0	−80	−54	7.30
Pip	+	130.0	84.0	50	20	1.98
		130.0	56.0	50	43	1.98
NHP	+	146.1	100.0	40	14	3.53
		146.1	70.0	40	24	3.53

SA salicylic acid, *SAG* salicylic acid 2-O-β-D-glucoside, Pip pipecolic acid, *NHP* N-hydroxypipecolic acid, *SM* scan mode. Q1, the precursor ion selected in the first quadrupole (Q1). Q3, the product ion selected in the third quadrupole (Q3). DP, the declustering potential. CE, the collision energy (V). RT, the retention time (min). For each compound, the first Q3 is used for quantification, while the second Q3 is used as a quality control

[a]Retention times listed in this table are obtained under HPLC and MS conditions mentioned in this protocol

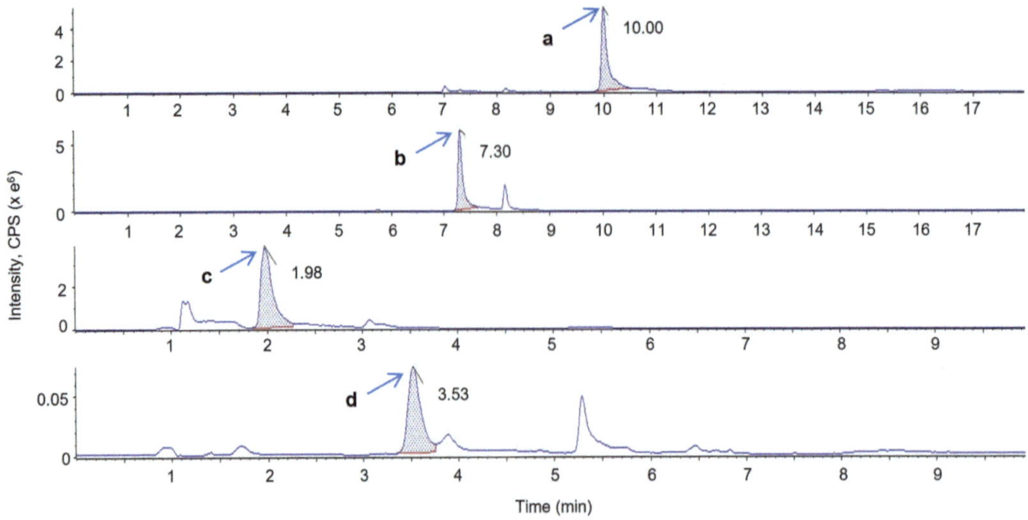

Fig. 3 The MRM chromatograms of detected plant hormones in fresh *Arabidopsis* leaf crude extracts. Four-week-old Col-0 plants (grown healthy) were infiltrated with *Pst* DC3000 at 1×10^6 cfu/mL and leaves were sampled 24 hours later. (**a**) MRM chromatogram of detected SA (137.0 > 93.0); (**b**) MRM chromatogram of detected SAG (299.0 > 137.0); (**c**) MRM chromatogram of detected Pip (130.0 > 84.0); (**d**) MRM chromatogram of NHP (146.1 > 100)

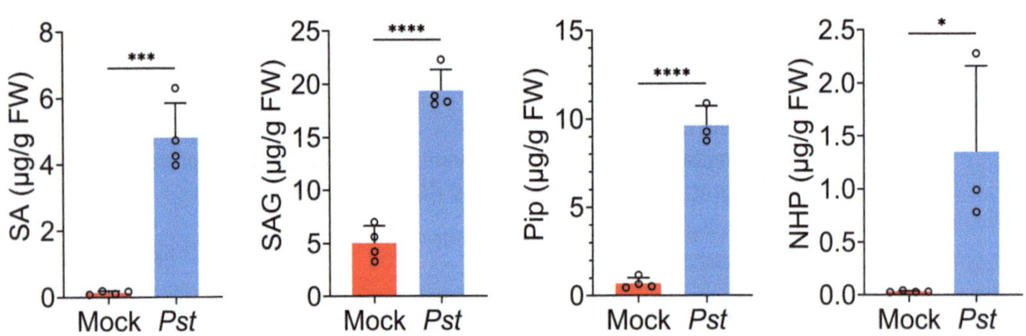

Fig. 4 Amounts of plant hormones/metabolites in fresh *Arabidopsis* leaves after pathogen infection determined by HPLC-MS/MS according to the protocol. *Arabidopsis thaliana* wild type (Col-0) plants were grown for 4 weeks as described in the protocol, leaves were infiltrated with *Pst* DC3000 at 1×10^6 cfu/mL, and leaves were sampled 24 hours later. Leaves infiltrated with sterile water were set as Mock. Leaves were processed and analyzed as described in the protocol. Data represent mean ± standard deviation ($n = 3$–4 biological replicates). Data was analyzed for normal distribution and subjected to parametric *t*-test using Prism. Statistical significance is indicated by asterisks (* for $p < 0.05$; ** for $p < 0.01$; *** for $p < 0.001$; **** for $p < 0.0001$)

10. Calculate the amount of the hormone in the plant sample according to the following formula: The plant hormone amount per fresh weight = (hormone concentration x dilution factor x hormone molar mass x the volume to dissolve powder)/ fresh weight of the sample. Examples showing typical results can be seen in Fig. 4.

4 Notes

1. The sterile water can be replaced with 10 mM $MgCl_2$.
2. Infiltrate sterile water or $MgCl_2$ solution into leaves as mock treatment.
3. This is to distinguish leaves injected with different bacteria or water.
4. Based on previous experience, it usually takes 24 h after the bacterial infection to detect SA, SAG, Pip, and NHP.
5. Two fully expanded *Arabidopsis* leaves will weigh ~0.1 g.
6. Wounding can induce changes in plant hormones; hence, it is important to perform **steps 5–7** quickly.
7. Keep EB on ice before performing the following process.
8. Ensure samples are in a liquid nitrogen environment throughout the procedure to prevent samples from thawing.
9. The 2 mL tubes comfortably hold 1.8 mL.
10. If the precipitate is accidentally aspirated, centrifuge again and aspirate the supernatant for subsequent operations.
11. Plant hormones are unstable at room temperature. Once concentration is complete, remove the tube from the concentrator and immediately store on ice. And if possible, it is better to use the vacuum concentrator system which can keep the temperature at 4 °C.
12. The concentrated powder may be stored at −20 °C for up to one month.
13. The supernatant may be stored at −20 °C for up to 1 week. Under our conditions, to detect SA, SAG, and NHP, use the original solution. However, to detect Pip, dilute it 10 times before testing. Record the dilution factor, which is essential for calculating the hormone amount.
14. The following example is to detect SAG, SA, Pip, and NHP in *Arabidopsis* plants after pathogen infection. Adjustments may be required to make sure the hormone concentrations are included in the standard serial dilution.
15. Compounds and concentrations can be adjusted according to your own experiments.
16. Mixed standards may be stored at −20 °C for up to one month.
17. To remove air bubbles from the liquid, mobile phases A and B should be sonicated for 5 min before use.
18. The parameters would need to be optimized in other triple quadrupole MS instruments.
19. If you use other triple quadrupole MS from another manufacturer, please use the related software for analysis.

Acknowledgments

We thank Yuanyuan Gao at the Metabolomics-Mass Spectrometry facility at the CAS Center for Excellence in Molecular Plant Sciences for the help with LC-MS/MS performance and analysis. Dr. Lingya Yao is supported by the Youth Program of Natural Science Foundation of Shanghai Municipality (project number: 25ZR1402413). Dr. Christian Danve M. Castroverde is supported by the Natural Sciences and Engineering Research Council of Canada (NSERC), Canada Foundation for Innovation, Ontario Research Fund and Wilfrid Laurier University. We would like to acknowledge that Wilfrid Laurier University is located on the shared traditional territory of the Neutral, Anishinaabe and Haudenosaunee peoples.

References

1. Jones JDG, Staskawicz BJ, Dangl JL (2024) The plant immune system: from discovery to deployment. Cell 187(9):2095–2116. https://doi.org/10.1016/j.cell.2024.03.045
2. Kessler A (2015) The information landscape of plant constitutive and induced secondary metabolite production. Curr Opinion Insect Sci 8:47–53. https://doi.org/10.1016/j.cois.2015.02.002
3. Zhou J-M, Zhang Y (2020) Plant immunity: danger perception and signaling. Cell 181(5):978–989. https://doi.org/10.1016/j.cell.2020.04.028
4. van der Burgh AM, Joosten MHAJ (2019) Plant immunity: thinking outside and inside the box. Trends Plant Sci 24(7):587–601. https://doi.org/10.1016/j.tplants.2019.04.009
5. Ngou BPM, Jones JDG, Ding P (2022) Plant immune networks. Trends Plant Sci 27(3):255–273. https://doi.org/10.1016/j.tplants.2021.08.012
6. Yuan M, Ngou BPM, Ding P, Xin X-F (2021) PTI-ETI crosstalk: an integrative view of plant immunity. Curr Opin Plant Biol 62(102030):102030. https://doi.org/10.1016/j.pbi.2021.102030
7. Yuan M, Jiang Z, Bi G, Nomura K, Liu M, Wang Y et al (2021) Pattern-recognition receptors are required for NLR-mediated plant immunity. Nature 592(7852):105–109. https://doi.org/10.1038/s41586-021-03316-6
8. Ngou BPM, Ahn H-K, Ding P, Jones JDG (2021) Mutual potentiation of plant immunity by cell-surface and intracellular receptors. Nature 592(7852):110–115. https://doi.org/10.1038/s41586-021-03315-7
9. Durrant WE, Dong X (2004) Systemic acquired resistance. Annu Rev Phytopathol 42(1):185–209. https://doi.org/10.1146/annurev.phyto.42.040803.140421
10. Zeier J (2021) Metabolic regulation of systemic acquired resistance. Curr Opin Plant Biol 62(102050):102050. https://doi.org/10.1016/j.pbi.2021.102050
11. Kachroo A, Kachroo P (2020) Mobile signals in systemic acquired resistance. Curr Opin Plant Biol 58:41–47. https://doi.org/10.1016/j.pbi.2020.10.004
12. Vlot AC, Sales JH, Lenk M, Bauer K, Brambilla A, Sommer A et al (2021) Systemic propagation of immunity in plants. New Phytol 229(3):1234–1250. https://doi.org/10.1111/nph.16953
13. Maldonado AM, Doerner P, Dixon RA, Lamb CJ, Cameron RK (2002) A putative lipid transfer protein involved in systemic resistance signalling in Arabidopsis. Nature 419(6905):399–403. https://doi.org/10.1038/nature00962
14. Shields A, Shivnauth V, Castroverde CDM (2022) Salicylic acid and N-hydroxypipecolic acid at the fulcrum of the plant immunity-growth equilibrium. Front Plant Sci 13:841688. https://doi.org/10.3389/fpls.2022.841688
15. Huang W, Wang Y, Li X, Zhang Y (2020) Biosynthesis and regulation of salicylic acid and N-Hydroxypipecolic acid in plant immunity. Mol Plant 13(1):31–41. https://doi.org/10.1016/j.molp.2019.12.008

16. Hartmann M, Zeier J (2019) N-hydroxypipecolic acid and salicylic acid: a metabolic duo for systemic acquired resistance. Curr Opin Plant Biol 50:44–57. https://doi.org/10.1016/j.pbi.2019.02.006
17. Peng Y, Yang J, Li X, Zhang Y (2021) Salicylic acid: biosynthesis and signaling. Annu Rev Plant Biol 72(1):761–791. https://doi.org/10.1146/annurev-arplant-081320-092855
18. Spoel SH, Dong X (2024) Salicylic acid in plant immunity and beyond. Plant Cell 36(5): 1451–1464. https://doi.org/10.1093/plcell/koad329
19. Kumar S, Zavaliev R, Wu Q, Zhou Y, Cheng J, Dillard L et al (2022) Structural basis of NPR1 in activating plant immunity. Nature 605(7910):561–566. https://doi.org/10.1038/s41586-022-04699-w
20. Lim G-H, Liu H, Yu K, Liu R, Shine MB, Fernandez J et al (2020) The plant cuticle regulates apoplastic transport of salicylic acid during systemic acquired resistance. Sci Adv 6(19): eaaz0478. https://doi.org/10.1126/sciadv.aaz0478
21. Wang C, Liu R, Lim G-H, de Lorenzo L, Yu K, Zhang K et al (2018) Pipecolic acid confers systemic immunity by regulating free radicals. Sci Adv 4(5):eaar4509. https://doi.org/10.1126/sciadv.aar4509
22. Li Q, Zhou M, Chhajed S, Yu F, Chen S, Zhang Y, Mou Z (2023) N-hydroxypipecolic acid triggers systemic acquired resistance through extracellular NAD(P). Nat Commun 14(1):6848. https://doi.org/10.1038/s41467-023-42629-0
23. Hartmann M, Zeier T, Bernsdorff F, Reichel-Deland V, Kim D, Hohmann M et al (2018) Flavin monooxygenase-generated N-hydroxypipecolic acid is a critical element of plant systemic immunity. Cell 173(2):456–469.e16. https://doi.org/10.1016/j.cell.2018.02.049
24. Yildiz I, Mantz M, Hartmann M, Zeier T, Kessel J, Thurow C et al (2021) The mobile SAR signal N-hydroxypipecolic acid induces NPR1-dependent transcriptional reprogramming and immune priming. Plant Physiol 186(3):1679–1705. https://doi.org/10.1093/plphys/kiab166
25. Chen Y-C, Holmes EC, Rajniak J, Kim J-G, Tang S, Fischer CR et al (2018) N -hydroxy-pipecolic acid is a mobile metabolite that induces systemic disease resistance in Arabidopsis. Proc Natl Acad Sci USA 115(21): E4920–E4929. https://doi.org/10.1073/pnas.1805291115
26. Holmes EC, Chen Y-C, Sattely ES, Mudgett MB (2019) An engineered pathway for N-hydroxy-pipecolic acid synthesis enhances systemic acquired resistance in tomato. Sci Signal 12(604):eaay3066. https://doi.org/10.1126/scisignal.aay3066
27. Cai J, Jozwiak A, Holoidovsky L, Meijler MM, Meir S, Rogachev I, Aharoni A (2021) Glycosylation of N-hydroxy-pipecolic acid equilibrates between systemic acquired resistance response and plant growth. Mol Plant 14(3): 440–455. https://doi.org/10.1016/j.molp.2020.12.018
28. Holmes EC, Chen Y-C, Mudgett MB, Sattely ES (2021) Arabidopsis UGT76B1 glycosylates N-hydroxy-pipecolic acid and inactivates systemic acquired resistance in tomato. Plant Cell 33(3):750–765. https://doi.org/10.1093/plcell/koaa052
29. Mohnike L, Rekhter D, Huang W, Feussner K, Tian H, Herrfurth C et al (2021) The glycosyltransferase UGT76B1 modulates N-hydroxy-pipecolic acid homeostasis and plant immunity. Plant Cell 33(3):735–749. https://doi.org/10.1093/plcell/koaa045
30. Bauer S, Mekonnen DW, Hartmann M, Yildiz I, Janowski R, Lange B et al (2021) UGT76B1, a promiscuous hub of small molecule-based immune signaling, glucosylates N-hydroxypipecolic acid, and balances plant immunity. Plant Cell 33(3):714–734. https://doi.org/10.1093/plcell/koaa044
31. Park S-W, Kaimoyo E, Kumar D, Mosher S, Klessig DF (2007) Methyl salicylate is a critical mobile signal for plant systemic acquired resistance. Science (NY) 318(5847):113–116. https://doi.org/10.1126/science.1147113
32. Gong Q, Wang Y, He L, Huang F, Zhang D, Wang Y et al (2023) Molecular basis of methyl-salicylate-mediated plant airborne defence. Nature 622(7981):139–148. https://doi.org/10.1038/s41586-023-06533-3
33. Mohnike L, Huang W, Worbs B, Feussner K, Zhang Y, Feussner I (2023) N-Hydroxy pipecolic acid methyl ester is involved in Arabidopsis immunity. J Exp Bot 74(1):458–471. https://doi.org/10.1093/jxb/erac422
34. Huang X-X, Zhu G-Q, Liu Q, Chen L, Li Y-J, Hou B-K (2018) Modulation of plant salicylic acid-associated immune responses via glycosylation of dihydroxybenzoic acids. Plant Physiol 176(4):3103–3119. https://doi.org/10.1104/pp.17.01530
35. Zhang Y, Zhao L, Zhao J, Li Y, Wang J, Guo R et al (2017) S5H/DMR6 encodes a salicylic acid 5-hydroxylase that fine-tunes salicylic acid homeostasis. Plant Physiol 175(3):

1082–1093. https://doi.org/10.1104/pp.17.00695

36. Kim JH, Castroverde CDM, Huang S, Li C, Hilleary R, Seroka A et al (2022) Increasing the resilience of plant immunity to a warming climate. Nature 607(7918):339–344. https://doi.org/10.1038/s41586-022-04902-y
37. Engelberth J, Schmelz EA, Alborn HT, Cardoza YJ, Huang J, Tumlinson JH (2003) Simultaneous quantification of jasmonic acid and salicylic acid in plants by vapor-phase extraction and gas chromatography-chemical ionization-mass spectrometry. Anal Biochem 312(2):242–250. https://doi.org/10.1016/s0003-2697(02)00466-9
38. Loba VC, Pollmann S (2017) Highly sensitive salicylic acid quantification in milligram amounts of plant tissue. Method Mol Biol (Clifton, N.J.) 1497:221–229. https://doi.org/10.1007/978-1-4939-6469-7_18
39. Yao L, Jiang Z, Wang Y, Hu Y, Hao G, Zhong W et al (2023) High air humidity dampens salicylic acid pathway and NPR1 function to promote plant disease. EMBO J 42(21):e113499. https://doi.org/10.15252/embj.2023113499
40. Huang WE, Huang L, Preston GM, Naylor M, Carr JP, Li Y et al (2006) Quantitative in situ assay of salicylic acid in tobacco leaves using a genetically modified biosensor strain of Acinetobacter sp. ADP1. Plant J 46(6):1073–1083. https://doi.org/10.1111/j.1365-313X.2006.02758.x
41. Defraia CT, Schmelz EA, Mou Z (2008) A rapid biosensor-based method for quantification of free and glucose-conjugated salicylic acid. Plant Methods 4(1):28. https://doi.org/10.1186/1746-4811-4-28
42. Yu K, Liu H, Kachroo P (2020) Pipecolic acid quantification using gas chromatography-coupled mass spectrometry. Bio-Protocol 10(23):e3841. https://doi.org/10.21769/BioProtoc.3841

Chapter 3

Comprehensive Evaluation of Abscisic Acid Content in Plants Through Quantitative and Qualitative Techniques

Estefanía Rodríguez-Dobreva, Silvia Martínez-Fenoll, Adrián González Ortega-Villaizán, and Stephan Pollmann

Abstract

Global agricultural productivity is currently challenged by a range of emerging environmental stressors associated with the ongoing climate change scenario. The future of food security is critically reliant on the capacity to mitigate and manage these stresses on crops and trees. Over recent decades, substantial progress has been made in elucidating plant responses to various stresses, including heat, drought, and soil salinity, and in understanding how plant acclimatization to these responses is regulated by a limited number of signalling molecules, known as phytohormones. In response to abiotic stresses, abscisic acid (ABA) plays a pivotal role as it regulates, among other physiological processes, stomatal aperture and, consequently, the gas exchange and water uptake of plants. Therefore, methods for the comprehensive monitoring of changes in ABA levels are of particular significance. In this chapter, we provide a detailed description of a highly sensitive and accurate method for the quantitative analysis in plants using gas chromatography-coupled tandem mass spectrometry (GC-MS/MS). Additionally, we describe an indirect method for estimating ABA contents in *Arabidopsis thaliana* using ABA signalling reporter lines.

Key words Abscisic acid, Abscisic acid signalling, Gas chromatography-electron impact-tandem mass spectrometry (GC-EI-MS/MS), Solid-phase extraction, Stable isotopes, Derivatization, Plant hormone analysis, Reporter lines

1 Introduction

Plant growth and developmental processes are meticulously regulated by a limited number of small signalling molecules, referred to as plant hormones. These compounds also play a pivotal role in adjusting the intrinsic growth program of plants to align with prevailing environmental conditions. Typically, plant hormones are categorized into growth-promoting hormones, such as auxins,

Estefanía Rodríguez-Dobreva, Silvia Martínez-Fenoll and Adrián González Ortega-Villaizán contributed equally with all other contributors.

Shahid Mukhtar (ed.), *Plant Hormones: Methods and Protocols*, Methods in Molecular Biology, vol. 3026,
https://doi.org/10.1007/978-1-0716-5214-5_3,

cytokinins, and gibberellins, and growth-inhibiting hormones, including jasmonic acid, salicylic acid, and abscisic acid (ABA). ABA is instrumental in regulating a wide array of developmental and growth characteristics in plants, such as leaf abscission and the inhibition of fruit ripening. It is widely acknowledged as one of the most important stress hormones, involved in plant responses to a diverse range of environmental stresses, encompassing both biotic and abiotic stressors [1].

From a chemical point of view, ABA is classified as a sesquiterpene, characterized by a nonplanar configuration and multiple functional moieties. Its de novo synthesis occurs during desiccation, while its degradation takes place during rehydration following dehydration [2]. ABA is predominantly synthesized in plant roots and terminal buds. The C-15 skeleton of the ABA backbone is prevalent in biosynthetic precursors such as xanthoxin, abscisic aldehyde, and abscisic alcohol, as well as in oxidized catabolites including phaseic acid, dihydrophaseic acid, and 8′-hydroxy-ABA. In response to environmental stresses, endogenous ABA levels increase rapidly, thereby activating specific signalling pathways and altering downstream gene expression levels [3–5]. The stress-induced endogenous production of ABA involves the activation of genes encoding enzymes that synthesize ABA from β-carotene, with 9-*cis* epoxy carotenoid dioxygenases (NCEDs) catalyzing the rate-limiting step in ABA biosynthesis [6].

ABA plays a pivotal role in various cellular processes, including seed development, vegetative growth, seed maturation and germination, and responses to environmental stress [7]. In relation to the induced stress responses in plants, ABA regulates the synthesis of enzymes crucial for cellular protection against dehydration [8, 9], thermal stress [10], and other processes such as water transport [11, 12], as well as salinity and high light stress [13]. It is commonly accepted that plant responses to variations in microclimatic parameters, such as temperature, moisture, and radiation, can be modulated by alterations in ABA content. Herein, elevated ABA levels are supposed to substantially contribute to cold resistance, as demonstrated by increased ABA levels in bean plant exudates following brief heat stress [14]. Furthermore, ABA is involved in stomatal closure during periods of low CO_2 demand or drought conditions, thereby preventing excessive water loss through transpiration. ABA is directly regulating stomatal aperture [11, 15], tissue hydraulic conductivity [12, 16], and root and shoot growth [17]. At the whole-plant level, ABA serves as a crucial mediator of communication between roots and shoots during water and salt stress, interacting with other plant-derived signals for organ-to-organ communication [18]. A further primary function of ABA is the inhibition of seed germination, as it is believed to prevent germination immediately after seeds are sown [19].

In the context of plant stress responses to environmental challenges, the detailed monitoring of ABA levels in small tissue samples, such as plant roots or leaves, or even at the cellular level, is of considerable interest. However, the quantitative analysis of ABA is technically challenging due to its presence in only trace amounts within these tissues. Advanced and highly sensitive technologies, such as tandem mass spectrometry and confocal laser scanning microscopy, are particularly well-suited to address this analytical challenge. The methodologies delineated in this chapter facilitate both quantitative and qualitative analyses of ABA in small tissue samples. These methods have been rigorously established and validated, demonstrating successful application in *Arabidopsis thaliana*. Furthermore, the mass spectrometric analysis has been employed for the analysis of ABA from several other plant species, including potato, barley, tobacco, tomato, corn, and rice. For those readers seeking a comprehensive multiplex method for the concurrent analysis of all acidic plant hormones, we recommend consulting the protocol detailed by Pérez-Alonso et al. [20] for further information.

2 Materials

2.1 Abscisic Acid Extraction, Pre-purification, Derivatization, and GC-MS/MS Analysis

To prevent unwanted interferences, it is essential that all solvents utilized are prepared using reagents of pro analysis (p.a.) grade or, at a minimum, high-performance liquid chromatography (HPLC) grade. If pure reagents are not used directly, it is advisable to freshly prepare the solvents and employ them at room temperature, unless specified otherwise. All reagents should be stored at room temperature.

1. Acetone p.a.
2. Methanol p.a.
3. Diethyl ether p.a.
4. Stable isotope-labelled ABA ([d_6]-ABA) (*see* **Note 1**).
5. Acetone/methanol solution [9:1, (v/v)].
6. 2 M (Trimethylsilyl)diazomethane solution in hexanes (dry, free of peroxides).
7. Washing solvent for solid-phase extraction: Chloroform/2-propanol [2:1, v/v].
8. Elution solvent for solid-phase extraction: diethyl ether +2% acetic acid [v/v].
9. Thermomixer.
10. Aminopropyl solid-phase extraction columns (Bond Elut NH2 50 mg, 1 mL, Agilent, Santa Clara, USA).
11. 600 μL Crimp-top conical micro-vials.

12. 8 mm Crimp seal with PTFE/rubber septa.
13. Centrifuge and corresponding rotor.
14. Speed-vac concentrator.
15. Extraction vacuum manifold.
16. Membrane vacuum pump stand.
17. Ball mill.
18. Stainless steel balls, 3 mm diameter.
19. Magnet.
20. Ultrasonic bath.
21. 30 m 0.25 mm i.d. fused silica capillary column with a chemical bond 0.25 μm ZB35 stationary phase.
22. GC triple-quadrupole MS.

2.2 Growth and Analysis of Abscisic Acid Reporter Plants

To mitigate the adverse effects of biotic stressors, it is imperative to implement a rigorously sterile cultivation approach.

1. 70% ethanol (v/v).
2. 50% (v/v) sodium hypochlorite from a commercial bleach solution (4% w/v).
3. Sterile 6-well, flat bottom, tissue culture plates.
4. Sterile 0.5 × Murashige & Skoog medium [21] (*see* **Note 2**).
5. Sterile 0.5 × Murashige & Skoog medium solidified with 1% (w/v) agar on square (120 × 120 mm) Petri dishes.
6. ABA signalling reporter lines 6 × ABRE_A::erGFP (NASC ID: N71619) and 6 × ABRE_R::erGFP (NASC ID: N71620) [22] are available from the Nottingham Arabidopsis Stock Centre (NASC, https://arabidopsis.info/).
7. Horizontal laminar flow cabinet.
8. Breathable medical tape.
9. Plant growth chamber.
10. Microscope glass slides and cover slips.
11. Confocal laser scanning microscope equipped with a 488 nm Argon laser and a 494–596 nm broadband filter.

3 Methods

3.1 Quantitative Analysis of Abscisic Acid by GC-MS/MS

3.1.1 Extraction of Plant Tissue

1. Weigh between 50 mg and 100 mg of plant tissue (fresh weight) into a 1.5 mL disposable reaction tube (*see* **Note 3**). Add 1 mL of methanol, 30 pmol of the [d_6]-ABA internal standard (*see* **Note 4**), and three steel balls.
2. Utilize a thermomixer to heat the sample to 60 °C, maintaining gentle agitation for 3 min, followed by homogenization

through vigorous vortexing or shaking for several minutes. Ideally, a vibrating-ball micro-mill may be employed, subjecting the sample to two cycles of 3 min each at 30 Hz (*see* **Note 5**). Subsequent extraction should occur at room temperature for a minimum of 1 h. During the incubation period, the sample should be vortexed or inverted intermittently.

3. Utilize a magnet to remove the steel balls, then centrifuge the sample using a desktop centrifuge at approximately 21,000× *g* for 1 min to sediment cell wall debris and other suspended particles.
4. Carefully transfer the supernatant to a new reaction tube, ensuring that no portions of the pellet are included. Evaporate the sample to complete dryness using a speed-vac set to 45 °C (*see* **Note 6**) and maximum vacuum. It is important to note that, once dried, the sample can be stored at −80 °C until further processing.

3.1.2 Pre-purification and Enrichment of Abscisic Acid by Solid-Phase Extraction

1. Thoroughly dissolve the crude residue in 50 μL of methanol, followed by the addition of 200 μL of diethyl ether. Subject the sample to sonication for 5 min in an ultrasonic bath, and subsequently centrifuge it for 5 min at 21,000× *g*.
2. Concurrently, initiate the equilibration of the aminopropyl solid-phase extraction column (*see* **Note 7**). This is achieved by washing the column with 2 × 500 μL of diethyl ether.
3. Transfer the entire sample onto the conditioned column, allowing it to pass through the matrix by gravity flow without the application of vacuum to the solid-phase extarction (SPE) manifold (*see* **Note 8**).
4. Following the sample passage, wash the column twice with 500 μL of freshly prepared washing solvent. Clean the tip of the plastic insert holding the column with white laboratory paper.
5. Elute the sample from the matrix into a fresh reaction tube by adding 2 × 500 μL of the elution solvent. Finally, dry the combined eluates using a speed-vac concentrator (10 mbar, 45 °C) and remove residual acetic acid with a gentle stream of nitrogen.

3.1.3 Derivatization of the Sample Prior to GC-MS/MS Analysis

The contained ABA within the sample is analyzed using gas chromatography-mass spectrometry (GC-MS) in the form of its methyl ester (MeABA). The corresponding ester is characterized by a significantly lower melting point, thereby enhancing its vaporization properties during subsequent GC-MS analysis (*see* **Note 9**).

1. To prepare the derivatization solvent for the methylation of ABA, combine 190 μL of the acetone/methanol solution (*see* **Note 10**) with 30 μL of diethyl ether and 30 μL of the (trimethylsilyl)diazomethane solution (*see* **Note 11**). The resulting 250 μL of solvent is adequate for the chemical modification of ten samples (*see* **Note 12**).

2. Dissolve the sample in 25 μL of methanol and transfer the solution into a 600 μL crimp-top conical micro-vial. Dry the sample completely using a gentle stream of nitrogen.
3. Add 20 μL of the freshly prepared derivatization solvent to the sample. Immediately seal the vial with an appropriate 8 mm crimp seal equipped with PTFE/rubber septa. Allow the sample to rest for 30 min at room temperature before proceeding with GC-MS analysis.

From the derivatized sample, inject a 1 μL aliquot into the GC-MS system for gas chromatographic separation followed by mass spectrometric analysis.

3.1.4 Gas Chromatographic-Mass Spectrometric Assessment

(A) GC Setup

1. Inject the sample in splitless mode. Apply a pressure pulse of 25 psi for 1 min to facilitate the transfer of compounds from the injector into the column.
2. Subsequently, fully open the split for 1.5 min before adjusting it to a split ratio of 20% for the remainder of the run time.
3. Set the injector temperature to 250 °C and maintain the column temperature at 50 °C for 1.2 min.
4. Thereafter, increase the temperature at a rate of 30 °C/min to 120 °C, followed by a further increase to 325 °C at a rate of 10 °C/min.
5. Finally, hold the temperature at 325 °C for an additional 4 min. For the selection of the capillary column, *see* **Note 13**. Utilize helium at a flow rate of 1 mL/min as the mobile phase.

(B) MS Setup.

The method delineated herein is optimized for triple-quadrupole mass spectrometers operating in positive polarity. The mass spectrometer should be configured to function in EI-MRM mode.

1. The transfer line temperature should be set to 250 °C, while the ion source temperature should be maintained at 200 °C.
2. The ion source should be adjusted to 70 eV with a filament emission current of 80 μA, and the dwell time should be set to 175 ms.
3. Argon is to be utilized as the collision gas, with a setting of 2.0 mTorr.
4. For the target compounds, which include derivatized endogenous ABA and the deuterated internal standard, a minimum of two transitions should be recorded. One transition serves as a quantifier ion, while the others function as qualifier ions, providing additional information about the analyte and

Table 1
Characteristic precursor and product ions used for the detection of methylated abscisic acid (MeABA) and its internal standard [d_6]-MeABA

Compound	Retention time (min)	Retention time window (min)	Precursor ion (m/z)	Product ion (m/z)	Collision energy (eV)	Scan time (%)	Quantifier ion
MeABA	16.20	1	190	161	15	50	
	16.20	1	162	133	10	50	X
[d_6]-MeABA	16.20	1	196	167	15	50	
	16.20	1	168	139	10	50	X

The retention time for the target compound may exhibit moderate variation due to parameters specific to the system. It is advisable to employ reference compounds to ascertain retention times for the specific setup prior to conducting the analysis (*see* **Note 13**)

indicating the presence of potential impurities. The selected precursor ions and their corresponding diagnostic product ions are enumerated in Table 1 (*see* **Note 13**).

5. The quantity of the endogenous compound is subsequently calculated from the signal ratio of the unlabelled to the stable isotope-containing mass fragment, as observed in parallel measurements (Fig. 1).

3.2 Comparative Analysis of Changes in Abscisic Acid Signalling in Arabidopsis Thaliana

The fluorescent intensity of ABA-signalling reporter constructs can be used as a proxy to assess changes in cellular ABA levels. Suitable reporter lines harboring a green fluorescent protein (GFP) reporter gene under the control of a synthetic ABA-responsive promoter [22] are publicly available. For sterile *in vitro* plant culture, both on plates containing solidified medium or in liquid medium, seeds must be surface sterilized.

3.2.1 Sterilization and Growth of Abscisic Acid Reporter Lines

1. Transfer between 50 and 100 reporter line seeds into a sterile reaction tube. Introduce 1 mL of 70% ethanol (v/v) and allow the mixture to incubate for 2 min. Subsequently, centrifuge the mixture for 1 min at 14,000× *g*. Carefully remove the supernatant using a pipette, ensuring that the seeds are not inadvertently aspirated. Next, add 1 mL of sodium hypochlorite solution and incubate for 5 min. Repeat the centrifugation process and discard the sodium hypochlorite solution. Finally, wash the seeds five times with autoclaved water. Store the seeds in 1 mL sterile water.
2. Enclose the tubes in aluminium foil or place them within an opaque container. Stratify the imbibed seeds by storing them in darkness for 48 h at 4 °C to synchronize germination.
3. Individual seeds should be placed on plates containing sterile 0.5× MS medium, ensuring they are evenly spaced along a line, 2 cm from the top edge. The transfer of seeds from the tube to

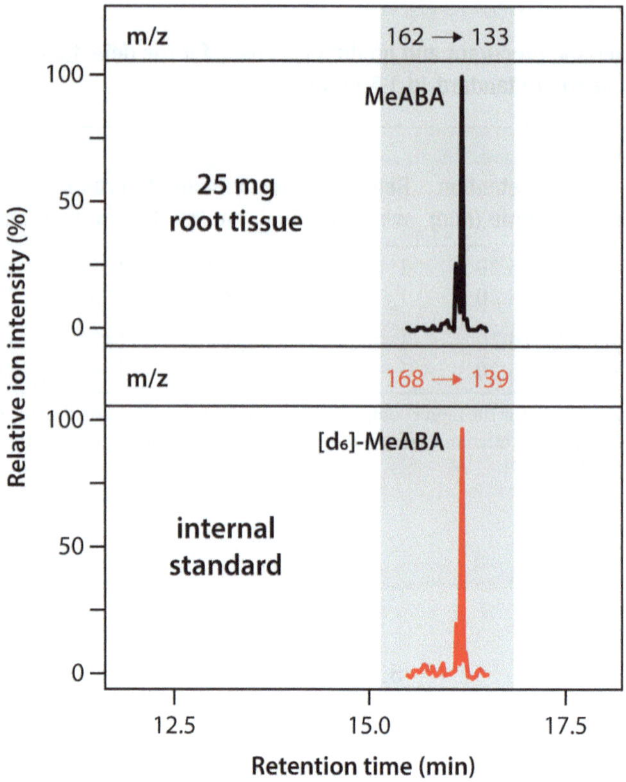

Fig. 1 Representative chromatograms from a parallel GC-MS/MS analysis conducted on *Arabidopsis thaliana* roots, utilizing a total of 25 mg of tissue derived from 10- to 14-days-old plants. The chromatograms depict the intensities of characteristic ion transitions selected for both endogenous ABA (black) and the stable isotope-labelled internal standard (red). The gray box delineates the scanned retention time window for ABA and its corresponding internal standard

the plate can be accomplished using a pipette equipped with a sterile yellow tip. It is recommended to plate the seeds under a laminar flow hood and not to place more than 4–5 seeds per square Petri dish. Thereafter, the plates should be sealed with medical tape. If a subsequent test in liquid medium is anticipated, up to 24 seeds may be placed on round 9 cm Petri dishes, maintaining a consistent distance between the seeds.

4. Germinate and grow the seedlings for 4–6 days in a plant growth chamber under controlled conditions (16 h light/8 h dark cycle, at 22 °C) with a photosynthetically active radiation of 100 $\mu mol \cdot m^{-2} \cdot s^{-1}$. The square plates should be grown vertically.

5. To achieve the desired treatment, transfer a subset of the plates to a chamber with an altered temperature or relocate the seedlings to plates containing NaCl or mannitol to induce salt and

osmotic stress responses, respectively. Alternatively, the transfer may be performed into liquid medium in 6-well tissue culture plates to assess short-term responses. Based on our experience, an incubation period of 2–4 h with chemical compounds, e.g., with a 10 μM ABA solution (positive control), is sufficient to initiate the ABA-dependent expression of the GFP reporter gene (*see* **Note 14**). Ensure that corresponding controls of similarly handled seedlings grown under control conditions are available.

3.2.2 Microscopic Assessment of Abscisic Acid Reporter Lines

1. Transfer individual plant roots to microscope glass slides. Add a drop of sterile water and cover the samples with a glass slip.
2. Mount the specimen in a confocal laser scanning microscope and focus on the primary root tip region.
3. Excite the GFP reporter protein with an Argon laser using the 488 nm line. Capture the emitted GFP fluorescence using a 494–596 nm broadband filter. If possible, also capture a bright-field image of the sample.
4. As illustrated in Fig. 2, the images captured with the microscope facilitate the visual inspection of ABA-induced GFP fluorescence (ensure identical laser energy and detector gain settings across all samples). However, the fluorescence intensity in the images can also be quantified by using suitable software (*see* **Note 15**) to obtain statistically assessable results.

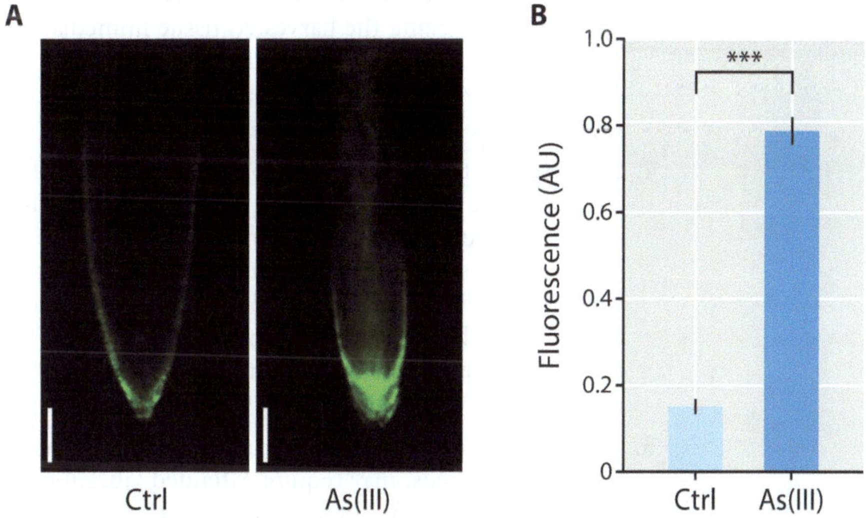

Fig. 2 Qualitative and quantifive analysis of GFP fluorescence in the root tips of 14-days-old 6xABRE_R:erGFP seedlings grown under control conditions or on plates additionally containing 10 μM As(III). (**a**) Confocal microscopy images of primary root tips under the different test conditions. Scale bar = 100 μm. (**b**) The bars show means ± SE of $n = 6$ independent measurements. Asterisks mark significantly altered fluorescence levels. Student's *t*-test: $***p \leq 0.001$. AU, arbitrary units

4 Notes

1. Stable isotope-labelled ABA, intended for use as internal standards in mass spectrometric analysis, is commercially available and can be procured from suppliers such as ICON Isotopes (www.iconisotopes.com) or OlChemIm Ltd. (www.olchemim.cz). It is permissible to utilize standards from alternative vendors, provided that the isotopic enrichment of the substance exceeds 96%, and the compound neither decays nor undergoes detectable exchange of heavy isotopes during the sample preparation and analysis process. It is strongly advised that the mass difference between the internal standard and the analyte is not less than two atomic mass units (AMU).
2. Murashige & Skoog medium, along with the corresponding agar for plant growth media, is available for purchase from suppliers such as Sigma-Aldrich (http://www.sigma-aldrich.com/) and Duchefa (https://www.duchefa-biochemie.com/). For experiments involving, e.g., nutrient limitation conditions, alternative media, such as Gamborg medium [23] or custom-made media with extra stressors such as NaCl (salt stress) or mannitol (osmotic stress), may be utilized, provided that sterility criteria are strictly maintained.
3. The production of abscisic acid is significantly induced under abiotic stress conditions. When conducting organ-specific plant hormone analysis, which requires the dissection of plants, it is essential to rapidly quench the metabolism. We strongly recommend freezing the harvested tissue immediately in reaction tubes containing liquid nitrogen. This procedure effectively minimizes biases in the data obtained by preventing self-induced ABA production.
4. It is essential to ascertain the appropriate quantity of the internal [d_6]-ABA standard for use in preliminary experiments. The optimal concentration of the standard may vary depending on the specific tissues and developmental stages under investigation. It is advisable that the amount of standard added to the sample should not exceed five times the endogenous content of the endogenous ABA, nor should it be less than one-fifth of the endogenous ABA levels in given tissues.
5. Plant structures with increased rigidity, such as lignified tissues like stems or seeds, may require extended vibration durations. In the absence of a vibrating-ball micro-mill, the tissue can alternatively be pulverized in liquid nitrogen using micro-pistils and standard reaction tubes. It is imperative that the resultant powder remains frozen until methanol is introduced.
6. Ensuring the complete desiccation of the sample, with no residual moisture, is crucial. However, excessive processing in

the speed-vac concentrator should be avoided, as it may lead to analyte loss due to over-drying. Given that the included ABA has a melting point of 163 °C, the risk of analyte loss through evaporation is minimal. Based on our experience, processing durations of up to 2 h do not adversely impact analyte content.

7. Commercially available SPE columns or microtips from other vendors, such as Phenomenex or Chromabond, may be utilized for solid-phase extraction. However, it is important to note that the protocol presented herein has not been optimized for these alternative columns and pipet microtips and will require adaptation if their use is intended.
8. Loading the sample slowly via gravity flow is likely to enhance the association of acidic compounds with the NH_2 matrix.
9. As an alternative to methylation using the (trimethylsilyl)-diazomethane solution, other reagents for derivatization may be employed. However, in our experience, trimethyl silylation of ABA using N,O-bis(trimethylsilyl)trifluoroacetamide (BSTFA) combined with tri-methylsilyl chloride (TMCS), in a 99:1 [v/v] ratio (Supelco), has not proven to be effective.
10. Prepare 10 mL of the acetone/methanol solution in a 9:1 (v/v) ratio. When stored in darkness and within an airtight container, this mixture can be preserved at room temperature for an extended duration.
11. Alternatively, ethereal diazomethane may be utilized for methylation. Ethereal diazomethane can be synthesized from N-nitrosomethylurea, which is recrystallized from methanol. N-nitrosomethylurea is available from Sigma-Aldrich. It is essential that safety precautions are strictly adhered to when preparing and handling diazomethane-containing solutions. All procedures must be conducted within a well-ventilated laboratory fume hood.
12. It is advisable to prepare only the necessary amount of derivatization solution for immediate use. The solvent should be freshly prepared prior to use.
13. GC capillary columns with an intermediate polarity, like the ZB-35 column (35% phenyl–65% dimethylpolysiloxane, low bleeding) employed here, are standard for acidic phytohormone profiling. However, other stationary phases, such as those containing 5% or 50% phenyl, which offer either lower or higher polarity, may also be utilized. It should be noted that such columns may require a slightly different temperature program than the one reported here to achieve satisfactory separation of the analytes.
14. A variety of abiotic stresses can be assessed by administering chemical compounds at specific concentrations directly onto

the plates. However, the infiltration of the medium at precise concentrations can be challenging due to potential variations in the absolute volume of the medium between plates. Therefore, we recommend conducting short-term treatments in a liquid medium under gentle agitation (40 rpm). The shaker should be operated within a plant growth chamber to prevent bias resulting from altered light and temperature conditions. Additionally, young seedlings must be handled with care when transferring them either from one plate to another or from the initial plate into the liquid medium to prevent bias from plant wounding.

15. Fluorescence levels within the region of interest (ROI) can be evaluated using the specialized software provided by the microscope vendor. Alternatively, fluorescence levels across various images can be analyzed using open-source software, such as ImageJ2 [24], which is available for download on the ImageJ website (https://imagej.net/software/imagej2/). For comparative and statistical analyses of fluorescence levels across different sample classes, it is essential to maintain consistent laser energy and detector gain settings throughout the entire experiment.

Acknowledgments

This work obtained financial support by grant PID2023-151327OB-I00 funded by MCIN/AEI/10.13039/501100011033 and, as appropriate, by "ERDF A way of making Europe," by the "European Union" or by the "European Union NextGenerationEU/PRTR" to S.P.

References

1. Zhang D-P (2014) Abscisic acid: metabolism, transport and signaling. Springer, Dordrecht. https://doi.org/10.1007/978-94-017-9424-4
2. Roychoudhury A, Paul S, Basu S (2013) Cross-talk between abscisic acid-dependent and abscisic acid-independent pathways during abiotic stress. Plant Cell Rep 32(7):985–1006. https://doi.org/10.1007/s00299-013-1414-5
3. Seki M, Narusaka M, Ishida J et al (2002) Monitoring the expression profiles of 7000 *Arabidopsis* genes under drought, cold and high-salinity stresses using a full-length cDNA microarray. Plant J 31(3):279–292. https://doi.org/10.1046/j.1365-313x.2002.01359.x
4. Rabbani MA, Maruyama K, Abe H et al (2003) Monitoring expression profiles of rice genes under cold, drought, and high-salinity stresses and abscisic acid application using cDNA microarray and RNA gel-blot analyses. Plant Physiol 133(4):1755–1767. https://doi.org/10.1104/pp.103.025742
5. Zeller G, Henz SR, Widmer CK et al (2009) Stress-induced changes in the *Arabidopsis thaliana* transcriptome analyzed using whole-genome tiling arrays. Plant J 58(6):1068–1082. https://doi.org/10.1111/j.1365-313X.2009.03835.x
6. Felemban A, Braguy J, Zurbriggen MD et al (2019) Apocarotenoids involved in plant development and stress response. Front Plant

Sci 10:1168. https://doi.org/10.3389/fpls.2019.01168

7. Chen K, Li GJ, Bressan RA et al (2020) Abscisic acid dynamics, signaling, and functions in plants. J Integr Plant Biol 62(1):25–54. https://doi.org/10.1111/jipb.12899
8. Finkelstein RR, Gibson SI (2002) ABA and sugar interactions regulating development: cross-talk or voices in a crowd? Curr Opin Plant Biol 5(1):26–32. https://doi.org/10.1016/S1369-5266(01)00225-4
9. Li C, Yue J, Wu X et al (2014) An ABA-responsive DRE-binding protein gene from *Setaria italica*, *SiARDP*, the target gene of SiAREB, plays a critical role under drought stress. J Exp Bot 65(18):5415–5427. https://doi.org/10.1093/jxb/eru302
10. Huang X, Shi H, Hu Z et al (2017) ABA is involved in regulation of cold stress response in Bermudagrass. Front Plant Sci 8:1613. https://doi.org/10.3389/fpls.2017.01613
11. Assmann SM (2003) *OPEN STOMATA1* opens the door to ABA signaling in *Arabidopsis* guard cells. Trends Plant Sci 8(4):151–153. https://doi.org/10.1016/S1360-1385(03)00052-9
12. Parent B, Hachez C, Redondo E et al (2009) Drought and abscisic acid effects on aquaporin content translate into changes in hydraulic conductivity and leaf growth rate: a trans-scale approach. Plant Physiol 149(4):2000–2012. https://doi.org/10.1104/pp.108.130682
13. Segarra-Medina C, Alseekh S, Fernie AR et al (2023) Abscisic acid promotes plant acclimation to the combination of salinity and high light stress. Plant Physiol Biochem 203:108008. https://doi.org/10.1016/j.plaphy.2023.108008
14. Itai C, Benzioni A (1974) Regulation of plant response to high temperature. Bulletin, The Royal Society of New Zealand (No.12), pp 477–482
15. Christmann A, Weiler EW, Steudle E et al (2007) A hydraulic signal in root-to-shoot signalling of water shortage. Plant J 52(1):167–174. https://doi.org/10.1111/j.1365-313X.2007.03234.x
16. Hose E, Steudle E, Hartung W (2000) Abscisic acid and hydraulic conductivity of maize roots: a study using cell- and root-pressure probes. Planta 211(6):874–882. https://doi.org/10.1007/s004250000412
17. Saab IN, Sharp RE, Pritchard J et al (1990) Increased endogenous Abscisic acid maintains primary root growth and inhibits shoot growth of maize seedlings at low water potentials. Plant Physiol 93(4):1329–1336. https://doi.org/10.1104/pp.93.4.1329
18. Kuromori T, Fujita M, Takahashi F et al (2022) Inter-tissue and inter-organ signaling in drought stress response and phenotyping of drought tolerance. Plant J 109(2):342–358. https://doi.org/10.1111/tpj.15619
19. Sano N, Marion-Poll A (2021) ABA metabolism and homeostasis in seed dormancy and germination. Int J Mol Sci 22(10). https://doi.org/10.3390/ijms22105069
20. Pérez-Alonso MM, Ortiz-García P, Moya-Cuevas J et al (2021) Mass spectrometric monitoring of plant hormone cross talk during biotic stress responses in potato (*Solanum tuberosum* L.). Methods Mol Biol 2354:143–154. https://doi.org/10.1007/978-1-0716-1609-3_7
21. Murashige T, Skoog F (1962) A revised medium for rapid growth and bio assays with tobacco tissue cultures. Physiol Plant 15(3):473–497. https://doi.org/10.1111/j.1399-3054.1962.tb08052.x
22. Wu R, Duan L, Pruneda-Paz JL et al (2018) The 6xABRE synthetic promoter enables the spatiotemporal analysis of ABA-mediated transcriptional regulation. Plant Physiol 177(4):1650–1665. https://doi.org/10.1104/pp.18.00401
23. Gamborg OL, Miller RA, Ojima K (1968) Nutrient requirements of suspension cultures of soybean root cells. Exp Cell Res 50(1):151–158. https://doi.org/10.1016/0014-4827(68)90403-5
24. Rueden CT, Schindelin J, Hiner MC et al (2017) ImageJ2: ImageJ for the next generation of scientific image data. BMC Bioinformatics 18(1):529. https://doi.org/10.1186/s12859-017-1934-z

Chapter 4

Characterization and Quantification of Strigolactones in Root Exudates

Jian You Wang, Guan-Ting Erica Chen, Kawthar F. Alashoor, and Salim Al-Babili

Abstract

This chapter details the methodology for collecting and processing plant root exudates using a C_{18} solid-phase extraction column, followed by the isolation, identification, and quantification of strigolactones (SLs) through liquid chromatography-mass spectrometry (LC-MS).

Key words Strigolactones, Root exudates, C_{18} solid-phase extraction column, Liquid chromatography-mass spectrometry (LC-MS), Liquid chromatography with tandem mass spectrometry (LC-MS/MS), Rice, Sorghum, Pearl millet, Rhizosphere signals

1 Introduction

Strigolactones (SLs) are plant hormones that inhibit shoot branching/tillering [1, 2] and regulate root development, stem thickness, and leaf senescence [3]. In addition, SLs released by plant roots act as key signaling molecules that promote the establishment of arbuscular mycorrhizal (AM) symbiosis. However, released SLs are exploited by root parasitic weeds, such as Striga, which use them to trigger seed germination, thereby paving the way for host infestation [4, 5].

There are more than 35 natural SLs with different structures [5]. All SLs are characterized by the presence of a butenolide ring (D-ring; Fig. 1), which is essential for their biological activity [6]. In natural SLs, this ring is coupled with a second moiety through an enol ether bridge in *R* configuration. According to the presence or absence of the BC-ring in the second moiety (Fig. 1), natural SLs are classified as canonical and noncanonical

Jian You Wang and Guan-Ting Erica Chen have equally contributed to this chapter.

Shahid Mukhtar (ed.), *Plant Hormones: Methods and Protocols*, Methods in Molecular Biology, vol. 3026, https://doi.org/10.1007/978-1-0716-5214-5_4,

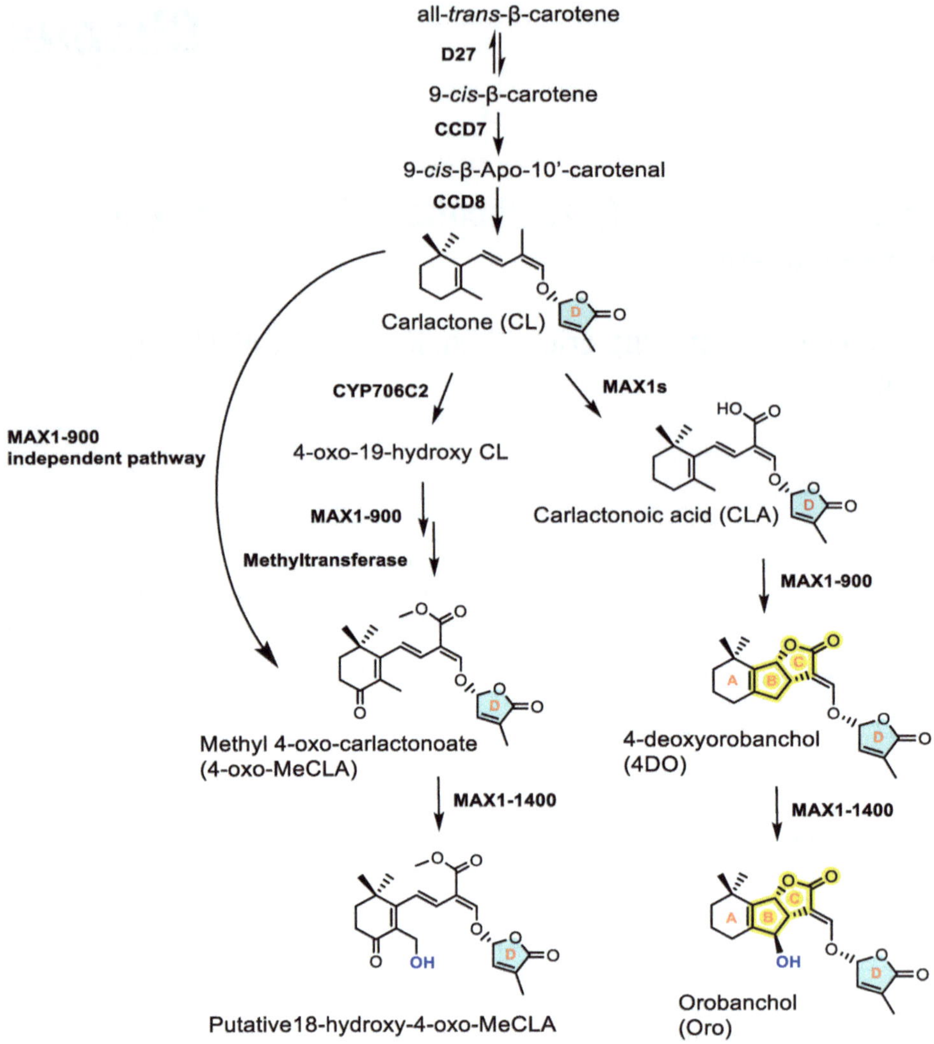

Fig. 1 Scheme of SL biosynthesis in rice. The biosynthesis of SLs begins with the reversible isomerization of all-*trans*-β-carotene into 9-*cis*-β-carotene, catalyzed by the isomerase DWARF27 (D27). This is followed by cleavage and rearrangement reactions mediated by CAROTENOID CLEAVAGE DIOXYGENASE 7 (CCD7) and CCD8, resulting in the formation of carlactone (CL). In rice, the cytochrome P450 enzyme MAX1–900 then oxygenates CL to produce carlactonoic acid (CLA), which is further converted into the canonical SL, 4-deoxyorobanchol (4DO). Another MAX1 homolog, MAX1–1400, catalyzes the conversion of 4DO into orobanchol (Oro). In addition to this pathway, CL undergoes metabolism by CYP706C2, yielding a noncanonical SL, 4-oxo-19-hydroxy-CL. MAX1–900, in conjunction with methyltransferases, further converts 4-oxo-19-hydroxy-CL into the noncanonical SL methyl 4-oxo-carlactonoate (4-oxo-MeCLA), previously identified as a methoxy-5-deoxystrigol isomer. The presence of 4-oxo-MeCLA in *max1–900* rice mutants suggests an additional, MAX1–900-independent minor pathway for its biosynthesis. Moreover, the rice MAX1–1400 hydroxylates 4-oxo-MeCLA, producing the tentative compound 18-OH-4-oxo-MeCLA. Abbreviation: *D27* Dwarf27, *CCD* Carotenoid Cleavage Dioxygenase, *MAX1* More Axillary Growth 1

SLs, respectively [5]. The SL biosynthetic pathway starts with the reversible conversion of all-*trans*-β-carotene into 9-*cis*-β-carotene by the isomerase DWARF27 (D27) [7]. Then, two carotenoid cleavage dioxygenases (CCDs), CCD7 and CCD8, cleave successively 9-*cis*-β-carotene into carlactone (CL), the core intermediate of SL biosynthesis [8–12]. The structural diversity of SLs arises from modifications of CL, which are mediated by cytochrome P450 monooxygenases (CYPs), including CYP711A subfamily members represented by the Arabidopsis More Axillary Growth 1 (MAX1), CYP722A/B/C in Arabidopsis and tomato, CYP728B in sorghum, and CYP706C in maize and rice [13–17]. A key function of MAX1 orthologue enzymes, i.e., CYP711A1 enzymes, is to catalyze the conversion of CL into carlactonoic acid (CLA) [13]. In rice, OsMAX1–900 (CYP711A2) sequentially oxygenates CL to generate the canonical SL 4-deoxyorobanchol (4DO), which is subsequently hydroxylated into orobanchol (Oro) by OsMAX1–1400 (CYP711A3) [6, 18–21]. In addition, OsCYP706C2 converts CL into 4-oxo-19-OH-CL, a precursor for methyl 4-oxo-carlactonoate (4-oxo-MeCLA) [16, 22] (Fig. 1). Beyond CYPs, several other enzyme families contribute to SL diversity, including CLA methyltransferase (CLAMT), Lateral Branching Oxidoreductase (LBO), and Low Germination Stimulant 1 (LGS1) [23–26].

Given their structural diversity and instability, the characterization of SLs requires a detailed, standardized, and reproducible protocol. Furthermore, because SLs are exuded at extremely low concentrations (~pM per liter of root exudates), even under phosphate-deficient conditions, which enhances their production and release, their extraction and quantification remain highly challenging [27].

In this chapter, we describe a comprehensive, reproducible, step-by-step methodology for extracting SLs from the root exudates of rice [19–22, 28–31], sorghum [32], tomato [36], and pearl millet [26] cultivated in hydroponic and/or sand culture systems. The extracted SLs are subsequently analyzed by liquid chromatography-mass spectrometry (LC-MS)-based detection and quantification. In addition, their bioactivity can be assessed through germination assays with the root parasitic plant *Striga hermonthica* using software SeedQuant [33].

2 Materials

2.1 Key Materials and Equipment

1. Methanol (LC-MS grade, ≥99.9%).
2. Water (LC-MS grade).
3. Milli-Q Water.
4. Ethyl Acetate (HPLC grade, ≥99.8%).

5. Acetonitrile (LC-MS grade, ≥99.9%).
6. Formic acid (LC-MS grade, ≥98%).
7. 4-deoxyorobanchol (Standard, ≥99%).
8. Orobanchol (Standard, ≥99%).
9. GR24 (Standard, ≥99%).
10. Ammonium nitrate (≥99%).
11. Magnesium sulfate heptahydrate (≥99%).
12. Iron(II) sulfate heptahydrate, plant cell culture tested (≥99%).
13. Ethylenediaminetetraacetic acid disodium salt dihydrate (EDTA; ≥99%).
14. Calcium chloride dihyd (99% acs).
15. Potassium nitrate (≥99%).
16. Potassium phosphate dibasic trihydrate ReagentPlus (R) (≥99%).
17. Boric acid (≥99%).
18. Manganese chloride tetrahydrate (≥99%).
19. Copper sulfate pentahydrate (≥99%).
20. Zinc chloride (≥99%).
21. Sodium molybdate (≥99%).
22. Murashige & Skoog Basal Medium: Dissolve 2.21 g Murashige & Skoog Basal Medium (without vitamins), fill up to 1 L with Milli-Q water and adjust the pH to 5.8 before use. Autoclave ½ MS medium for further use (*see* **Note 1**).

    Preparation of stock for 500X modified Hoagland nutrient solution: (a) Dissolve 224.1 g NH_4NO_3 fill up to 1 L with Milli-Q water (Stock No. 1); (b) Dissolve 50.0 g $K_2HPO_4 \cdot 3H_2O$ in Milli-Q water and make up to 1 L (Stock No. 2); (c) Dissolve 98.6 g $MgSO_4 \cdot 7H_2O$ and 70.0 g K_2SO_4 in Milli-Q water and make up to 1 L (Stock No. 3); (d) Dissolve 2.55 g $FeSO_4 \cdot 7H_2O$ and 1.86 g $Na_2EDTA \cdot 2H_2O$ in Milli-Q water and make up to 1 L (Stock No. 4); (e) Dissolve 117.6 g $CaCl_2 \cdot 2H_2O$ in Milli-Q water and make up to 1 L (Stock No. 5); (f) Dissolve 40.4 g KNO_3 in Milli-Q water and make up to 1 L (Stock No. 6); (g) For micronutrients, dissolve 0.711 g H_3BO_3, 0.445 g $MnCl_2 \cdot 4H_2O$, 0.037 g $CuSO_4 \cdot 5H_2O$, 0.102 g $ZnCl_2$, and 0.012 g $Na_2MoO_4 \cdot 2H_2O$ in Milli-Q water and make up to 1 L (Stock No. 7). All the stock solutions are stored at 4 °C for up to 3 months.
23. Preparation of half-strength modified Hoagland nutrient solution: (a) To make +Pi solution (normal growth condition), add 1 mL of each stock solution (listed in 22; No1 to No.7), except use 10 mL of $FeSO_4 \cdot 7H_2O$ and $Na_2EDTA \cdot 2H_2O$, to 900 mL

of Milli-Q water, then fill with Milli-Q water to a final volume of 1 L, and adjust the pH to 5.8 before use; (b) For low Pi (−Pi) or No Pi conditions, prepare the solution with either 0.01 mL/L of 0.4 mM K_2HPO_4 or omit K_2HPO_4 entirely, and adjust the pH to 5.8 before use.

24. Grade 4 qualitative filter papers, Whatman™, 90 mm, 1 cm thickness.
25. Plastic round Petri dish (PETRI DISH 90X16.2 MM).
26. Silver sand.
27. Hypersil GOLD™ C18 Selectivity HPLC Columns.
28. SPE C18-Fast (500 mg/3 mL).
29. UHPLC-Orbitrap ID-X Tribrid Mass Spectrometer.
30. HPLC-triple quadrupole/linear ion trap instrument (QTRAP5500).
31. UHPLC-Triple-Stage Quadrupole Mass Spectrometer (TSQ-AltisTM).

2.2 Software and Algorithms

1. Xcalibur™ Software (Thermo Fisher Scientific; OPTON-30965).
2. SeedQuant [33].
3. MultiQuant 2.1 (SCIEX).

3 Methods

3.1 Plant Growth: Rice (Oryza Sativa L.) as an Example: 21 Days for Hydroponic; 42 Days for Sand

1. Surface sterilize the seed in a laminar flow cabinet (*see* **Note 2**).
2. After imbibition, place 10 seeds on a 100 mm × 15 mm Petri dish, containing two filter papers moistened with 5 mL of ½ MS medium (*see* **Note 3**).
3. Seal the Petri dishes with parafilm, and wrap them in aluminum foil to avoid exposure to light to the seeds.
4. Keep the Petri dishes in a 30 °C incubator for 2 days.
5. Remove the aluminum foil and the parafilm, and transfer the germinated seeds to a growth chamber with a day/night temperature of 28/22 °C and a 12 h photoperiod, 200 μmol photons m^{-2} s^{-1} for 5 days. Check the seeds on the third day and add ½ MS medium if needed.
6. To grow rice seedlings under hydroponic conditions, make one hole in the center of the cap of the 50 mL black tube. Cut the bottom of a 1.5 mL black Eppendorf tube, remove the lid if it has one, and place it in the hole of the 50 mL black tube's cap (Fig. 2a). Fill the 50 mL black tube with +Pi Hoagland nutrient solution. Transfer One or two one-week-old rice

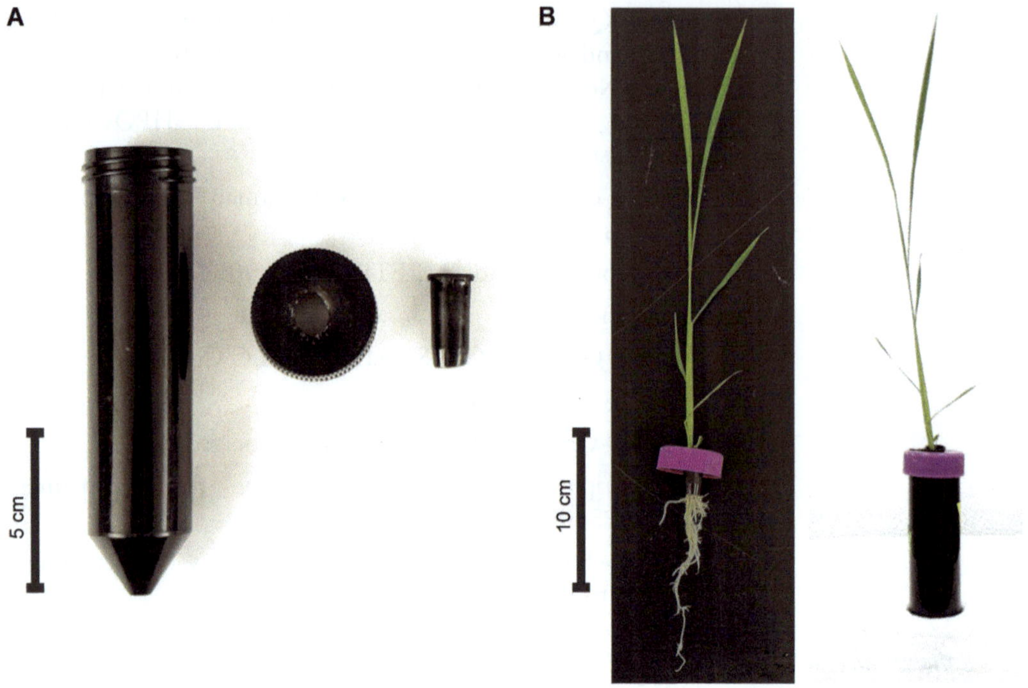

Fig. 2 The setup of the hydroponic system using a 50 mL black Falcon tube. (**a**) Drill a hole in the center of the cap of the 50 mL Falcon tube. Cut off the bottom of a 1.5 mL black Eppendorf tube (lid removed) and insert it into the hole in the cap. (**b**) Representative image of rice seedlings grown in the hydroponic system

seedling(s) through the 1.5 mL bottomless Eppendorf tube in the center of each cap and fix the cap into 50 mL black tubes (Fig. 2b). Keep the tubes with seedlings in a growth chamber for 1 week (+Pi) and refresh the nutrient solution two times per week (*see* **Note 4**).

7. Trigger the SL production and exudation by applying −Pi half-strength modified Hoagland nutrient solution for another 1 week [28]. Rice seedlings can be directly grown under low Pi conditions for 2 weeks, which generally showed higher SL content compared to −Pi conditions [34].

3.2 SL Collection and Extraction

Try to avoid direct light exposure during the extraction process (Fig. 3). Plant roots continuously release SLs into the medium when hydroponically grown under phosphate starvation. Therefore, we recommend letting SLs accumulate in the medium for 6 h before collecting it [28]: we usually refresh the solution in the early morning and collect the medium for SL extraction 6 h later (Fig. 3) (*see* **Note 5**).

1. To collect root exudates, prepare fresh −Pi/low Pi half-strength modified Hoagland solution with pH adjusted to ~5.8. Refresh the medium (50 mL) of each tube containing

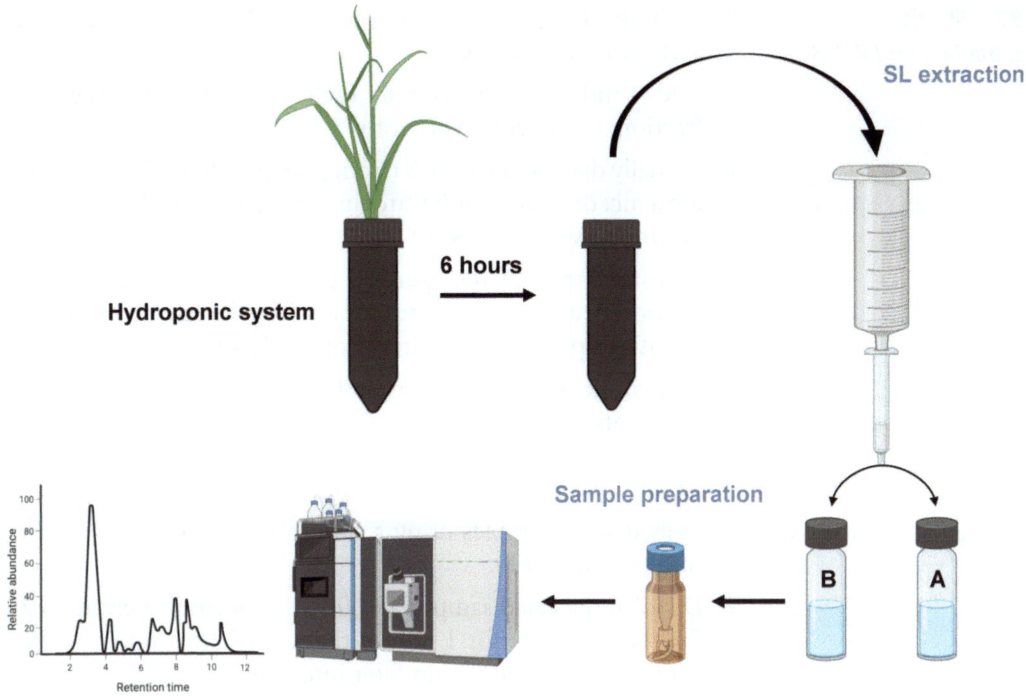

Fig. 3 A simplified workflow of SL collection and extraction from root exudates

rice seedlings in the early morning and wait for 6 h. After 6 h, remove the plants from the tubes and store the collected medium with 50 mL tubes on ice (*see* **Note 6**).

2. To extract SL from root exudates, keep the collected root exudates on ice. In the lab, spike each sample, collected both from hydroponic and sand, with labeled SLs, e.g., of D_6–5-deoxystrigol, used as internal standard (IS); alternatively, the internal standard can be replaced by using the SL analog GR24 (*see* **Note 7**). Add 3 mL of methanol, followed by 3 mL of Milli-Q water to precondition the Fast SPE C_{18} column (500 mg/3 mL; GracePure®), connected with SPE Vacuum Manifold. Attach a 50 mL syringe to the preconditioned C18 column and load the collected root exudate samples (**step 1**) onto it. After enrichment, wash the C18 column with 3 mL Milli-Q water to remove impurities. Elute SLs twice, into two separate 8 mL brown glass vials: collect the first fraction (A) using 2 mL of acetone, followed by a second fraction (B) with 3 mL of acetone (*see* **Note 8**).
3. Cap the collected samples in the brown glass vials and keep the acetone extracted samples in −20 °C refrigerator for 2-week storage or at −80 °C for long-term storage.

3.3 SLs Re-extraction for LC-MS

1. Fully dry the fraction B (from **step 3.2.2**; containing mostly acetone) under vacuum.
2. Add 1 mL ethyl acetate to each vial containing the dried fraction, then gently vortex for 5 s.
3. Partially dry the fraction A (from previous step 3.2.2); containing a mix of acetone and water under vacuum until ~300 μL left (acetone-water) (*see* **Note 9**).
4. Transfer 1 mL of resuspended fraction B in ethyl acetate to fraction A, and gently vortex for 5 s. Two layers will form: ethyl acetate (top layer) and water (bottom layer).
5. Centrifuge the samples at 3800 rpm for 2 min at room temperature.
6. Carefully transfer 750 μL from the upper layer (SLs enriched organic phase) to a new 1.5 mL Eppendorf tube (*see* **Note 10**). Fully dry the samples using a speed vacuum at room temperature (*see* **Note 11**).
7. Dissolve the dried samples in 100 μL of acetonitrile:water (25: 75, v:v), sonicate for 1 min in the water bath, and filter them gently through a 0.22 μm filter into a glass autosampler vial.
8. Close the vial with the cap before LC-MS/MS analysis. For better detection, gently tap the bottom of the vial to release any trapped gases in the sample (*see* **Note 12**).

3.4 SL Identification by LC-MS/MS

Although SLs are divided into canonical and noncanonical subfamilies [5], they are generally traced and identified using the same methods. In our case study, we use the IDX-Orbitrap high-resolution Mass Spectrometry instrument, including the MS and MS/MS scans. The full scan MS analysis revealed the presence of the SL 4DO at *m/z* 331.15402 (RT: 14.95 min) (Fig. 4a). The MS/MS scan further confirmed the presence of the common D-ring moiety fragment ion at *m/z* 97.02823, which is a conserved diagnostic product ion of SLs (Fig. 4b) (*see* **Note 13**).

1. Separate the metabolites using a Hypersil GOLD™ C_{18} Selectivity HPLC column (150 × 4.6 mm; 3 μm), and maintain at 35 °C.
2. Use a gradient elution method as described in Table 1. The mobile phases consisted of (A) 100% LC-MS grade water with 0.1% formic acid and (B) 100% Acetonitrile with 0.1% formic acid.
3. Perform the identification of SLs using a UHPLC system coupled with an Orbitrap ID-X Tribrid Mass Spectrometer, which runs in positive and negative modes. The MS parameters are listed in Table 2 and the parameters MS/MS are listed in Table 3.

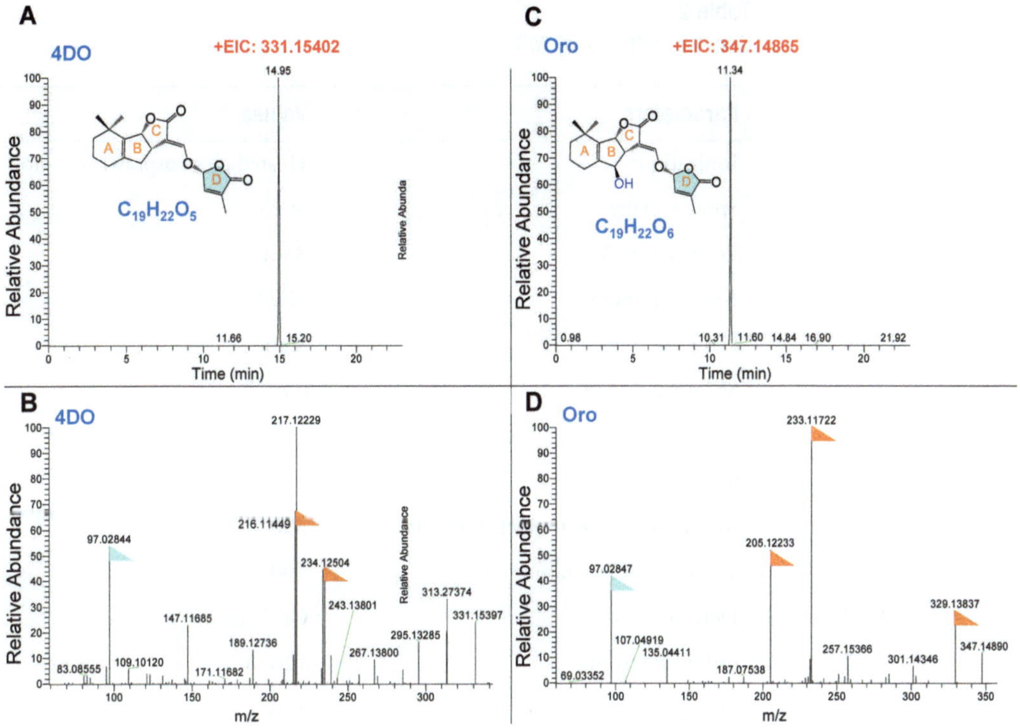

Fig. 4 The chromatography of SL identification from authentic standards. (**a**) Accurate mass of 4-deoxyorobanchol (4DO). (**b**) MS/MS fragmentation of 4DO. (**c**) Accurate mass of Orobanchol (Oro). (**d**) MS/MS fragmentation of Oro. The colored flags point out the major fragments. +EIC = extracted ion chromatogram of positive mode

Table 1
Solvent composition and LC gradient condition for SL identification

Reagent	Final concentration	Amount	
(A) Water	N/A	500 mL	
(B) Acetonitrile	N/A	500 mL	
Formic acid	0.1% (V/V)		
Time	Flow (mL/min)	%A	%B
0	0.5	75	25
15	0.5	0	100
20	0.5	0	100
21	0.5	75	25
23	0.5	75	25

Table 2
General settings of MS

Parameters	Values
Ionization	Heated electrospray ionization
Spray voltage	Static
Positive ion (V)	3500
Negative ion (V)	2500
Gas mode	Static
Sheath gas (Arb)	60
Aux gas (Arb)	15
Sweep gas (Arb)	2
Ion transfer tube temperature (°C)	350
Vaporized temperature (°C)	400
Detector type	Orbitrap
Orbitrap resolution	120,000
Use quadrupole isolation	TRUE
Scan range (m/z)	50–500
RF lens (%)	60
AGC target	Standard
Microscans	1
Data type	Centroid

4. Prepare solvents as described in Table 1.
5. Install the running solvent lines into the UHPLC solvent reservoirs.
6. Purge the solvent lines for 5 min.
7. Equilibrate the Orbitrap ID-X Tribrid system as shown in Tables 2 and 3.
8. Create a batch table.
9. Analyze samples in both positive and negative ion modes.
10. Injection volumes can vary from 5 μL to 10 μL, depending on sample concentrations.
11. Process the data using Xcalibur™ Software.
12. Use the elemental composition to calculate the theoretical accurate mass, e.g., 331.15400 for 4DO and 347.14865 for Oro (Fig. 4a–c), and search from the mass data using Xcalibur™ Software with 5 ppm mass tolerance.

Table 3
General settings of MS/MS

Parameters	Values
Isolation mode	Quadrupole
Isolation window (m/z)	1.6
Isolation offset	Off
Activation type	HCD
Collision energy mode	Assisted
HCD assisted collision energies (%)	10, 20, 30, 40
Detection type	Orbitrap
Orbitrap resolution	60,000
Scan range mode	Auto
AGC target	Standard
Microscans	1
Data type	Centroid

13. Check MS/MS data to make sure to identify with the authentic standard(s) (Fig. 4b, d). The MS/MS spectrum of authentic 4DO, Oro, 4-oxo-19-OH-CL, 4-oxo MeCLA, GR24, putative Oxo-CL, putative 18-hydroxy-4-oxo-MeCLA, putative pennilactones can be found in Fig. 4 and in our published literature [16, 19, 20, 22, 26, 34]. Besides, the Oro and Orobanchyl acetate showed the same MS/MS, which could be distinguished by retention time [34].

3.5 SL Quantification by LC-MS/MS and Statistical Analysis

A highly specific and sensitive method has been accomplished to selectively quantify compounds within complex mixtures by quadrupole MS analyzer. Quantification of SLs was performed by the UHPLC-TSQ-Altis Mass spectrometer using Multiple Reaction Monitoring (MRM) experiment (Fig. 5) (*see* **Note 14**).

1. Separate the metabolites using a Hypersil GOLD™ C_{18} Selectivity HPLC column (150 × 4.6 mm; 3 μm), maintained at 35 ° C. The mobile phases consist of (A) 100% LC-MS grade water with 0.1% formic acid and (B) 100% Acetonitrile with 0.1% formic acid are used for gradient elution (Table 4).
2. Perform the identification of SLs using an UHPLC system coupled with a Triple-Stage Quadrupole Mass Spectrometer (TSQ-Altis™), which runs in positive mode. The general MS parameters of MRM are listed in Table 5, and the ion transitions (MRM) of SLs are listed in Table 6.

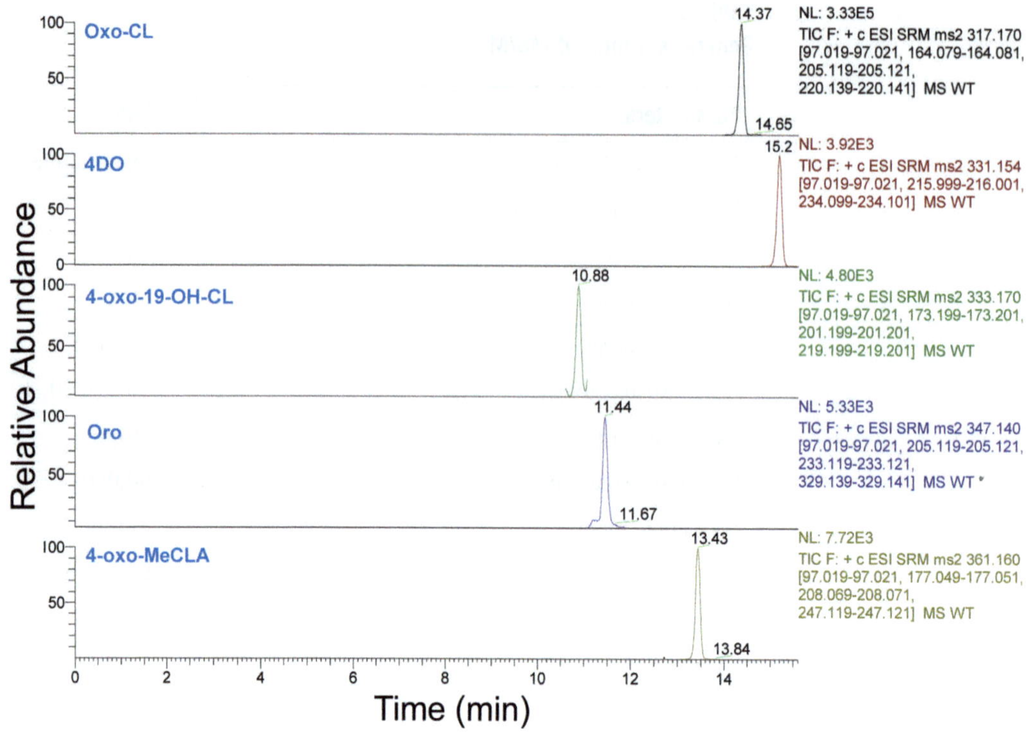

Fig. 5 Overview of MRM chromatography from TSQ-Altis Mass spectrometer. Oxo-CL; oxo-carlactone; 4DO, 4-deoxyorobanchol; 4-oxo-19-OH-CL, 4-oxo-19-hydroxyl-carlactone; Oro, Orobanchol; 4-oxo-MeCLA, methyl 4-oxo-carlactonoate

Table 4
Solvent composition and LC gradient condition for MRM analysis

Reagent	Final concentration	Amount	
(A) Water	N/A	500 mL	
(B) Acetonitrile	N/A	500 mL	
Formic acid	0.1% (V/V)		
Time	Flow (mL/min)	%A	%B
0	0.5	75	25
15	0.5	0	100
20	0.5	0	100
21	0.5	75	25
25	0.5	75	25

Table 5
MS parameter settings of SRM analysis

Parameters	Values
Ionization	Heated electrospray ionization
Spray voltage	Static
Positive ion (V)	4000
Negative ion (V)	3500
Sheath gas (Arb)	40
Aux gas (Arb)	15
Sweep gas (Arb)	5
Ion transfer tube temperature (°C)	350
Vaporized temperature (°C)	350
Cycle time (Sec)	0.8
Q1 resolution (FWHM)	0.2
Q3 resolution (FWHM)	0.2
CID gas (mTorr)	2
Chromatographic peak width (Sec)	6

3. Prepare solvents as described in Table 4.
4. Install the running solvent lines into the UHPLC solvent reservoirs.
5. Purge the solvent lines for 5 min.
6. Equilibrate the Triple-Stage Quadrupole Mass Spectrometer (TSQ-Altis™) system as shown in Table 5.
7. Create a batch table and insert pooled QC samples once every ten samples (*see* **Note 15**).
8. Analyze samples in positive ion modes. Injection volumes can be changed from 5 μL to 10 μL, depending on the sample concentrations (*see* **Note 16**).
9. Analyze data in Xcalibur™ Software.
10. Create a processing method based on the MRM results before batch analysis. Output automatic integrative peak area results as a CSV file for calculation (*see* **Note 17**).
11. In this case study, the quantification of analytes is summarized in Table 7, while the anticipated outcomes are detailed in **Note 18**.

Table 6
The MRM detection parameters of SLs

Strigolactones	Monitoring transitions (m/z; $[M + H]^+$) Precursor ion	Diagnostic product ion	Confirming product ion
D_6–5-Deoxystrigol (D_6–5DS)	337.19	97.028	240.16 | 222.15
GR24	299.08	97.028	157.06 | 185.05
4-Deoxyorobanchol (4DO)	331.15	97.028	234.1 | 216.0
Orobanchol (Oro)	347.14	97.028	329.14 | 233.12 | 205.12
Methyl 4-oxo-carlactonoate (4-oxo-MeCLA)	361.16	97.028	247.12 | 208.07 | 177.05
4-oxo-19-hydroxyl-carlactone (4-oxo-19-OH-CL)	333.17	97.028	219.2 | 201.2 | 173.2
Putative oxo-carlactone (Oxo-CL)	317.17	97.028	220.14 | 205.12 |164.08
Putative Pennilactone1	359.15	97.028	345.13 | 299.12 | 249.1 | 203.1
Orobanchyl acetate (Oro-acetate)	389.15 411.1 $[M + Na]^+$	97.028 97.028	329.14 | 233.12 | 205.12 351.1 | 254.1

4 Notes

1. The pH of the half-strength modified Hoagland nutrient solution should be readjusted to 5.8 every time before use. The growth medium (½ MS medium and half-strength modified Hoagland nutrient solution) can be kept at room temperature for up to 1 month.
2. (a) Sterilize the rice or Pearl millet seeds in a 50 mL falcon tube with 50% commercial bleach (sodium hypochlorite) solution (Milli-Q water + commercial bleach; 1:1 v/v) solution. (b) Invert the tube in a rotator for 10 min continuously to sterilize the surface of the seeds. Optional: For pearl millet (*Pennisetum glaucum*) seeds, sterilize for only 5 min. (c) In a sanitized laminar flow cabinet, wash the seeds successively 5–6 times with sterilized Milli-Q water. (d) For imbibition, keep the seeds in the falcon tube with 30 mL sterilized Milli-Q water overnight in an incubator, in the dark at 30 °C.

Table 7
Example of MRM raw data from LC-MS/MS (Sample collected volume 50 mL)

Sample	Analyte name	Analyte peak area (counts)	Analyte retention time (min)	IS peak name	IS peak area (counts)	IS retention time (min)	Content (pg/mL)
d17-1	4-DeoxyOrobanchol	N/A	N/A	GR24	6.23E+06	12.04	N/A
d17-2	4-DeoxyOrobanchol	N/A	N/A	GR24	6.79E+06	12.06	N/A
d17-3	4-DeoxyOrobanchol	N/A	N/A	GR24	7.14E+06	12.02	N/A
d17-4	4-DeoxyOrobanchol	N/A	N/A	GR24	7.56E+06	12.04	N/A
WT-1	4-DeoxyOrobanchol	5.44E+05	15.21	GR24	6.36E+06	12.03	34.21
WT-2	4-DeoxyOrobanchol	5.68E+05	15.19	GR24	7.68E+06	12.02	29.58
WT-3	4-DeoxyOrobanchol	5.89E+05	15.20	GR24	7.24E+06	12.03	32.54
WT-4	4-DeoxyOrobanchol	5.90E+05	15.21	GR24	7.34E+06	12.05	32.15
d17-1	Orobanchol	N/A	N/A	GR24	6.23E+06	12.04	N/A
d17-2	Orobanchol	N/A	N/A	GR24	6.79E+06	12.06	N/A
d17-3	Orobanchol	N/A	N/A	GR24	7.14E+06	12.02	N/A
d17-4	Orobanchol	N/A	N/A	GR24	7.56E+06	12.04	N/A
WT-1	Orobanchol	1.12E+06	11.44	GR24	6.36E+06	12.03	70.44
WT-2	Orobanchol	1.56E+06	11.41	GR24	7.68E+06	12.02	81.25
WT-3	Orobanchol	1.67E+06	11.42	GR24	7.24E+06	12.03	92.27
WT-4	Orobanchol	1.83E+06	11.44	GR24	7.34E+06	12.05	99.73
d17-1	4-oxo MeCLA	N/A	N/A	GR24	6.23E+06	12.04	N/A
d17-2	4-oxo MeCLA	N/A	N/A	GR24	6.79E+06	12.06	N/A
d17-3	4-oxo-MeCLA	N/A	N/A	GR24	7.14E+06	12.02	N/A
d17-4	4-oxo-MeCLA	N/A	N/A	GR24	7.56E+06	12.04	N/A
WT-1	4-oxo-MeCLA	1.56E+06	13.43	GR24	6.36E+06	12.03	98.11
WT-2	4-oxo-MeCLA	1.78E+06	13.43	GR24	7.68E+06	12.02	92.71
WT-3	4-oxo-MeCLA	1.50E+06	13.45	GR24	7.24E+06	12.03	82.87
WT-4	4-oxo-MeCLA	1.67E+06	13.41	GR24	7.34E+06	12.05	91.01

3. For pearl millet, place around 20 seeds per 100 mm × 15 mm Petri dish, containing two filter papers moistened with 5 mL of ½ MS medium.
4. For pearl millet, transfer seven germinated seedlings to a 2 L pot filled with silver sand and apply 1 L + Pi half-strength modified Hoagland nutrient solution twice per week for 4 weeks.

5. For plants grown in the sand system, we recommend collecting the exudate 24 h after pre-washing the pots with 1 L −Pi half-strength modified Hoagland solution.
6. If working with the sand system, we recommend collecting 500 mL to 1000 mL root exudates by adding −Pi/low Pi half-strength modified Hoagland solution on the top of each pot and collecting the flow-through, from the bottom of the pot, in 1 L plastic bottle. During the collection, protect the collected root exudates from light to avoid SL degradation. We also recommend extracting the collected samples as soon as possible without any delay, as SLs degrade quickly in the water when exposed to light and high temperatures.
7. If the collected exudate is to be applied to parasitic seeds, do not add internal standards. The latter will trigger high germination.
8. Following preconditioning, ensure the columns remain wetted with water and do not let them dry before loading the sample. This step prepares the column for effective analyte adsorption, and maintaining the adsorbent in a wetted state with an appropriate solvent ensures reproducibility. We recommend pooling at least 1000 mL root exudate for better MS resolution and sensitivity for SL identification. If the volume of root exudate exceeds 200 mL, we strongly advise adding a pre-step, using glass fiber filter paper through a vacuum assembly, to remove impurities and plant debris.
9. We recommend vacuum-drying Fraction A and B together if space allows.
10. CRITICAL: Do not touch the lower layer to avoid water contamination.
11. Pause point: Keep the extracts in −20 °C refrigerator if you do not want to process them immediately.
12. Pause point: Keep the filtered samples in a − 20 °C refrigerator for storage.
13. The IDX Orbitrap Mass Spectrometer is an analytical technique that contains three mass analyzers used for determining the m/z of small and big molecules. The Orbitrap IDX spectrometer could reach a high resolution (> 120,000) and reliable mass accuracy (<3 ppm mass error). The Mass spectrometer was calibrated using a purchasable "Calibration Mix ESI (Thermo Scientific)" and by following the manufacturer's guidelines. Electrospray ionization in positive mode (ESI+) was applied for the studied compounds, using the following parameters: vaporized temperature = 100 °C, voltage = 3500 V, sheath gas = 30, auxiliary gas: 15, ion source fragmentation = 35 V, capillary temperature = 300 °C. 10 μL

of each sample were injected through a loop injection to a C_{18} column using an independent UPLC pump.

14. The TSQ Altis is a triple-stage quadrupole Mass Spectrometer that determines the m/z of small molecules and provides the best transmission and peak shape. This instrument has an electrospray ionization source and is capable of scan functions, such as Selective Ion Monitoring (SIM), Selective/Multiple Reaction Monitoring (S/MRM), and precursor ion scanning. The Mass spectrometer was calibrated using a purchasable "Calibration Mix ESI (Thermo Scientific)" by following the manufacturer's guidelines.
15. Optional: Insert authentic standard samples, if having any, at the beginning and end of running.
16. CRITICAL: To prevent retention time-shifting, the same batch of samples should be analyzed within 48 h in the same run.
17. Alternatively, if no IS is used, simply measure the peak area of the selected analyte and compare it to the total average peak area across all the studied genotypes.

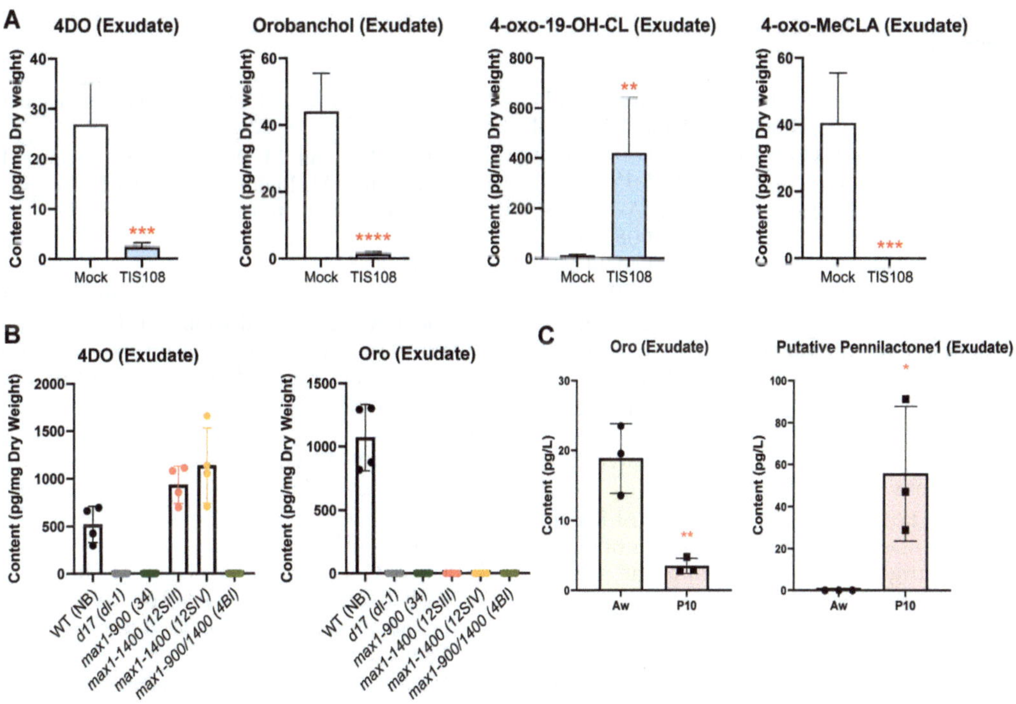

Fig. 6 Expected results obtained from this protocol. (**a**) Evaluation of the CYP450 inhibitor TIS108 effect by SL quantification (Adapted from Ito et al., 2022 [19]). (**b**) SL quantification among different rice SL mutants (Adapted from Chen et al., 2023 [21]). (**c**) SL quantification between two pearl millet genotypes (Adapted from Kuijer et al., 2024 [26]). WT, wild type; *d17*, *dwarf17*; *max1*, *MORE AXILLARY GROWTH1*; Aw, 29AW; P10, SOSAT-C88 P10

18. Expected outcomes: This protocol enables the unequivocal identification of endogenous SLs by comparing MS/MS patterns with authentic standards. Additionally, it allows for the evaluation of synthetic compounds, such as zaxinone mimics (MiZax) and TIS108, which regulate SL biosynthesis and release [19, 30, 35] (Fig. 6a). The protocol also facilitates the quantification of SL levels across different plant species and genotypes [21, 26] (Fig. 6b, c). Furthermore, by applying collected root exudates to Striga seeds—a method significantly more sensitive than LC-MS detection—we can directly correlate bioassay results (detailed in [33, 34]) with LC-MS findings. This approach further helps confirm the bioactivity of released SLs in the rhizosphere.

Acknowledgments

This work was supported partly by Academia Sinica to J. Y. W and by baseline funding and the Competitive Research Grants (CRG2024) Program from King Abdullah University of Science and Technology (KAUST) to S. A-B.

References

1. Gomez-Roldan V, Fermas S, Brewer PB, Puech-Pagès V, Dun EA, Pillot J-P et al (2008) Strigolactone inhibition of shoot branching. Nature 455(7210):189–194
2. Umehara M, Hanada A, Yoshida S, Akiyama K, Arite T, Takeda-Kamiya N et al (2008) Inhibition of shoot branching by new terpenoid plant hormones. Nature 455(7210):195–200
3. Al-Babili S, Bouwmeester HJ (2015) Strigolactones, a novel carotenoid-derived plant hormone. Annu Rev Plant Biol 66:161–186
4. Fiorilli V, Wang JY, Bonfante P, Lanfranco L, Al-Babili S (2019) Apocarotenoids: old and new mediators of the arbuscular mycorrhizal symbiosis. Front Plant Sci 10:1186
5. Wang JY, Chen G-TE, Braguy J, Al-Babili S (2024) Distinguishing the functions of canonical strigolactones as rhizospheric signals. Trends Plant Sci
6. Yoneyama K, Xie X, Yoneyama K, Kisugi T, Nomura T, Nakatani Y et al (2018) Which are the major players, canonical or non-canonical strigolactones? J Exp Bot 69(9):2231–2239
7. Abuauf H, Haider I, Jia K-P, Ablazov A, Mi J, Blilou I et al (2018) The Arabidopsis DWARF27 gene encodes an all-trans–/9-cis-β-carotene isomerase and is induced by auxin, abscisic acid and phosphate deficiency. Plant Sci 277:33–42
8. Alder A, Jamil M, Marzorati M, Bruno M, Vermathen M, Bigler P et al (2012) The path from β-carotene to carlactone, a strigolactone-like plant hormone. Science 335(6074):1348–1351
9. Bruno M, Hofmann M, Vermathen M, Alder A, Beyer P, Al-Babili S (2014) On the substrate-and stereospecificity of the plant carotenoid cleavage dioxygenase 7. FEBS Lett 588(9):1802–1807
10. Seto Y, Sado A, Asami K, Hanada A, Umehara M, Akiyama K et al (2014) Carlactone is an endogenous biosynthetic precursor for strigolactones. Proc Natl Acad Sci U S A 111(4):1640–1645
11. Chen G-TE, Wang JY, Jamil M, Braguy J, Al-Babili S (2022) 9-cis-β-Apo-10′-carotenal is the precursor of strigolactones in planta. Planta 256(5):1–7
12. Wang JY, Chen GTE, Balakrishna A, Jamil M, Berqdar L, Al-Babili S (2024) Strigolactone biosynthesis in rice can occur via a 9-cis-3-OH-10′-apo-β-carotenal intermediate. FEBS Lett 598(5):571–578
13. Yoneyama K, Mori N, Sato T, Yoda A, Xie X, Okamoto M et al (2018) Conversion of

carlactone to carlactonoic acid is a conserved function of MAX 1 homologs in strigolactone biosynthesis. New Phytol 218(4):1522–1533

14. Wakabayashi T, Hamana M, Mori A, Akiyama R, Ueno K, Osakabe K et al (2019) Direct conversion of carlactonoic acid to orobanchol by cytochrome P450 CYP722C in strigolactone biosynthesis. Sci Adv 5(12): eaax9067
15. Li C, Dong L, Durairaj J, Guan J-C, Yoshimura M, Quinodoz P et al (2023) Maize resistance to witchweed through changes in strigolactone biosynthesis. Science 379(6627):94–99
16. Li C, Haider I, Wang JY, Quinodoz P, Suarez Duran HG, Méndez LR et al (2024) OsCYP706C2 diverts rice strigolactone biosynthesis to a noncanonical pathway branch. Sci Adv 10(35):eadq3942
17. Zhou A, Kane A, Wu S, Wang K, Santiago M, Ishiguro Y et al (2025) Evolution of interorganismal strigolactone biosynthesis in seed plants. Science 387(6731):eadp0779
18. Zhang Y, Van Dijk AD, Scaffidi A, Flematti GR, Hofmann M, Charnikhova T et al (2014) Rice cytochrome P450 MAX1 homologs catalyze distinct steps in strigolactone biosynthesis. Nat Chem Biol 10(12):1028–1033
19. Ito S, Braguy J, Wang JY, Yoda A, Fiorilli V, Takahashi I et al (2022) Canonical strigolactones are not the major determinant of tillering but important rhizospheric signals in rice. Sci Adv 8(44):eadd1278
20. Wang JY, Braguy J, Chen GTE, Jamil M, Balakrishna A, Berqdar L et al (2022) Perspectives on the metabolism of strigolactone rhizospheric signals. Front Plant Sci 13:1062107
21. Chen G-TE, Wang JY, Votta C, Braguy J, Jamil M, Kirschner GK et al (2023) Disruption of the rice 4-deoxyorobanchol hydroxylase unravels specific functions of canonical strigolactones. Proc Natl Acad Sci 120(42): e2306263120
22. Wang JY, Balakrishna A, Martínez C, Chen GTE, Sioud S, de Lera AR et al (2024) The rice orobanchol synthase catalyzes the hydroxylation of the noncanonical strigolactone methyl 4-oxo-carlactonoate. New Phytol 244(6):2121
23. Brewer PB, Yoneyama K, Filardo F, Meyers E, Scaffidi A, Frickey T et al Lateral Branching Oxidoreductase acts in the final stages of strigolactone biosynthesis in Arabidopsis. Proc Natl Acad Sci USA 113(22):6301–6306
24. Yoda A, Mori N, Akiyama K, Kikuchi M, Xie X, Miura K et al (2021) Strigolactone biosynthesis catalyzed by cytochrome P450 and sulfotransferase in sorghum. New Phytol 232(5):1999–2010
25. Mashiguchi K, Seto Y, Onozuka Y, Suzuki S, Takemoto K, Wang Y et al (2022) A carlactonoic acid methyltransferase that contributes to the inhibition of shoot branching in Arabidopsis. Proc Natl Acad Sci U S A 119(14): e2111565119
26. Kuijer HN, Wang JY, Bougouffa S, Abrouk M, Jamil M, Incitti R et al (2024) Chromosome-scale pearl millet genomes reveal CLAMT1b as key determinant of strigolactone pattern and Striga susceptibility. Nat Commun 15(1):6906
27. Boutet-Mercey S, Perreau F, Roux A, Clavé G, Pillot JP, Schmitz-Afonso I et al (2018) Validated method for strigolactone quantification by ultra high-performance liquid chromatography–electrospray ionisation tandem mass spectrometry using novel deuterium labelled standards. Phytochem Anal 29(1): 59–68
28. Wang JY, Haider I, Jamil M, Fiorilli V, Saito Y, Mi J et al (2019) The apocarotenoid metabolite zaxinone regulates growth and strigolactone biosynthesis in rice. Nat Commun 10(1): 1–9
29. Shrestha JK, Lin C, Wang JY, Tang I, Hu C, Tsednee M, Zhang Y, Jamil M, Berqdar L, Blilou I, Wang C, Yeh K (2026) Carotenoid biosynthesis drives root plasticity through aerenchyma and iron plaque formation in rice. Nature Plants, 1–12. https://doi.org/10.1038/s41477-025-02170-y
30. Wang JY, Jamil M, Lin P-Y, Ota T, Fiorilli V, Novero M et al (2020) Efficient mimics for elucidating zaxinone biology and promoting agricultural applications. Mol Plant 13(11): 1654–1661
31. Wang JY, Chen G-TE, Braguy J, Jamil M, Berqdar L, Al-Babili S (2023) Disruption of the cytochrome CYP711A5 gene reveals MAX1 redundancy in rice strigolactone biosynthesis. J Plant Physiol:154057
32. Mutinda S, Jamil M, Wang JY, Berqdar L, Ateka E, Bellis ES et al (2025) Strigolactone biosynthesis lgs1 mutant alleles mined from the sorghum accession panel are a promising resource of resistance to witchweed (Striga) parasitism. Plants People Planet 7(2):382–395
33. Braguy J, Ramazanova M, Giancola S, Jamil M, Kountche B A, Zarban R, Felemban A, Wang JY, Lin P, Haider I, Zurbriggen M, Ghanem, B (2021) SeedQuant: A deep learning-based tool for assessing stimulant and inhibitor activity on root parasitic seeds. Plant Physiol 186 (3):1632–1644. https://doi.org/10.1093/plphys/kiab173

34. Wang JY, Chen GE, Jamil M, Braguy J, Sioud S, Liew KX, Balakrishna A, Al- Babili S (2022) Protocol for characterizing strigolactones released by plant roots. STAR Protoc 3 (2):101352. https://doi.org/10.1016/j.xpro.2022.101352

35. Jamil M, Lin P-Y, Berqdar L, Wang JY, Takahashi I, Ota T et al (2023) New series of Zaxinone mimics (MiZax) for fundamental and applied research. Biomolecules 13(8):1206

36. Stra A, Moreno JC, Wang JY et al (2025) LYCOPENE β-CYCLASE overexpression improves growth, modulates hormone content, and affects rhizospheric interactions in tobacco and tomato roots. Plant Cell Rep 44:240. https://doi.org/10.1007/s00299-025-03623-6

Chapter 5

LC-MS Analysis and Quantification of Plants Apocarotenoids

Mohamed Salem, Kawthar F. Alashoor, Kit Xi Liew, and Salim Al-Babili

Abstract

Carotenoid oxidative cleavage products (apocarotenoids; APOs) represent an important class of specialized metabolites that exert diverse and important biological functions. Apocarotenoids are precursors of the plant hormones abscisic acid and strigolactones, pigments such as crocin and picrocrocin, and volatiles such as β-ionone and β-cyclocitral. Recently, carotenoid-derived bioactive metabolites such as zaxinone, anchorene, β-cyclocitral, and retinal have been identified as plant root growth regulators. Analysis of plant APOs is crucial for elucidating their biological function and metabolism as well as for determining their content in food. In this chapter, we describe an efficient protocol for a simultaneous analysis of plant APOs utilizing ultra-high performance liquid chromatographic (UPLC) separation and tandem mass spectrometry (MS). Additionally, we introduce practical details to assist researchers in setting up their experiments, extraction, and analysis of APOs.

Key words Apocarotenoids, Liquid chromatography, Mass spectrometry, Qualitative and quantitative analysis, Rice, Zaxinone

1 Introduction

Carotenoids are C_{40} terpenoid pigments with an extended conjugated double bond system. They contribute to different aspects of plant's life, including growth and development, stress response and biotic interactions, and essential components of the photosynthetic apparatus [1–3]. The carotenoid's conjugated double bond system makes them susceptible to oxidative cleavage via carotenoid cleavage dioxygenases (CCDs) or by nonenzymatic means, in some cases [4, 5]. This process produces bioactive compounds known as apocarotenoids, which include the precursors of as the phytohormones abscisic acid (ABA) and strigolactones (SLs), as well as compounds like β-cyclocitral, β-ionone, and 3-OH-β-apo-13-carotenone (Zaxinone) that regulate plant growth, development, and response to biotic and abiotic stress (Fig. 1) [1, 4]. Several pigments such as crocins, bixins, and citraulin belong also to the apocarotenoid class

Shahid Mukhtar (ed.), *Plant Hormones: Methods and Protocols*, Methods in Molecular Biology, vol. 3026,
https://doi.org/10.1007/978-1-0716-5214-5_5,

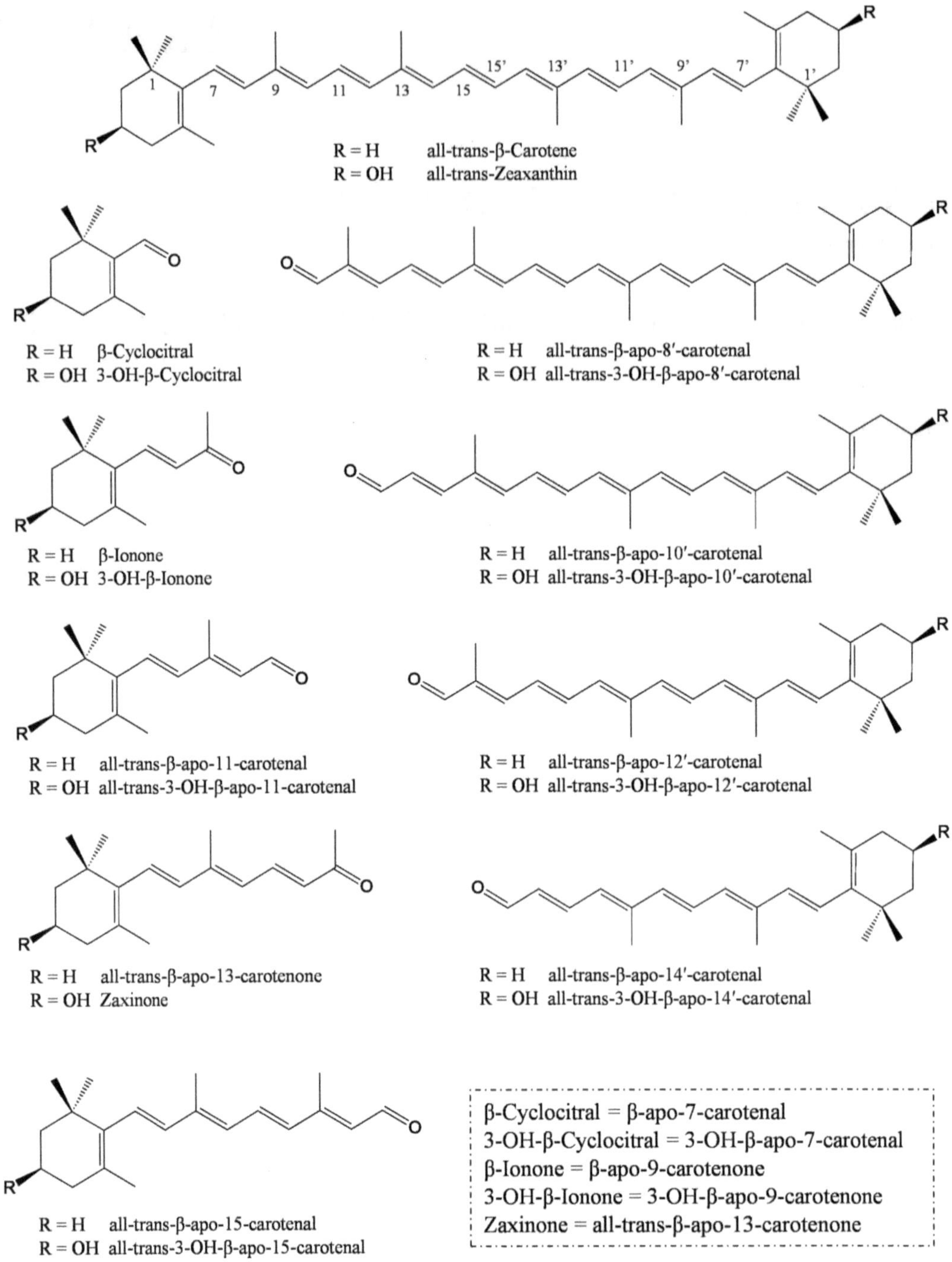

Fig. 1 Representative examples and chemical structures of apocarotenoids (APOs) that can be identified and quantified using the described protocol

of compounds. Moreover, carotenoids and long-chain apocarotenoids are sources of provitamin A and thus, apocarotenoids are not only important for biological processes in plants but are also important for human health and nutrition [1, 6–8].

Due to their diverse roles, the analysis and quantification of apocarotenoids in plants and crops helps in understanding their biological functions and aids in the development of more nutritional crops [4, 9]. Ultra-high performance liquid chromatography (UHPLC) paired with tandem mass-spectrometry (MS) is a flexible analytical tool for the identification and quantification of these compounds [10–13]. In the case of apocarotenoids in plant extracts, this quantification is highly dependent on the extraction method, as apocarotenoids are present in low amounts. A reliable and reproducible method of extraction and analysis of apocarotenoids is thus necessary.

In this chapter, we describe an optimized protocol for the extraction, identification, and quantification of apocarotenoids from plant extracts [10–13] and present a list of apocarotenoids detected from rice tissue.

2 Materials

Use ultrapure water and high-quality chemicals (HPLC grade or above) for the extraction. Perform all the extraction process in a fume hood and wear all suitable protective equipment (*see* **Note 1**). Keep all the solvent bottles, measuring cylinders and pipettes free from dust and detergents.

2.1 Plant Material Harvest and Sampling

1. Plant material of interest.
2. Clean scissors and forceps.
3. Liquid nitrogen.
4. Harvesting tubes (e.g., 2 mL amber color microcentrifuge tubes).
5. Metal balls.
6. Tissue grinder.
7. −80 °C freezer.
8. Analytical balance.
9. Freeze dryer.

2.2 Apocarotenoids Extraction

1. Extraction solution of methanol—0.1% (w/v) butylated hydroxytoluene (BHT).
2. Extraction tubes (e.g., 4 mL amber color glass vials with screw caps).
3. Vortexer.

4. Sonicator.
5. Centrifuge.
6. Nitrogen drying apparatus or vacuum concentrator.

2.3 LC-MS Analysis

1. LC-MS system (e.g., Vanquish™ Duo UHPLC System coupled to TSQ Altis™ triple quadrupole, with a heated electrospray ionization source, Thermo Scientific, USA).
2. HPLC C18 column (e.g., ACQUITY UPLC BEH C18 column, 100 x 2.1 mm, 1.7 mm with an UPLC BEH C18 guard column, 5 x 2.1 mm, 1.7 mm).
3. HPLC glass vial and screw caps.
4. Resuspension solvent (90% aqeuous acetonitrile).
5. Mobile Phase eluent (A): water, acetonitrile and 0.1% formic acid (80, 20, 0.1, v/v/v) (*see* **Note 2**).
6. Mobile Phase eluent (B): acetonitrile, isopropanol and 0.1% formic acid (60, 40, 0.1, v/v/v) (*see* **Note 3**).
7. Computer and software for LC/MS data analysis (e.g., Xcalibur by Thermo Scientific).

3 Methods

This chapter introduces an optimized protocol for the qualitative and quantitative analysis of plant apocarotenoids (Fig. 2). We here

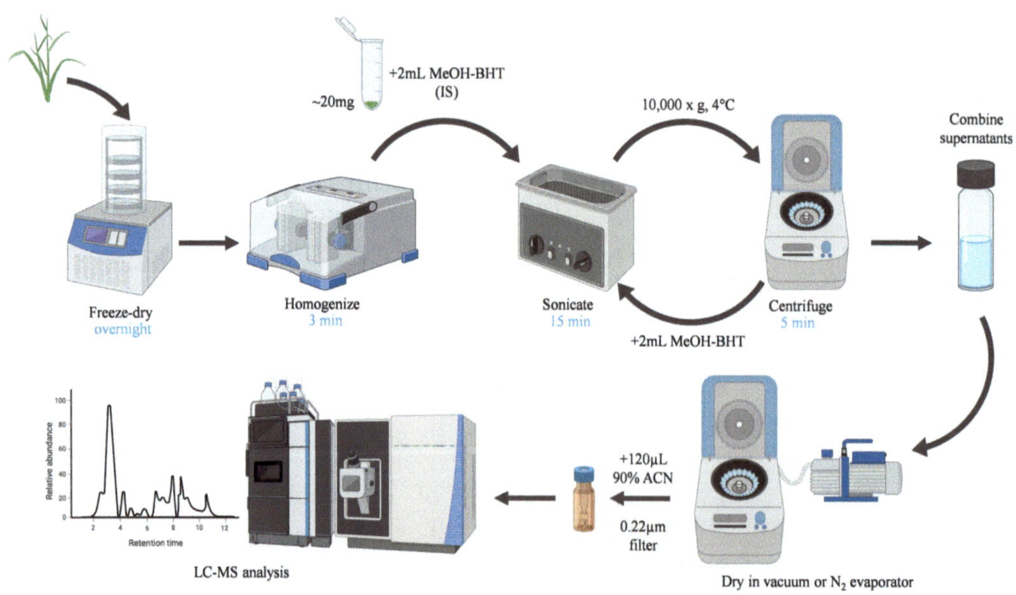

Fig. 2 Schematic representation of the workflow for the extraction and subsequent LC/MS analysis of apocarotenoids (APOs) from plant tissues (Partially created in BioRender)

utilize rice leaves; however, the protocol works for other common model plants such as Arabidopsis, tomato, and other crops. Apocarotenoids can be influenced by growth conditions and are largely different qualitatively and quantitatively between tissue types. Perform all the extraction procedures under cold conditions on ice, unless otherwise specified. Whenever possible, use amber-colored glassware for extraction.

3.1 Plant Cultivation and Sampling

1. Grow rice plants under required growth conditions (*see* **Note 4**).
2. Harvest leaves using a clean scissor and forceps.
3. Place the collected leaves into 2.0-mL microcentrifuge tubes and freeze the tubes immediately in liquid nitrogen (*see* **Note 5**).
4. Store the frozen harvested leaves at −80 °C until further use.
5. Freeze-dry the samples overnight to ensure complete removal of water.
6. Put two or three metal balls into each microcentrifuge tube and close the tube properly.
7. Grind the dry leaves with tissue grinder for 3 min at 25 Hz.
8. Aliquot and transfer 20 ± 2 mg the dry leaf powder into the extraction tubes, record the sample weight.

3.2 Extraction of Apocarotenoids

1. Add 2 mL extraction solvent to each extraction tube (*see* **Note 6**).
2. Prepare the internal standard mixture and spike it into each sample prior to extraction (*see* **Note 7**).
3. Vortex the samples for 1 min and put the samples into ice-cooled ultrasonic bath for 15 min.
4. Centrifuge the tubes for 5 min at 10,000× *g* at 4 °C.
5. Transfer the supernatant to a new tube.
6. Resuspend the pellet with 2 mL extraction solvent.
7. Vortex the samples for 1 min and put the samples into ice-cooled ultrasound bath for 15 min.
8. Centrifuge the tubes for 5 min at 10,000× *g* at 4 °C.
9. Combine the supernatants into one tube.
10. Dry the samples using nitrogen drying apparatus or vacuum concentrator.

3.3 LC-MS Analysis of Apocarotenoids

1. Resuspend samples in 0.12 mL of 90% aqueous acetonitrile and filter through a 0.22 mm filter.
2. Transfer the resuspended samples to LC glass vials.
3. Put all samples in the autosampler tray.

4. Set up the LC-MS system for apocarotenoid analysis (*see* **Note 8**).
5. Maintain the column temperature at 35 °C.
6. Use the following gradient program: 0–1 min, 100% A; 1–3 min, 100% A to 60% A; 3–8 min, 60% A to 20% A; 8–14 min, 20% A to 10% A; 14–15 min, 10% A to 0% A; followed by washing with 0% A and equilibration with 100% A.
7. Set the flow rate was 0.2 mL/min and the injection volume to 5 μL.
8. Use the following MS parameters: sheath gas flow rate of 40 arbitrary units, auxiliary gas flow rate of 20 arbitrary units, positive electrospray ionization mode, spray voltage of 4.0 kV, capillary temperature of 350 °C, auxiliary gas heater temperature of 250 °C (*see* **Note 9**).

3.4 Data Analysis

1. Check the quality of the LC/MS obtained data with software provided by the MS vendor.
2. Analyze the LC/MS data with software provided by the MS vendor. Identify plant APOs by coelution profiling and mass fragmentation with the running authentic standard compound. In Table 1, we present an example of apocarotenoids that can be detected. An exemplary description is shown in Fig. 3.
3. Export the signal intensity data of the detected apocarotenoid peaks.
4. Perform quantifications for the detected APOs as follows:

$$\text{Amount}_{[\text{target compound}]} = \text{Area}_{[\text{target compound}]} / \text{Area}_{[\text{spiked IS}]} \times \text{Amount}_{[\text{spiked IS}]}.$$

4 Notes

1. Organic solvents like methanol, acetonitrile, and isopropanol are toxic and should be handled with care under the fume hood. Formic acid is corrosive and should be handled under the fume hood.
2. Combine 200 mL of acetonitrile with 1 mL of formic acid, water should be added in increments to reach 1000 mL final solution. This solution should be freshly prepared before analysis.
3. Combine 400 mL of isopropanol with 1 mL of formic acid, acetonitrile should be added in increments to reach 1000 mL final solution. This solution should be freshly prepared before analysis.

Table 1
Liquid chromatography-mass spectrometry (LC-MS) analysis of representative APOs from plant samples [11]

Name	RT (min)	Chemical formula	$(M + H)^+$	MRM transitions
3-OH-β-ionone	5.54	$C_{13}H_{20}O_2$	209.1	109.0, 135.0, 191.1
3-OH-β-apo-11-Carotenal	6.32	$C_{15}H_{22}O_2$	235.2	119.1, 161.1, 217.2
3-OH-β-apo-13-Carotenone	7.14	$C_{18}H_{26}O_2$	275.2	135.1159.1, 173.2, 257.2
3-OH-β-apo-15-Carotenal	7.95	$C_{20}H_{28}O_2$	301.2	161.1, 173.2, 191.1
β-Cyclocitral	8.10	$C_{10}H_{16}O$	153.1	67.0, 95.1, 109.1
β-Ionone	8.98	$C_{13}H_{20}O$	193.2	95.1, 109.1, 175.1
3-OH-β-apo-12′-Carotenal	9.88	$C_{25}H_{34}O_2$	367.3	105.1, 119.1, 161.1
β-Apo-11-Carotenal	9.99	$C_{15}H_{22}O$	219.2	91.1, 109.1, 201.2
3-OH-β-apo-10′-Carotenal	10.10	$C_{27}H_{36}O_2$	393.3	91.1, 145.1, 175.1
β-Apo-13-Carotenone	10.98	$C_{18}H_{26}O$	259.2	119.1, 175.1, 201.2
3-OH-β-apo-8′-Carotenal	11.30	$C_{30}H_{40}O_2$	433.3	105.1, 119.1, 145.1
β-Apo-15-Carotenal	11.77	$C_{20}H_{28}O$	285.2	119.1, 161.1, 175.1
β-Apo-14′-Carotenal	12.53	$C_{22}H_{30}O$	311.2	95.0, 121.1, 159.1
β-Apo-12′-Carotenal	14.08	$C_{25}H_{34}O$	351.3	119.1, 161.1, 209.1
β-Apo-10′-Carotenal	14.38	$C_{27}H_{36}O$	377.3	105.1, 121.1, 147.1
β-Apo-8′-Carotenal	15.72	$C_{30}H_{40}O$	417.3	119.1, 131.1, 145.1, 157.1

4. For this chapter, three-week-old wild-type rice (*Oryza sativa* "Nipponbare") seedlings were hydroponically grown in 50 mL black tubes supplemented with low phosphate modified Hoagland nutrient solution under controlled conditions (day/night temperature of 28/22 °C with a 12/12 h photoperiod, 200 μmol photons m^{-2} s^{-1}, and 85% relative humidity) [11, 14].
5. Close all the tubes properly ensuring that liquid nitrogen does not enter the tubes. Perform this step as quickly as possible to prevent samples from thawing.
6. Butylated hydroxytoluene (BHT, 0.1%) is incorporated into the extraction solvent to protect apocarotenoids from oxidation.
7. We use internal standards (IS) stock solution containing deuterium-labeled APO standards (1 mg/mL of each standard in ethyl acetate). Working solutions at concentrations of 20 ng/mL can be prepared by diluting the stock solution with MeOH containing 0.1% BHT. For quantitative analysis,

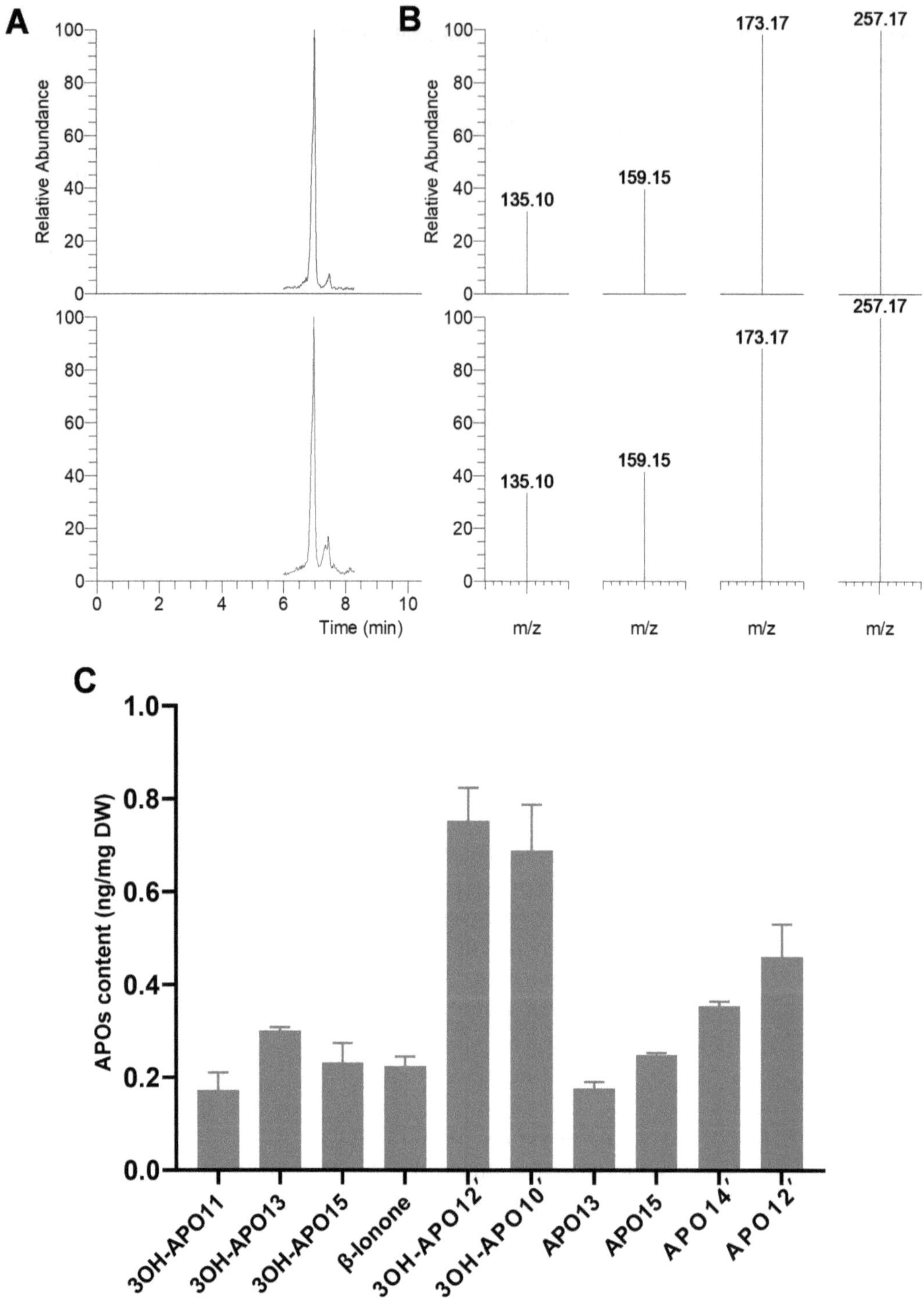

Fig. 3 Detection of apocarotenoids (APOs) from rice leaves by LC/MS. (A) Extracted ion chromatograms (EICs) of standard (top) and rice endogenous (bottom) 3-OH-β-apo-13-carotenone (zaxinone). (B) Mass spectra of 3-OH-β-apo-13-carotenone (zaxinone) standard (top) and rice endogenous compound (bottom). The levels of APOs detected in the leaves of three-week-old rice grown hydroponically under low phosphate conditions. Error bars indicate means ± SDs for three biological replicates

the internal standard mixture is prepared and spiked into each sample prior to extraction, yielding a final concentration equivalent to 2 ng per sample. We use a mix of D1-β-cyclocitral, D3-β-ionone, D3-β-apo-8′-carotenal, D3–βapo-10′-carotenal, D3-β-apo-11-carotenal, D3-β-apo-12′-carotenal, D3-β-apo-13-carotenone, D3–3-OH-β-apo-13-carotenone, and D3-β-apo-15-carotenal.

8. Equilibrate the LC-MS system by measuring proper method blanks.
9. Different MS instruments can be used, so optimal MS conditions must be determined experimentally. For QQQ mass spectrometer, selected ion monitoring (SIM) or multiple reaction monitoring (MRM) modes can be used. For a Quadrupole-Orbitrap (Q-Orbitrap) mass spectrometer, parallel reaction monitoring (PRM) mode can be applied.

Acknowledgments

This work was supported by baseline funding and Competitive Research Grant 2024 (CRG2024) from King Abdullah University of Science and Technology given to S. A.-B.

References

1. Sun T, Rao S, Zhou X, Li L (2022) Plant carotenoids: recent advances and future perspectives. Mol Hortic 2(1):3. https://doi.org/10.1186/s43897-022-00023-2
2. Zheng X, Yang Y, Al-Babili S (2021) Exploring the diversity and regulation of Apocarotenoid metabolic pathways in plants. Front Plant Sci 12. https://doi.org/10.3389/fpls.2021.787049
3. Li L, Rodríguez-Concepción M, Al-Babili S (2025) Advances in carotenoid and apocarotenoid metabolisms and functions in plants. Plant Physiol 198(4):kiaf304. https://doi.org/10.1093/plphys/kiaf304
4. Moreno JC, Mi J, Alagoz Y, Al-Babili S (2021) Plant apocarotenoids: from retrograde signaling to interspecific communication. Plant J 105(2):351–375. https://doi.org/10.1111/tpj.15102
5. Wang JY, Lin P-Y, Al-Babili S (2021) On the biosynthesis and evolution of apocarotenoid plant growth regulators. Semin Cell Dev Biol 109:3–11. https://doi.org/10.1016/j.semcdb.2020.07.007
6. Felemban A, Braguy J, Zurbriggen MD, Al-Babili S (2019) Apocarotenoids involved in plant development and stress response. Front Plant Sci 10. https://doi.org/10.3389/fpls.2019.01168
7. Fiorilli V, Wang JY, Bonfante P, Lanfranco L, Al-Babili S (2019) Apocarotenoids: old and new mediators of the Arbuscular Mycorrhizal Symbiosis. Front Plant Sci 10. https://doi.org/10.3389/fpls.2019.01186
8. Ke D, Guo J, Li K, Wang Y, Han X, Fu W, Miao Y, Jia K-P (2022) Carotenoid-derived bioactive metabolites shape plant root architecture to adapt to the rhizospheric environments. Front Plant Sci 13:2022. https://doi.org/10.3389/fpls.2022.986414
9. Wang JY, Haider I, Jamil M, Fiorilli V, Saito Y, Mi J, Baz L, Kountche BA, Jia K-P, Guo X, Balakrishna A, Ntui VO, Reinke B, Volpe V, Gojobori T, Blilou I, Lanfranco L, Bonfante P, Al-Babili S (2019) The apocarotenoid metabolite zaxinone regulates growth and strigolactone biosynthesis in rice. Nat Commun 10(1):810. https://doi.org/10.1038/s41467-019-08461-1
10. Mi J, Jia K-P, Balakrishna A, Wang JY, Al-Babili S (2019) An LC-MS profiling method reveals a route for apocarotene glycosylation and shows

its induction by high light stress in Arabidopsis. Analyst 144(4):1197–1204. https://doi.org/10.1039/C8AN02143K

11. Mi J, Jia K-P, Wang JY, Al-Babili S (2018) A rapid LC-MS method for qualitative and quantitative profiling of plant apocarotenoids. Anal Chim Acta 1035:87–95. https://doi.org/10.1016/j.aca.2018.07.002
12. Mi J, Liew KX, Al-Babili S (2022) Ultrahigh-performance liquid chromatography–mass spectrometry analysis of carotenoid-derived hormones and apocarotenoids in plants. Curr Protoc 2(2):e375. https://doi.org/10.1002/cpz1.375
13. Mi J, Moreno JC, Alagoz Y, Liew KX, Balakrishna A, Zheng X, Al-Babili S (2022) Chapter nine—ultra-high performance liquid chromatography-mass spectrometry analysis of plant apocarotenoids. In: Wurtzel ET (ed) Methods in enzymology, vol 670. Academic Press, pp 285–309. https://doi.org/10.1016/bs.mie.2021.10.012
14. Berqdar L, Salem MA, Wang JY, Alrudayan A, Jamil M, Al-Babili S (2025) Development of a rapid and efficient protocol for seed germination and seedling establishment of Oryza coarctata. Bio-protocol 15(4):e5203. https://doi.org/10.21769/BioProtoc.5203

Chapter 6

Extraction and Analysis of Cytokinins from Mammalian Tissues by HPLC–HRMS

Muhammad Naseem, Kenesha Wilson, Feras Lafi, Elena Bencurova, Khalid Muhammad, Thomas Dandekar, Alfonso Albacete, and Eman M. Othman

Abstract

Cytokinins (CKs) are adenine-derived plant hormones regulating almost every aspect of plant growth and development. But CK-like compounds may also occur in microbial as well as mammalian systems. Detecting CKs in animal matrices is challenging due to low abundance and complex backgrounds. We describe a high-performance liquid chromatography–high-resolution mass spectrometry (HPLC–HRMS) protocol for CK extraction, purification, and quantification. Samples are frozen or lyophilized, homogenized, and extracted with methanol: water (80:20, v/v) containing stable isotope labeled CKs as internal standards. Solid-phase extraction (SPE) with C18 cartridges removes proteins and lipids, and purified eluates are dried, reconstituted, and analyzed by reversed-phase HPLC coupled to an Orbitrap mass spectrometer. Accurate-mass detection (≤5 ppm) and diagnostic fragmentation confirm identity, while isotope-dilution calibration enables picomolar sensitivity. This method provides high recovery (>75%) and selectivity, enabling investigation of CK occurrence in ex-planta cellular systems.

Key words Plant hormones, Cytokinins, Two-component systems, Active form, Storage form, Signalling molecules, Non-plant systems

1 Introduction

Cytokinins (CKs) are adenine-derived plant hormones central to the regulation of growth and development in plants [1, 2]. While classically studied in a botanical context, evidence suggests that CK-like molecules may occur beyond plants [3]. In animals, potential sources may include endogenous purine metabolism [4], dietary intake [5], or microbial production [6]. However, no endogenous biosynthetic mechanism has yet been identified to account for the source of CKs in mammalian systems. Determining

Shahid Mukhtar (ed.), *Plant Hormones: Methods and Protocols*, Methods in Molecular Biology, vol. 3026, https://doi.org/10.1007/978-1-0716-5214-5_6,

their presence in animal systems is therefore relevant not only to evolutionary biology but also to understanding their implications for nutrigenomics.

Detecting CKs in animal tissues is technically demanding. Concentrations are expected to be extremely low, and animal matrices contain abundant interfering metabolites [7]. Traditional immunoassays or bioassays lack the specificity to distinguish CKs from related purines. For definitive analysis, high-performance liquid chromatography coupled with high-resolution mass spectrometry (HPLC–HRMS) is required, as it combines separation of complex extracts with accurate-mass identification [8].

Extraction is the critical first step. Animal tissues, cells, or biofluids are homogenized under cold conditions to suppress enzymatic activity. Aqueous–organic solvents containing weak acid stabilize adenine derivatives and prevent oxidative loss [9]. Stable isotope-labeled CKs are spiked into each sample to serve as internal standards, correcting for recovery losses and matrix effects throughout the workflow [10].

Purification is performed by solid-phase extraction (SPE). Mixed-mode sorbents remove proteins and lipids while retaining CKs [9]. The eluate is concentrated and injected into the HPLC system, where reversed-phase gradients separate canonical CKs, such as trans-zeatin, cis-zeatin, and isopentenyladenine, from other nucleoside derivatives [8]. Chromatographic resolution minimizes ion suppression and improves downstream mass spectrometric detection.

The HRMS stage provides the selectivity required for confident CK identification. Accurate-mass detection distinguishes CKs from isobaric purine metabolites common in animal tissues, with mass errors typically below 5 ppm. Tandem MS produces diagnostic fragmentation spectra, yielding adenine- and ribose-related ions characteristic of CKs [6, 8]. Together, exact mass and fragmentation profiles establish compound identity with high certainty.

Quantification is achieved through isotope-dilution calibration, where standard curves are generated with authentic CKs alongside isotopically labeled analogs [10, 11]. This approach enables reliable detection at picomolar concentrations—sensitivity necessary for animal matrices. Properly applied, it allows researchers to distinguish between trace dietary residues and possible endogenous metabolites.

The method's robustness depends on stringent controls. Samples from animals maintained on CKs-depleted diets help exclude dietary artifacts, while spiking experiments validate extraction recovery. Negative controls and blank runs reduce the risk of contamination. Even when CKs are detected, their physiological meaning requires careful interpretation, as no definitive biosynthetic or receptor pathways have yet been confirmed in animals [4, 5].

Nevertheless, HPLC–HRMS offers the most powerful tool for addressing whether CKs exist in animal systems and at what levels. It provides the sensitivity, selectivity, and quantitative accuracy needed to resolve long-standing questions on the cross-kingdom distribution of these hormones [2, 6]. Beyond detection, such analyses may open new perspectives on the nutrigenomics of CKs, expanding our view of their biology from plants into broader biological contexts [12].

2 Materials

The solvents utilized in this protocol are of HPLC-MS grade. Water used is de-ionized, using a Milli-Q system (18.2 MΩ·cm at 25 °C). All chemicals and solvents are used as received, without further purification.

2.1 CK Extraction and Purification

1. Solution of CKs hormones: This solution contains the following CKs at a concentration of 0.4 μg/mL in methanol:water (80:20, v/v): *trans*-Zeatin (Z), Isopentenyladenine (iP), Dihydrozeatin (DHZ), Isopentenyladenine 9-glucoside (iP9G), *trans*-Zeatin riboside (ZR), *trans*-Zeatin 9-glucoside (Z9G) (*see* Fig. 1, **Note 1**).
2. Solution of internal standard: This solution contains the following deuterated versions of the CKs at a concentration of 0.4 μg/mL: [$^2H_5$]*trans*-Zeatin (D-tZ), [$^2H_6$]N^6Isopentenyladenine (D-iP), [$^2H_3$]Dihydrozeatin (D-DHZ), [$^2H_6$]N^6-Isopentenyladenine 9-glucoside (D-iP9G), [$^2H_5$]*trans*-Zeatin riboside (D-ZR), [$^2H_5$]*trans*-Zeatin 9-glucoside (D-Z9G) (*see* **Note 2**).
3. Extraction solvent: The solvent used for CK extraction is methanol:water (80:20, v/v) (*see* **Note 3**).
4. Ready-to-use solid phase extraction (SPE) C_{18} cartridges (*see* **Note 4**).
5. Equilibrating solution: The solution used for the equilibration of the SPE C_{18} cartridges is the extraction solvent.
6. Eluting solution: The solvent used for eluting dry pellets prior HPLC-MS injection is methanol:water (20:80, v/v) (*see* **Note 5**, Fig. 1).
7. Tissue preparation: C57BL/6 mice were housed in a standard animal care facility under controlled conditions, including a 12-hour light/dark cycle and an ambient temperature of 24 ± 2 °C. All animals were euthanized, and samples were collected from key organs and tissues, including the heart, liver, ileum (gut), leg muscle, kidney, and intestinal stool content. Blood was collected via the retro-orbital sinus under appropriate anesthesia.

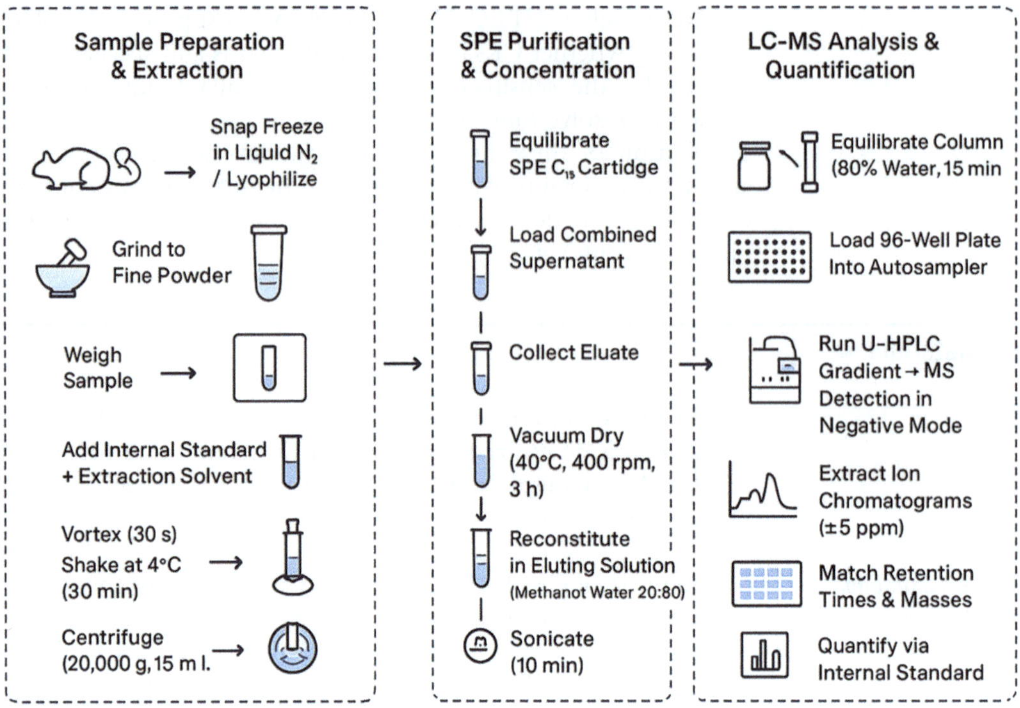

Fig. 1 Workflow for cytokinin (CK) extraction, purification, and quantification. We summarize the three main stages of the protocol. Left panel (Sample Preparation and Extraction): Tissue samples from mouse organs (heart, liver, ileum, muscle, kidney, stool, and blood) are collected, immediately frozen in liquid nitrogen or lyophilized, and ground to a fine powder using a ball mill. The weighed material is mixed with an internal standard solution and extraction solvent, vortexed for 30 s, shaken at 4 °C for 30 min, and centrifuged at 20,000 *g* for 15 min at 4 °C to collect supernatants. Middle panel (SPE Purification and Concentration): Supernatants are passed through equilibrated C18 solid-phase extraction (SPE) cartridges. The eluates are vacuum-dried (40 °C, 400 rpm, 3 h), reconstituted in methanol:water (20:80, v/v), sonicated for 10 min, and centrifuged again at 20,000 *g* for 15 min. Final extracts are transferred to 96-well plates and stored at −80 °C until analysis. Right panel (LC-MS Analysis & Quantification): The U-HPLC column is equilibrated with 80% water (15 min, 300 μL/min). Samples and calibration standards are analyzed using a pre-optimized U-HPLC gradient coupled to mass spectrometry in negative ion mode. Ion chromatograms are extracted at ±5 ppm mass tolerance, and CK concentrations are determined using the internal standard method

2.2 LC-MS Analysis and Quantification

1. The U-HPLC-MS system consists of an Accela Series U-HPLC (coupled with an Exactive mass spectrometer using a heated electrospray ionization (HESI) interface (*see* **Note 6**).
2. Accucore reversed-phase column (50 × 2.1 mm), which contains solid core particles, engineered to a diameter of 2.6 μm and a very narrow particle size distribution (*see* **Note 7**).
3. Calibration solutions: Six to eight calibration solutions should be prepared in order to adequately describe the calibration equations for each CK analyzed. These standard solutions should contain each hormone in equal concentration, in the

range of concentration expected for the samples (from 0.1 ng/mL to 100 ng/mL), along with 10 ng/mL of each internal standard (*see* **Note 8**).

3 Methods

This method is applicable to a variety of biological tissues and organs, and requires a minimum of 75 mg fresh weight or 20 mg dry weight of tissue. If the biological sample is liquid, it does not require any extraction, just filtration before analysis. The main advantage of this method is the simplicity of the extraction and purification protocol, which takes only a few hours (the most time-consuming step is the dryness process under vacuum), while the HPLC-MS analysis takes only 10 min. However, due to the number of CKs (and other potential analytes), an initial tuning of the mass spectrometer parameters is required, which could take some weeks to be accomplished. Another important advantage of this method is that it could be extended to the analysis of other plant-like hormone classes and derivatives, as a simple and robust protocol for the simultaneous analysis of complex tissue matrices (Fig. 1).

3.1 CK Extraction and Purification

1. Collect tissue samples and immediately freeze them by immersing them in liquid nitrogen, and store at −80 °C until analysis. Alternatively, samples could be lyophilized in a freeze-drier and stored in the fridge at 4 °C until analysis.
2. Grind the frozen or lyophilized tissue material into a fine and homogenous powder in a ball mill (*see* **Note 9**).
3. Weigh out the homogenized fresh or lyophilized biological material in a 1.5 mL tube. Add 4 mL of internal standard mix at a concentration of 10 ng/mL and 5 or 20 μL of extraction solution per mg of fresh or lyophilized material, respectively. Shake the tubes in a vortex for at least 30 s and place them on an orbital shaker at 4 °C for 30 min (*see* **Note 10**).
4. After extraction, shake the tubes in a vortex for 30 s, and centrifuge them at 20,000 *g* and 4 °C for 15 min. Transfer the supernatant to a glass test tube (13 × 100 mm), and keep at 4 °C.
5. Re-extract the pellets with the same amount of extraction solution, repeating the two previous steps, and mixing the two supernatants obtained per sample.
6. Equilibrate each SPE cartridge with 3 mL of equilibrating solution.
7. Pass each supernatant through a SPE C_{18} cartridge (*see* **Notes 11** and **12**), and collect them in glass test tubes (13 × 100 mm).
8. Evaporate the solvents of the samples in a vacuum dryer (*see* **Note 13**) at 40 °C and 400 rpm for 3 h (*see* Fig. 1, **Note 14**).

9. Elute the residue with 1 mL of eluting solution in a 1.5 mL tube.
10. Sonicate the tubes in an ultrasonic bath for 10 min at room temperature (*see* **Note 15**) to facilitate the resuspension of the dried residue in the eluting solution.
11. Centrifuge the samples at 20,000 *g* for 15 min (*see* **Note 16**).
12. Pipet the samples in a 96-wells plate together with the calibration solutions, cover with plastic film and store at −80 °C until analysis (*see* Fig. 1 and **Note 17**).

3.2 LC-MS Analysis and Quantification

1. Set the pump flow to 300 $\mu L{\cdot}min^{-1}$ in order to equilibrate the HPLC column during 15 min with the initial mobile phase (80% water) (*see* **Note 18**).
2. Put the 96-well plate in the auto-sampler and define the sequence with the Excalibur software.
3. Load the previously optimized HPLC method in the Excalibur software. We used a gradient with two mobile phases (aqueous and organic, *see* **Note 19**).
4. Load the previously optimized MS method in the Exactive Tune software. We perform the analyses in the negative mode $[M\text{-}H]^-$ because the method has been optimized for a vast range of hormones and derivatives, and not only for CKs (*see* **Notes 20** and **21**).
5. Run the sequence. The time of analysis per sample is 10 min (*see* **Note 22**).
6. Record the total ion chromatograms of the samples and extract the specific chromatogram of each compound with a mass tolerance of 5 ppm, using the Excalibur software (*see* **Note 23**).
7. Set up the retention times and exact masses of the different CKs and define the calibration levels of the analytes and internal standards, prior to the quantification of CK concentrations (*see* Fig. 1 and **Note 24**).
8. Concentrations and recovery rates of the different CKs in the samples are calculated according to the internal standard method (*see* **Note 25**).

4 Notes

1. This standard solution should be freshly prepared before use, from stock solutions of the different hormones at higher concentrations (typically 50 $mg{\cdot}l^{-1}$ or higher). These are the most representative CKs, but other CKs and derivatives could be also included in the analysis.

2. The internal standard mix should be also freshly prepared before use, from stock solutions of the different deuterated compounds at higher concentrations (typically 50 mg or higher).
3. Different percentages of organic solvent/water could be also used. The extraction solution could be slightly acidified with 1% of acetic acid (glacial) 99.9 + % purity.
4. Any other SPE C_{18} cartridge of similar characteristics could be used.
5. The percentage of organic solvent/water should be similar to that of the initial conditions of the (U)-HPLC mobile phase gradient.
6. Triple-quadrupole mass spectrometers are also widely used, but the molecules must be fractionated (MS/MS) because the resolution is lower than that of the Orbitrap spectrometer.
7. Other columns are also suitable depending on the (U)-HPLC system used (normally reversed-phase columns).
8. This is a typical calibration range, but can be adjusted to the characteristics of the samples analyzed, such as soft organs like colons, kidneys, heart tissue, or fecal material.
9. Samples can also be ground with a mortar.
10. It is desirable that the whole extraction and purification process is done under low light conditions, since some hormones are very sensitive to light.
11. This process should be carried out slowly. It removes proteins, lipids, and pigments, what may interfere in the analysis by HPLC-MS.
12. In the case of technical or biological replicates of animal tissue material, the cartridges can be re-used up to 3 times.
13. We use a Univapo 150 ECH vacuum dryer coupled to a Unicryo MC2L (−60 °C) unit, but any similar vacuum dryer of similar characteristics could be employed.
14. The time needed for total dryness depends on the amount of solvent and the sample characteristics.
15. The ultrasonic bath that we use to resuspend the samples is an Ultrasonics H-D system. Any other similar ultrasonic bath is also suitable for eluting the samples.
16. Alternatively, samples could be also filtrated through 13 mm diameter Millex filters with 0.22 μm pore size and nylon membrane.
17. The ACCELA auto-sampler allows the use of different plates, facilitating the final preparation of the samples, but normal HPLC vials could also be used.

18. These parameters could be modified and adjusted depending on the (U)-HPLC system employed.
19. The method should be optimized for the specific (U)-HPLC used.
20. In fact, the MS tune will vary depending on the analyte and the MS used, so it has to be optimized before. The typical instrument settings include: sheath gas flow rate = 35 mL·min^{-1}, auxiliary gas flow rate = 10 mL/min, spray voltage = 2.5 kV, capillary temperature = 275 °C, capillary voltage = −40 V, tube lens voltage = −110 V, skimmer voltage = −20 V.
21. Due to the high resolution of the Orbitrap, we record the total ion chromatogram of the samples and do not fragmentate the molecules. However, it is also possible to fragmentate to distinguish between two different CKs with the same m/z.
22. The time of analysis could vary depending on the LC system used. In the case of ultra-HPLC systems this time could be reduced up to less than 10 min. In the normal HPLC systems the time of analysis is 30 min or even higher.
23. We detect the ions by exact mass due to the high resolution of the Orbitrap (up to 100,000) and it is therefore possible to set a mass tolerance window of 5 ppm. With other lower resolution mass spectrometers (i.e., triple quadrupole), ion fragmentation is required.
24. This processing set up allows the determination of the limit of detection (LOD) and limit of quantification (LOQ) of each compound analyzed.
25. Acceptable recovery rates are those higher than 75%.

Acknowledgments

We gratefully acknowledge the research funding given to MN by Zayed University through the RIF grant (RIF23094) as well as the UAEU-ZU joint research grant (G-23219) to MN and KM (G-5310). TD acknowledges Deutsche Forschungsgemeinschaft (DFG) for funding, grant 492620490—SFB 1583 /INF.

References

1. Mok DW, Mok MC (2001) Cytokinin metabolism and action. Annu Rev Plant Physiol Plant Mol Biol 52:89–118
2. Naseem M, Wölfling M, Dandekar T (2014) Cytokinins for immunity beyond growth, galls and green islands. Trends Plant Sci 19:481–484. https://doi.org/10.1016/j.tplants.2014.04.001
3. Bruce NC, Rogers HJ, Burrell MM (1992) Cytokinin-like compounds in non-plant systems. Phytochemistry 31(5):1509–1512
4. Kasahara H (2016) Current aspects of cytokinin biosynthesis in *Arabidopsis*. J Exp Bot 67(2):491–498
5. Nicander B, Stenlid G, Ljung K, Olsson AS (2003) Dietary sources of plant hormones:

implications for mammalian exposure. Food Chem 82(3):293–300

6. Pertry I, Václavíková K, Depuydt S et al (2009) *Rhodococcus fascians* impacts plant development through the dynamic regulation of a cytokinin response throughout infection. Plant Physiol 150(3):1272–1285
7. Tarkowská D, Novák O (2018) Detection and quantification of cytokinins: advances and challenges. Plant Methods 14:77
8. Novák O, Hauserová E, Amakorová P, Doležal K, Strnad M (2008) Cytokinin profiling in plant tissues using ultra-performance liquid chromatography–electrospray tandem mass spectrometry. Phytochemistry 69(11):2214–2224
9. Kojima M, Kamada-Nobusada T, Komatsu H et al (2009) Highly sensitive and high-throughput analysis of plant hormones using MS after solid-phase extraction. Methods Mol Biol 495:39–54
10. Svačinová J, Novák O, Plačková L, Lenobel R, Holík J, Strnad M (2012) A new approach for cytokinin profiling in plants using ultra-performance liquid chromatography–electrospray tandem mass spectrometry. Plant Physiol 160(2):558–571
11. Albacete A, Ghanem ME, Martínez-Andújar C, Acosta M, Sánchez-Bravo J, Martínez V, Lutts S, Dodd IC, Pérez-Alfocea F (2008) Hormonal changes in relation to biomass partitioning and shoot growth impairment in salinized tomato (Solanum lycopersicum L.) plants. J Exp Bot 59:4119–4131
12. Naseem M, Sarukhanyan E, Dandekar T (2015) LONELY-GUY knocks every door: Crosskingdom microbial pathogenesis. Trends Plant Sci 20(12):781–783. https://doi.org/10.1016/j.tplants.2015.10.017

Chapter 7

In Vivo Detection of IAA Using IAA Nanosensor

Benny Jian Rong Sng, Duc Thinh Khong, Thomas K. Porter, Michael S. Strano, and In-Cheol Jang

Abstract

Indole-3-acetic acid (IAA) is an important auxin phytohormone that regulates development, directional growth, and stress responses in plants. Here, we provide a detailed method and protocol for use of a corona phase molecular recognition-based IAA nanosensor that allows for direct and real-time measurement of IAA in plants. The near-infrared signal of the IAA nanosensor also allows it to be used in green tissues, as it is not affected by chlorophyll content. Furthermore, the IAA nanosensor can be easily applied across various plant species and all tissues as the sensor is not genetically encoded.

Key words Corona phase molecular recognition, Single-walled carbon nanotube, Nanosensor, Auxin, Indole-3-acetic acid, Plant stress

1 Introduction

Indole-3-acetic acid (IAA) is a key auxin hormone that regulates growth and development in plants [1, 2]. Besides promoting growth, auxin concentration and auxin gradients are also crucial in determining plant development, such as the development of flowers and lateral roots, as well as the determining embryogenesis and leaf development [3–6]. Auxin also plays a large role in stress response in plants. IAA concentration and auxin response increase during responses to vegetative shade (shade avoidance) and high ambient temperature (thermomorphogenesis) [7–9]. In addition, the distribution of IAA in plants also affects directional growth like phototropism or development of plant structures such as apical hooks during skotomorphogenesis [10].

The amount and distribution of IAA in plants are regulated by a variety of processes, including biosynthesis, auxin transport, signal transduction, and inactivation of IAA. In the main IAA biosynthesis

Benny Jian Rong Sng and Duc Thinh Khong contributed equally.

Shahid Mukhtar (ed.), *Plant Hormones: Methods and Protocols*, Methods in Molecular Biology, vol. 3026, https://doi.org/10.1007/978-1-0716-5214-5_7,

pathway in plants, tryptophan is firstly converted to the intermediate indole-3-pyruvic acid (IPA) by TRYPTOPHAN AMINOTRANSFERASE family (TAA/TAR), followed by further conversion of IPA to IAA by YUCCA family (YUC) [11, 12]. IAA can then be inactivated, such as by GRETCHEN HAGEN 3 (GH3) family proteins into amino acid-IAA conjugates [13]. IAA can also undergo directional transport in plants, as PIN-FORMED (PIN) and ABC TRANSPORTER B (ABCB) family proteins regulate auxin efflux while AUXIN1/LIKE-AUX1 (AUX1/LAX) family controls auxin influx [14]. Lastly, as a plant hormone, IAA triggers downstream signaling and affects transcription of auxin-responsive genes. IAA binds to the TRANSPORT INHIBITOR RESPONSE 1 and AUXIN-SIGNALING F-BOX (TIR1/AFB) receptor complex, which triggers TIR1/AFB to interact with transcriptional repressors of the AUXIN/IAA (Aux/IAA) family and promote their degradation [15]. Protein degradation causes Aux/IAA to release their suppression of transcription factors AUXIN RESPONSE FACTORS (ARFs), thereby activating the transcription of auxin-responsive genes [15].

The multifaceted regulation of IAA in plants means that quantification of IAA needs to consider both amount and distribution. As such, various auxin reporter lines have been developed to visualize auxin in plants, such as DR5-GUS [16], DR5-GFP [17], DII-VENUS [18], R2D2 [19], C3PO [20], and AuxSen [21]. While each of these auxin reporter lines have their strengths, they all require the generation of genetically modified plants. Furthermore, auxin reporter lines that utilize fluorescent proteins tend to encounter interference when measuring green plant tissues due to strong chlorophyll autofluorescence [22, 23]. Lastly, some well-established auxin reporter lines do not directly measure auxin and instead infer auxin levels by measuring auxin response (e.g., DR5-GUS and DR5-GFP) or auxin signaling (e.g., DII-VENUS).

The IAA nanosensor was developed using the CoPhMoRe platform (Fig. 1) [24, 25]. The sensor construct entails the physical adsorption of a synthetic polymer onto the surface of single-wall carbon nanotubes (SWNT) [24], forming an aqueous corona phase that enables selective binding of the IAA analyte (Fig. 1). By utilizing carbon nanotubes as the signal transducer, the sensor provides highly photostable near-infrared (NIR) fluorescence, with wavelengths that typically avoid interference from the photo-absorption or autofluorescence of living tissues [26]. The core of IAA nanosensor was designed based on the interaction between IAA and TIR1, an F-box protein that is part of SKP-Cullin-F-box protein complex (SCF^{TIR1}) [25, 27]. Screening and optimization of the synthetic corona phase for IAA detection utilized a diverse library of synthetic polymers, resulting in the CoPhMoRe complex of water-soluble polyamic polymer-SWNT conjugates [25]. This complex offers a highly selective NIR fluorescent quenching

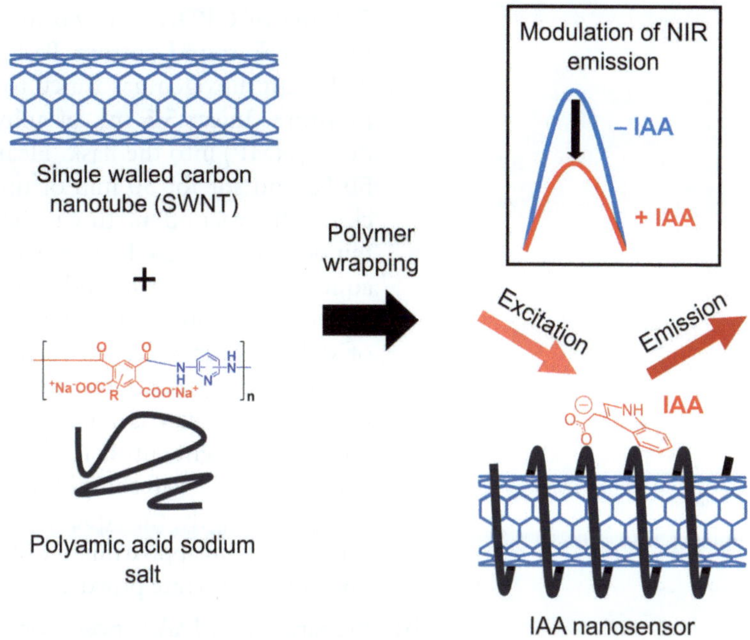

Fig. 1 Schematic diagram showing the design of the IAA nanosensor. The design of the IAA nanosensor is based on the interaction between IAA and TIR1 [25, 27]. The IAA nanosensor is generated by wrapping polyamic acid sodium salt polymers around single-walled carbon nanotubes (SWNTs). Subsequent binding of IAA to the nanosensor modulates the intensity of NIR emission, thereby allowing for measurement of IAA content. (Adapted from [25])

response to IAA, achieving a limit of detection as low as 0.1 μM in vivo [25].

In this chapter, we introduce the use of the IAA nanosensor for direct measurement and visualization of IAA in plants. The IAA nanosensor fluoresces in the NIR region and is thus unaffected by chlorophyll content and can be used in leaf tissues. Moreover, the IAA nanosensor can be easily applied across various species, as the sensor is not genetically encoded.

2 Materials

2.1 Reagents and Solutions

1. 2-(*N*-morpholino) ethanesulfonic acid (MES) buffer (10 mM, pH 5.5): Dissolve 2.13 g of MES in 1 L of milli-Q water. Adjust the pH of the MES buffer by adding 1 M potassium hydroxide dropwise until the pH is 5.5. Filter the MES buffer using a vacuum filtration unit with 0.2 μm membrane.
2. IAA nanosensor solution [25]:
   (a) Preparation of polymer 4,4'-Oxydiphthalic anhydride (OPDA)-2,2'-Bipyridine-6,6′-diamine (26DABP): Add

310 mg of OPDA and 186 mg of DABP into a 50 mL three-neck round-bottom flask. Vacuum and flush the flask with nitrogen gas thrice to remove residual air and moisture. Inject 5.5 mL of anhydrous *N*-methylpyrrolidone (NMP) into the flask. Heat the reaction mixture to 50 °C and stir for 30 min or until the solution becomes clear. The reaction mixture is then further stirred at room temperature for 24 h. Gradually add 50 mL of 0.1 N aqueous sodium hydroxide into the reaction mixture. White precipitates will be observed upon initial addition of sodium hydroxide, but further addition will gradually solubilize the precipitates. The reaction mixture is transferred into a dialysis membrane (regenerated cellulose, molecular weight cutoff 12–15 kDa) and dialyzed against deionized (DI) water until the polymer solution achieves pH of 6.5. Lastly, the aqueous polymer solution is freeze-dried to obtain approximately 300 mg of polymer OPDA-26DABP as a white powder.

(b) Preparation of IAA nanosensor using SWNT and OPDA-26DABP: Dissolve 40 mg of OPDA-26DABP with 4 mL of ultrapure water in a 15 mL tube. Add 4 mg of raw HiPCO® or CoMoCAT® SWNT into the same tube. Sonicate the resulting mixture using with a 1.6 mm probe tip sonicator for 1 h. Partially submerge the tube in an ice-bath while sonicating to prevent overheating. After sonication, ultracentrifuge the mixture at 153,100 *g* for 4 h at 22 °C to separate the aggregates and SWNT bundles. Carefully transfer about 75% volume of the supernatant to another tube as the stock IAA nanosensor. Estimate the sensor concentration using UV-Vis-NIR spectroscopy (*see* **Note 1**). Store the stock IAA nanosensor at 4 °C and away from light (*see* **Note 2**).

3. Reference nanosensor [25]: Combine 1 g of single-strand DNA (AT)15 and 0.25 mg of CoMoCAT® (6,5) SWNT in 1 mL of 50 mM sodium chloride in a tube. Sonicate the mixture using a 1.6 mm probe tip sonicator at 20% amplitude for 20 min. Keep the tube in an ice bath during sonication to prevent overheating. Centrifuge the sonicated mixture at 16,500 *g* for 90 min at room temperature to remove unsuspended SWNT bundles, then carefully discard the pellet. Repeat the centrifugation step (16,500 *g* for 90 min) to further remove aggregates. Transfer the supernatant to another tube as the stock reference sensor. Estimate the sensor concentration using UV-Vis-NIR spectroscopy (*see* **Note 1**). Store the stock reference nanosensor at 4 °C and away from light (*see* **Note 2**)

2.2 Plant Materials and Media

1. Rosette-stage Arabidopsis plants: Plants should have large and flat leaves for easy infiltration and detection.
2. Half-strength Murashige and Skoog with sucrose (1/2MS + suc) liquid media: Dissolve 2.2 g of basal MS salt, 10 g of sucrose, and 0.5 g of MES in 500 mL of DI water. Mix well using a stir bar and magnetic stirrer until all chemicals have fully dissolved. Adjust the pH of the liquid media by adding 1 M potassium hydroxide dropwise until the pH is 5.8. Make up to 1 L with DI water. Autoclave at 121 °C for 15 min. Keep at 4 °C for prolonged storage.
3. 1/2MS + suc agar media: Prepare 1/2MS + suc as above, until adjusting the pH to 5.8. Add 12 g of plant agar, then make up the volume to 1 L with DI water. Autoclave at 121 °C for 15 min, then cool the media to 70 °C. In an aseptic condition, pour the media into petri dishes and cool until the media solidifies completely. Store the media at 4 °C.

2.3 Consumables

1. Nitrile gloves (*see* **Note 3**).
2. Cytotoxic waste bags (*see* **Note 3**).
3. Needleless syringe (1 mL size).
4. Lint-free cleaning tissues.
5. Sterile forceps.
6. Microcentrifuge tube.
7. Sterile pipette tips.

2.4 Equipment

1. Stand-off NIR camera system, with a 785 nm excitation laser at 30 mW laser power and a 900 nm long pass filter for detecting NIR emission (*see* **Note 4**). The NIR images and videos are captured using a 2D InGaAs detector with a liquid nitrogen-cooled system. The NIR camera system is connected to a computer with Light Field® software (Teledyne Vision Solutions, version 6.15.2.2201) for experiment control and data collection.
2. Confocal microscope capable of detecting NIR emission wavelengths. We used the Olympus FV3000NIR inverted microscope, with a longpass filter for detecting NIR emission (*see* **Note 4**).
3. Laser safety goggles.

3 Methods

1. Nitrile gloves and cytotoxic waste bags are to be used for all subsequent steps and plant tissues (*see* **Note 3**).

Fig. 2 Infiltration of IAA nanosensor into Arabidopsis leaf. (**a**) A marker is used to indicate the leaf for application of IAA nanosensor (blue arrowhead). (**b**) Application of IAA nanosensor into Arabidopsis leaf using needleless syringe infiltration. (**c**) After nanosensor application, the area exposed to the nanosensor will appear water-soaked. IAA and reference nanosensors should be applied to different parts of the leaf, preferably separated by the midvein. Ref, reference nanosensor. IAA, IAA nanosensor

3.1 Application of IAA Nanosensor to Plants

3.1.1 Application of IAA Nanosensor to Mature Leaves

2. Indicate the leaves for infiltration with a felt-tip marker (Fig. 2a).
3. Dilute the IAA nanosensor solution to 10 mg/L using MES buffer.
4. Using a 1 mL needleless syringe, infiltrate the IAA nanosensor solution into the abaxial leaf surface (Fig. 2b). This process is similar to needleless syringe infiltration techniques such as infiltration of *Agrobacterium tumefaciens* into leaves for transient expression of genes.
5. Rinse the leaf surface with DI water and gently absorb the excess water on the leaf surface using lint-free cleaning tissues.
6. Using the same method, infiltrate the reference sensor into a nearby section of the same leaf (Fig. 2c).
7. Allow the infiltrated leaf to rest for at least 1 h before NIR imaging. The infiltrated leaf can rest overnight in regular growth condition (For Arabidopsis: long day, 21 °C, 100 μmol/m^2 s^1) before NIR imaging.

3.1.2 Application to Seedlings (e.g., for Study of Root Tip in Arabidopsis)

1. The following procedures should be done using aseptic techniques, such as in a laminar flow/biosafety cabinet and using sterile equipment/media.
2. Dilute concentrated IAA nanosensor solution to 10 mg/L using 1/2MS + suc liquid media in a microcentrifuge tube.
3. Pick up an Arabidopsis seedling growing on 1/2MS + suc agar (vertical agar media, such that the roots are growing on the agar surface).
4. To incubate IAA nanosensor into only the root tip, position the seedling on a pipette tip, such that the root tip faces the pointed

end of the pipette tip and the cotyledons faces the opposite direction. Insert the pipette tip into the IAA nanosensor solution.

5. For incubation of entire seedlings, use forceps to transfer and submerge seedlings in the IAA nanosensor solution.
6. Incubate the seedlings in IAA nanosensor solution for 2 days in regular growth condition (For Arabidopsis: long day, 21 °C, 100 μmol/m^2 s^1).
7. After incubation, carefully wash the Arabidopsis seedlings in MES buffer (10 mM, pH 5.5) thrice. There is no need to vortex or use other vigorous methods, as the washing step is meant to remove any nanosensors on the surface of the seedlings.
8. Seedlings can be immediately used for NIR detection.

3.2 Detection of IAA Nanosensor Signal

3.2.1 Nanosensor Imaging Using Stand-off NIR Camera System

The stand-off NIR camera system is designed to facilitate convenient and efficient observation of sensor responses in intact and living plants. This enables real-time monitoring with minimal disruption to the physiology of the plant. Figure 3 demonstrates an example of this setup for monitoring IAA in an Arabidopsis plant during shade avoidance response. The NIR camera system consists of: (1) 2D liquid nitrogen-cooled InGaAs detection system, (2) long pass filter to cut off any wavelength under 900 nm, (3) 785 nm laser module with divergent filter to increase the excitation area, and (4) a connected computer with LightField® software for experiment control and data collection (Fig. 3).

1. Position the sensor-functionalized Arabidopsis leaf in front of the camera. The leaf can be propped vertically, facing the camera lens at a distance of approximately 50 cm.
2. Turn on the white light fluorescent lamp to capture bright-field images using the LightField® software (Teledyne Vision Solutions, version 6.15.2.2201). Using the bright-field images, adjust lens of the detector system to optimize the focus. The leaf should be centered and in focus before capturing NIR signals. Turn off the lamp after capturing the bright-field images.
3. Turn on the laser probe and adjust its position so that the laser spot fully covers the leaf area containing the nanosensors. Always wear laser safety goggles during this process.
4. Turn off the laser probe. Turn on lamps for providing simulated shade condition (low intensity white light supplemented with far-red light). On the LightField® software, set the exposure time to 30 s. Perform background correction using the LightField® software.

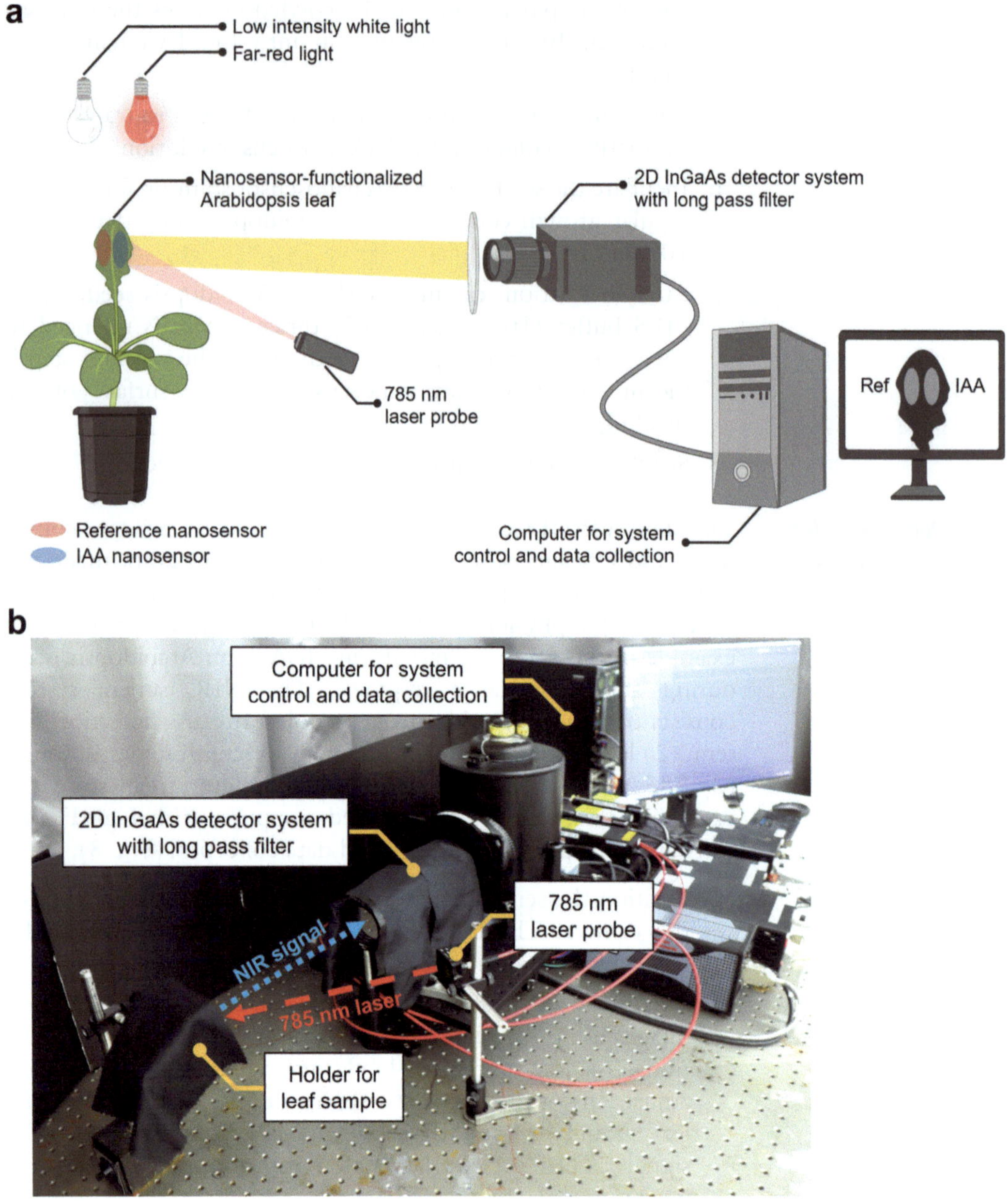

Fig. 3 Setup of stand-off NIR camera system. (**a**) Schematic diagram of the stand-off NIR camera system. Created in BioRender.com. (**b**) Photograph of NIR camera system

5. Turn on the laser probe. Continuously collect the NIR images for 3 h, with 30 s exposure time for each NIR image. The simulated shade condition should be provided throughout this process.
6. Save and export the image files in.tiff format.

7. The.tiff file can be further processed by image processing software such as ImageJ, MATLAB, etc.

3.2.2 Image Processing to Determine Average IAA Nanosensor Response

To analyze the IAA nanosensor signal as measured by the stand-off NIR camera system, we utilized a ratiometric analysis strategy to improve signal-to-noise ratio and minimize fluctuations in NIR signal caused by external factors such as variations in laser power or background lighting. This method requires the use of an inert nanosensor as a reference, alongside the IAA nanosensor. In brief, the ratiometric response is calculated by comparing the fluorescence intensity of the IAA nanosensor to the fluorescence intensity of the reference nanosensor. The following describes the basic steps for extracting average ratiometric response of IAA nanosensor using the image processing software ImageJ.

1. Open the ImageJ software.
2. Load the image file (.tiff) for processing into ImageJ. The.tiff file should consist of a stack of NIR fluorescence images, of which each frame corresponds to the 30 s exposure time.
3. Background correction: Use "Subtract background" function with rolling ball radius of 50.0 pixels in the "Process" menu to correct the background signal.
4. Image thresholding: Use "Threshold" function in the "Image -> Adjust" menu to manually define the total area of interest. Adjust the minimum and maximum sliders to select for the sensor-emitted NIR signal. The background values can be set to NaN (Do not click apply).
5. Define region of interests (ROI) for IAA nanosensor and reference nanosensor: Open "ROI Manager" window via "Analyze -> Tools." Use the shape tools in the main tool bar to designate the ROI and click "Add [t]" in "ROI Manager" to manually create the ROI for each nanosensor.
6. Extract mean gray value of each ROI: In "ROI Manager" window, click "More -> Multi Measure" (Ensure that the two ROIs are selected and that only "Mean gray value" is selected in the "Analyze -> Set Measurements" menu). The data in the new pop-up spreadsheet consists of three columns: (1) first column is the frame number, (2) all subsequent columns are the average mean gray value of each ROI. The spreadsheet can be exported in.csv format (via File - > Save As) to be further processed in other data analysis programs.
7. Processing of mean gray value in data analysis softwares. The following describes the general steps needed to obtain the ratiometric response of the IAA nanosensor:
   (a) Convert each frame number to time points by multiplying the frame number with the exposure time per frame (e.g.,

frame 10 in an experiment using 30 s exposure per frame = 10 × 30 s = 300 s = 5 min time point).

(b) Calculate the normalized response for each ROI, as denoted by the ratio I/I_o, where I represents the mean gray value (or intensity) of the sensor at any given time point and I_o is the mean gray value (or intensity) of the first time point (frame 1).

(c) Calculate the ratiometric response by taking the ratio of $(I/I_o)_{\_IAA\_sensor}$ to $(I/I_o)_{\_ref\_sensor}$. This normalizes the change in IAA nanosensor signal against the change in the reference nanosensor.

(d) Plot the ratiometric response against the corresponding time points as a line graph.

8. Prediction of change in auxin concentration can be calculated from the ratiometric response. This is done by assuming that the binding event between IAA and the IAA nanosensor (S) as a first-order reversible reaction:

$$IAA + S \underset{k_b}{\overset{k_f}{\rightleftharpoons}} S - IAA$$

where k_f and k_b are forward and backward rate constants, respectively. Derivation of the model to predict IAA level can be found in [25]. The following equation is used to calculate the IAA concentration:

$$[IAA] = \frac{k_b\left(1 - \frac{I}{I_0}\right) - \frac{d\left(I/I_o\right)}{dt}}{k_f \frac{I}{I_0}}$$

where k_f and k_b are forward and backward rate constants respectively (*see* **Note 5**), I/I_o is the normalized fluorescence intensity of the sensor, and $\frac{d(I/I_0)}{dt}$ is the differential of I/I_o over t.

Note that the steps outlined above provide a general guideline for image processing to derive the ratiometric profile of the IAA nanosensor and estimate IAA concentration levels from raw image data. However, this approach only yields temporal data, as spatial information is averaged out. To automate the entire process, it is recommended to code these steps in a programming software such as MATLAB or Python. This allows for the generation of both spatial and temporal distribution of auxin in two-dimensional space, thereby enhancing the analysis.

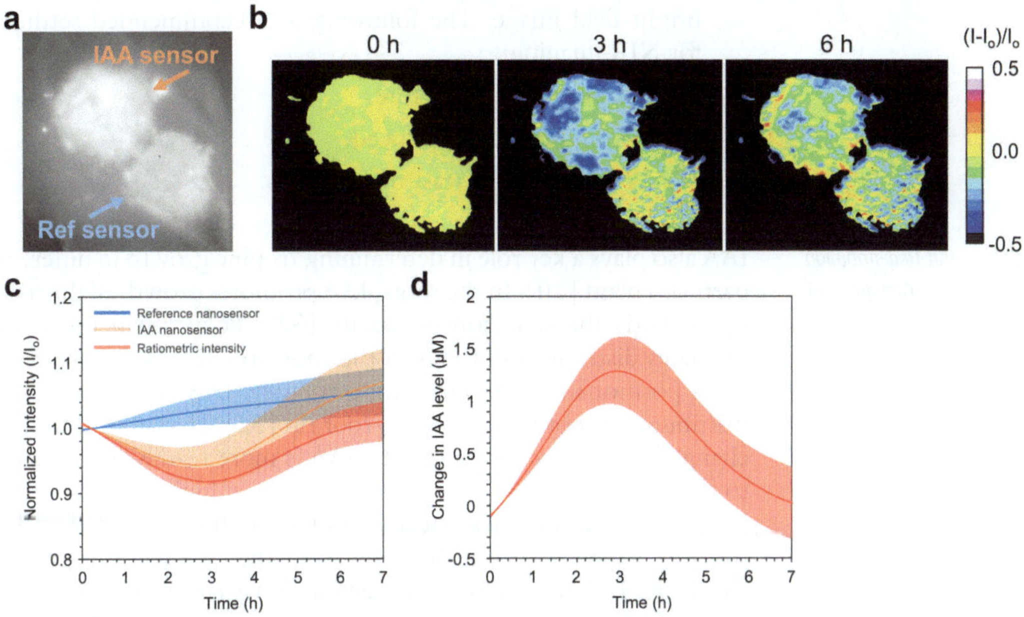

Fig. 4 Application of IAA nanosensor for measuring shade avoidance response. (**a**) NIR fluorescence image of IAA and reference nanosensors in *Nicotiana benthamiana* leaf. (**b**) False-color NIR fluorescence images of (**a**) at various time points of shade treatment. (**c**) Average normalized intensity (I/I_o) of IAA nanosensor, reference nanosensor, and corresponding ratiometric intensity. (**d**) Estimated average change in IAA concentration, based on average normalized intensities (I/I_o) in (**c**)

3.2.3 Measuring IAA Levels in Plants Undergoing Shade Avoidance Response

IAA is an important phytohormone for promoting shade avoidance response. Under vegetative shade, plants induce IAA biosynthesis in their leaves and cotyledons, which promote the development of shade-related phenotypes such as hypocotyl elongation and hyponasty [7, 28–30]. It is important to note that the increase in IAA content is transient and occurs shortly after the plant is exposed to shade condition [25, 29, 31]. Application of the IAA nanosensor to plant leaves enables real-time measurement and visualization of IAA content during shade avoidance response (Fig. 4). Quenching of the IAA nanosensor signal indicates an increase in IAA content, which peaks at 3 h shade treatment (Fig. 4). Subsequently, the IAA nanosensor signal returns to baseline at 3–6 h shade, highlighting the transient induction of IAA biosynthesis during shade treatment (Fig. 4). The same methodology can be used to measure IAA content during other plant stresses like high temperature [25].

3.2.4 Nanosensor Imaging Using Confocal Microscope with NIR Detection

1. Turn on the confocal microscope, excitation laser, and NIR detector.
2. Mount the seedling or root tip onto a glass slide using DI water and a cover slip.
3. Capture the bright-field and NIR fluorescence signals from the seedling sample, ensuring that the sample is in focus in the

bright-field image. The following are recommended settings for NIR imaging:

(a) Large confocal aperture.

(b) Minimum scanning speed.

(c) Averaging of at least three frames.

3.2.5 IAA Redistribution During Root Gravitropism

IAA also plays a key role in determining trophic growth in different parts of a plant [10]. In the roots, IAA promotes growth of the root tip towards the direction of gravity [32]. Polar auxin transport regulates distribution of IAA in the root tip [32]. When the root tip is positioned in the direction of gravity, IAA is symmetrically distributed in both halves of the root tip [32]. However, during a change in the direction of gravity, IAA is more readily transported to the bottom half of the root tip, resulting in higher IAA content [32]. This causes more elongative growth on the upper half of the root tip, which results in the bending of the root towards gravity [32]. The IAA nanosensor can be used to visualize the redistribution of IAA in root tips during gravitropic response (Fig. 5). Upon gravistimulation, the lower side displayed weaker NIR fluorescence than the upper side of the root tip (Fig. 5), indicating asymmetrical IAA content with higher IAA in the lower side of the root.

3.2.6 IAA Distribution in Cotyledons

IAA distribution is important for normal development in plants. IAA accumulates at the apex and border of the leaves and cotyledons, which influences leaf development such as leaf tooth formation [29, 33, 34]. Using the IAA nanosensor, we can visualize the natural distribution of IAA in cotyledons. Quenching of the NIR signal can be observed at the cotyledon apex and border (red and blue arrowheads, Fig. 6a), indicating high IAA content in these tissues, which corroborated with auxin-reporter line DR5-GFP (Fig. 6b) as well as other studies [29].

3.2.7 Modeling IAA Spatial Distribution and Transport

Based on NIR signals obtained from the IAA nanosensor, mathematical modeling can be used to describe the spatial distribution of IAA and obtain quantitative predictions regarding its transport. The following describes basic steps for developing a mathematical model that reflects the concentration profiles obtained from IAA nanosensors. In this example, we investigated IAA transport in *XVE::iaaM* Arabidopsis plants, in which IAA levels could be modulated via β-estradiol induction [25]. β-estradiol was topically applied to a small circular region of the nanosensor-functionalized leaf (red circle, Fig. 7), causing localized production of IAA. The nanosensors spanned a larger region of the leaf and were monitored for 18 h post-induction to visualize IAA distribution and transport.

1. Specify a model domain. In our example, a leaf strip centered around the induced spot was selected (blue rectangle, Fig. 7).

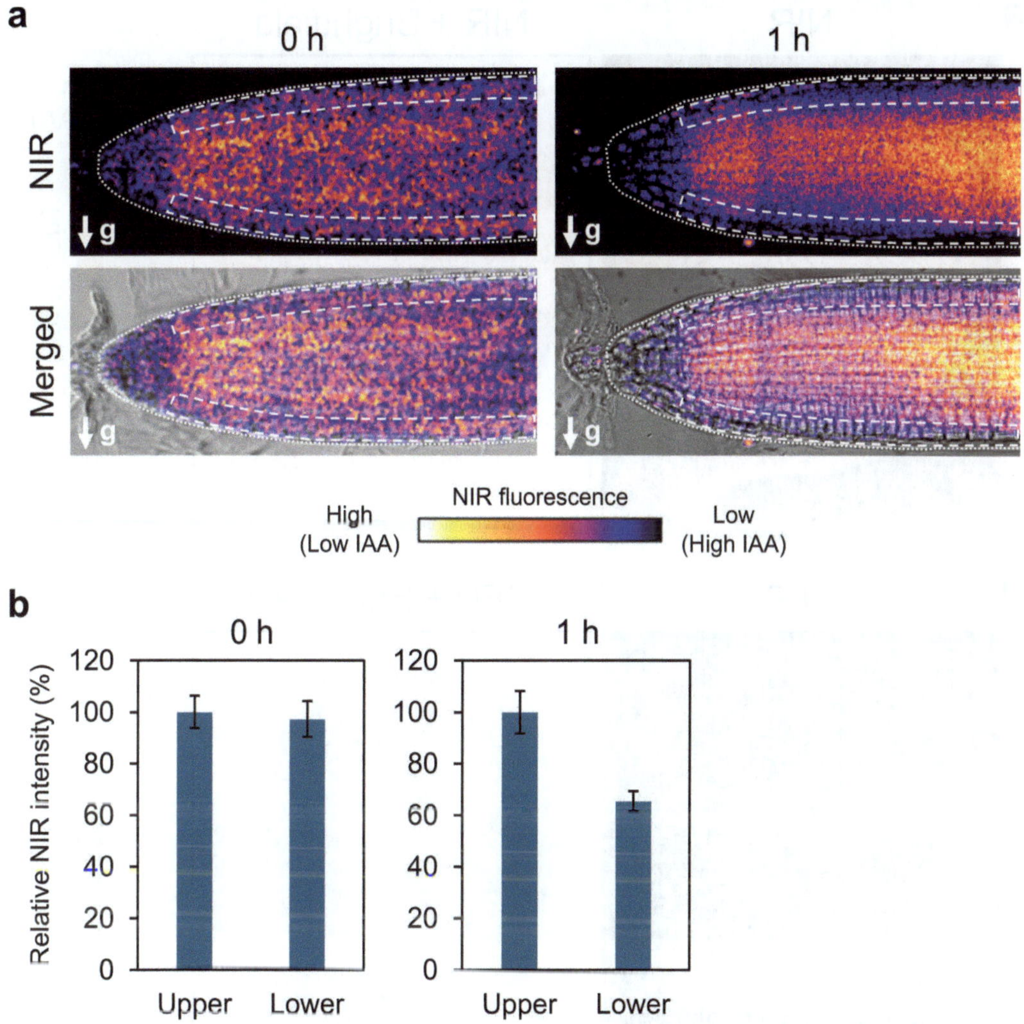

Fig. 5 IAA nanosensor reports redistribution of IAA during root gravitropic response. (**a**) Confocal images of Arabidopsis root tips during root gravitropic response. White dotted line indicates the outline of the root tip. Areas marked by white dashed lines indicate the upper and lower sides of the root tip. g, direction of gravity. (**b**) Relative NIR intensity in upper and lower sides of root tips in (**a**). Values indicate mean±SD. NIR fluorescence is symmetrically distributed in both sides of the root tip before inducing gravitropic response. During gravistimulation, lower NIR signal was found in the lower side of the root tip, indicating the accumulation of IAA

The concentration along the strip was averaged into one dimension for model simplicity.

2. Determine the relevant physical processes governing changes in IAA concentration. The change in IAA concentration over time ($\frac{d[IAA]}{dt}$) equals all sources of IAA generation subtracted by all sources of IAA depletion. These should include chemical reactions that involve the conversion of IAA to/from other

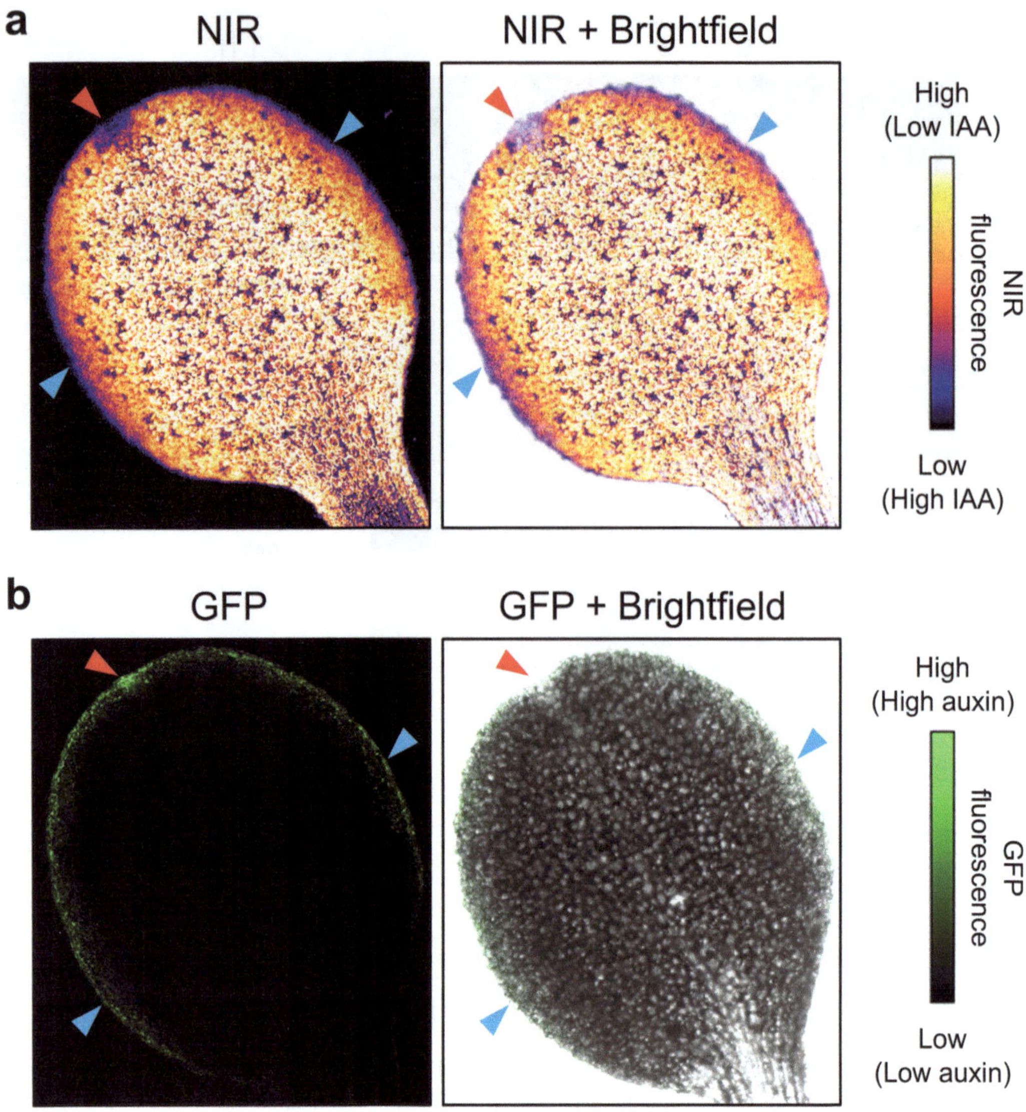

Fig. 6 Visualization of normal IAA distribution in cotyledons using the IAA nanosensor. (**a**) Cotyledon functionalized with the IAA nanosensor. Quenched NIR signals were found in the apex and border of the cotyledon, indicating high IAA content in these tissues. (**b**) Cotyledon of auxin-responsive reporter DR5-GFP. Strong GFP signals were found in the apex and border of the cotyledon, verifying the IAA distribution pattern as seen in (**a**). Red arrowhead, cotyledon apex. Blue arrowhead, cotyledon border

compounds, as well as physical processes like diffusion. We used the following model:

$$\frac{d\overline{C}}{dt} = \frac{d^2\overline{C}}{dx^2} + \overline{\gamma}\frac{1}{\overline{\sigma}\sqrt{2\pi}}\,exp\left[\frac{-(x-\overline{a})^2}{2\overline{\sigma}^2}\right] - \phi^2\overline{C}$$

where $\overline{C}$ represents IAA concentration, and the change in IAA concentration over time ($\frac{d\overline{C}}{dt}$) is governed by diffusion ($\frac{d^2\overline{C}}{dx^2}$

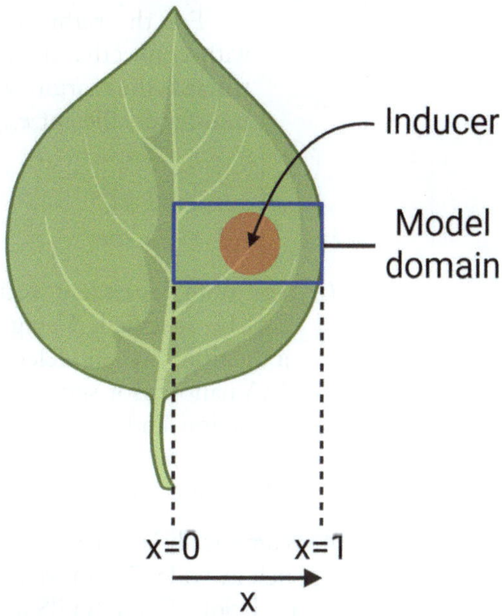

Fig. 7 Mathematical modeling of IAA distribution and transport in *XVE::iaaM* Arabidopsis plants [25]. β-estradiol inducer was topically applied to a small circular region (red circle) of a sensor-functionalized leaf, causing localized production of IAA. IAA nanosensors were applied across a large region of the leaf and were monitored for 18 h post-induction to visualize IAA transport. IAA concentrations calculated from the IAA nanosensor signals were used for modeling. The model domain (blue rectangle) was selected as a leaf strip centered around the inducer-treated region. The model domain was then spatially averaged into one dimension for model simplicity. The boundary conditions are defined at $x = 0$ and $x = 1$. (Created in BioRender.com. Adapted from [25])

), a Gaussian generative source from β-estradiol induction ($\overline{\gamma}\frac{1}{\overline{\sigma}\sqrt{2\pi}}\exp\left[\frac{-(x-\overline{a})^2}{2\overline{\sigma}^2}\right]$), and a reaction term for degradation ($-\phi^2\overline{C}$). x is the space variable, $\overline{a}$ and $\overline{\sigma}$ are Gaussian distribution parameters, $\overline{\gamma}$ is the magnitude of induced IAA production, and ϕ^2 is the degradation rate.

3. Determine boundary conditions based on the edges of the model domain. Generally, this will be equal to the order, or highest derivative, of the governing equation.

   (a) The diffusion term is second order, so two boundary conditions are needed to specify a solution. In this example, the left boundary is represented as an impermeable leaf midvein where IAA cannot cross (Fig. 7):

$$\left.\frac{d\overline{C}}{dx}\right|_{x=0} = 0$$

For the right boundary, the leaf margin is modeled with convective flux, representing a canal of auxin flow towards the margin, which acts as a sink governed by mass transfer coefficient k_{margin} (Fig. 7):

$$\left.\frac{d\overline{C}}{dx}\right|_{x=1} = -k_{margin}\overline{C}\ (x=1)$$

(b) For time-dependent equations, an initial condition should also be specified. Using *XVE::iaaM* plants with β-estradiol induction, we modeled concentration maps derived from IAA nanosensor signals measured at 6, 9, 14, and 18 h post-induction and assumed steady state ($\frac{dC}{dt}=0$) at each time point, which represents a slow change in concentrations [25]. Thus, an initial condition was not needed.

4. Numerically simulate the model. This can be done with solvers such as "ode15s" in MATLAB. Alternatively, commands such as "DSolve" and "FullSimplify" in Mathematica can be used to search for closed-form solutions.

5. Optimize model parameters to fit data. This involves setting initial conditions based on reasonable guesses and defining an objective function to minimize. Functions such as "fmincon" from the Optimization Toolbox in MATLAB can be used. Depending on model complexity, it can be helpful to use multiple different parameter initializations and parallel computing to search the parameter space faster (for example, using "MultiStart" with "UseParallel" set to true).

   (a) In our example, we used the "fmincon" solver with parallel "MultiStart" to solve for four parameters. The objective function was defined to minimize the sum of least squares of model vs. experimental data of IAA concentration, at each spatial location for a given time point.

   (b) The parameters should have similar order of magnitude values inside the optimizer. If necessary, scale the parameters before optimizing, then rescale later.

   (c) Some solvers like "fmincon" allow upper and lower bounds to be set. This can help restrict the searched parameter space to reasonable or physical values.

   (d) "fmincon" is a gradient-based solver. In some cases, such as noisy data or non-differentiable data, non-gradient-based solvers like genetic algorithm may yield better results.

   (e) Data smoothing may be necessary. Example methods include Savitzky-Golay filtering and cubic spline smoothing.

6. Perform downstream analysis and contextualize the model in terms of plant biology. This can include interpretation of the fitted parameter values and analysis of equation scaling.
7. In our model, we used non-dimensionalization to simplify the model. This reduces the number of parameters and facilitates scaling analysis, making it easier to interpret the relative importance of different variables. This enabled us to quantify the relationship between consumption and diffusion of IAA, both of which lowers IAA concentration in a given region. The model also quantitatively predicts an effective degradation rate of IAA and reflects how IAA content in the leaf margin saturates over time, which eventually results in a no-flux boundary.

4 Notes

1. The concentration of SWNT-based nanosensor in the aqueous solution can be estimated by UV-Vis-NIR spectroscopy, using its absorbance at 632 nm and extinction coefficient of 0.036 mg/L^1 cm^1.
2. When stored in a condition that is well-insulated, kept away from light, and at 4 °C, the IAA and reference nanosensors should be functional for at least one month. If the nanosensor solution was stored for more than one week, recentrifuge the solution at 16,500 *g* for 90 min, followed by an in vitro test to verify the sensor activity. We recommend testing the IAA nanosensor with 100 μM IAA, using a photoluminescence instrument such as Horiba Fluorolog. Otherwise, if visible aggregation is observed, the nanosensor suspension should be discarded.
3. Nitrile gloves should be worn when handling the IAA nanosensor solution and sensor-treated plant materials due to the human toxicity potential.
4. The excitation wavelength for the IAA nanosensor can vary from 550 nm to 800 nm. However, the NIR emission from the IAA nanosensor is stronger when the excitation is about 550–600 nm. While 785 nm excitation was used for the stand-off NIR camera system, the excitation wavelength used for the Olympus FV3000NIR inverted microscope was set at 561 nm to minimize integration time when capturing the signal.
5. The k_b and k_f parameter for IAA nanosensor were estimated using in vitro kinetic data [25].

Acknowledgments

This research was supported by the National Research Foundation (NRF), Prime Minister's Office, Singapore under its Campus for Research Excellence and Technological Enterprise (CREATE) program. Disruptive & Sustainable Technology for Agricultural Precision (DiSTAP) is an interdisciplinary research group of the Singapore-MIT Alliance for Research and Technology (SMART) Centre.

References

1. Zhao Y (2010) Auxin biosynthesis and its role in plant development. Annu Rev Plant Biol 61: 49–64. https://doi.org/10.1146/annurev-arplant-042809-112308
2. Sun P, Huang Y, Yang X et al (2023) The role of indole derivative in the growth of plants: a review. Front Plant Sci 13:1120613. https://doi.org/10.3389/fpls.2022.1120613
3. Du Y, Scheres B (2018) Lateral root formation and the multiple roles of auxin. J Exp Bot 69: 155–167. https://doi.org/10.1093/jxb/erx223
4. Xiong Y, Jiao Y (2019) The diverse roles of auxin in regulating leaf development. Plants (Basel) 8:243. https://doi.org/10.3390/plants8070243
5. Cucinotta M, Cavalleri A, Chandler JW et al (2021) Auxin and flower development: a blossoming field. Cold Spring Harb Perspect Biol 13:a039974. https://doi.org/10.1101/cshperspect.a039974
6. Verma S, Attuluri VPS, Robert HS (2021) An essential function for auxin in embryo development. Cold Spring Harb Perspect Biol 13: a039966. https://doi.org/10.1101/cshperspect.a039966
7. Ma L, Li G (2019) Auxin-dependent cell elongation during the shade avoidance response. Front Plant Sci 10:914. https://doi.org/10.3389/fpls.2019.00914
8. Burko Y, Willige BC, Seluzicki A (2022) PIF7 is a master regulator of thermomorphogenesis in shade. Nat Commun 13:4942. https://doi.org/10.1038/s41467-022-32585-6
9. Ai H, Bellstaedt J, Bartusch KS et al (2023) Auxin-dependent regulation of cell division rates governs root thermomorphogenesis. EMBO J 42:e111926. https://doi.org/10.15252/embj.2022111926
10. Michniewicz M, Brewer PB, Friml JÍ (2007) Polar auxin transport and asymmetric auxin distribution. Arabidopsis Book 5:e0108. https://doi.org/10.1199/tab.0108
11. Mashiguchi K, Tanaka K, Sakai T et al (2011) The main auxin biosynthesis pathway in Arabidopsis. Proc Natl Acad Sci USA 108:18512–18517. https://doi.org/10.1073/pnas.1108434108
12. Mano Y, Nemoto K (2012) The pathway of auxin biosynthesis in plants. J Exp Bot 63: 2853–2872. https://doi.org/10.1093/jxb/ers091
13. Luo P, Li TT, Shi WM et al (2023) The roles of GRETCHEN HAGEN3 (GH3)-dependent auxin conjugation in the regulation of plant development and stress adaptation. Plants (Basel) 12:4111. https://doi.org/10.3390/plants12244111
14. Geisler MM (2021) A retro-perspective on auxin transport. Front Plant Sci 12:756968. https://doi.org/10.3389/fpls.2021.756968
15. Lavy M, Estelle M (2016) Mechanisms of auxin signaling. Development 143:3226–3229. https://doi.org/10.1242/dev.131870
16. Ulmasov T, Murfett J, Hagen G et al (1997) Aux/IAA proteins repress expression of reporter genes containing natural and highly active synthetic auxin response elements. Plant Cell 9:1963–1971. https://doi.org/10.1105/tpc.9.11.1963
17. Ottenschläger I, Wolff P, Wolverton C et al (2003) Gravity-regulated differential auxin transport from columella to lateral root cap cells. Proc Natl Acad Sci USA 100:2987–2991. https://doi.org/10.1073/pnas.0437936100
18. Brunoud G, Wells DM, Oliva M et al (2012) A novel sensor to map auxin response and distribution at high spatio-temporal resolution. Nature 482:103–106. https://doi.org/10.1038/nature10791
19. Liao CY, Smet W, Brunoud G et al (2015) Reporters for sensitive and quantitative

measurement of auxin response. Nat Methods 12:207–210. https://doi.org/10.1038/nmeth.3279

20. Küpers JJ, Snoek BL, Oskam L et al (2023) Local light signaling at the leaf tip drives remote differential petiole growth through auxin-gibberellin dynamics. Curr Biol 33:75–85.e5. https://doi.org/10.1016/j.cub.2022.11.045
21. Herud-Sikimić O, Stiel AC, Kolb M et al (2021) A biosensor for the direct visualization of auxin. Nature 592:768–772. https://doi.org/10.1038/s41586-021-03425-2
22. Palmer AE, Qin Y, Park JG et al (2011) Design and application of genetically encoded biosensors. Trends Biotechnol 29:144–152. https://doi.org/10.1016/j.tibtech.2010.12.004
23. Donaldson L (2020) Autofluorescence in plants. Molecules 25:2393. https://doi.org/10.3390/molecules25102393
24. Zhang J, Landry MP, Barone PW et al (2013) Molecular recognition using corona phase complexes made of synthetic polymers adsorbed on carbon nanotubes. Nat Nanotechnol 8:959–968. https://doi.org/10.1038/nnano.2013.236
25. Khong DT, Vu KV, Sng BJR et al (2025) A near-infrared fluorescent nanosensor for direct and real-time measurement of indole-3-acetic acid in plants. ACS Nano 19:15302–15321. https://doi.org/10.1021/acsnano.4c13556
26. Boghossian AA, Zhang J, Barone PW et al (2011) Near-infrared fluorescent sensors based on single-walled carbon nanotubes for life sciences applications. ChemSusChem 4:848–863. https://doi.org/10.1002/cssc.201100070
27. Tan X, Calderon-Villalobos LI, Sharon M et al (2007) Mechanism of auxin perception by the TIR1 ubiquitin ligase. Nature 446:640–645. https://doi.org/10.1038/nature05731
28. Procko C, Crenshaw CM, Ljung K et al (2014) Cotyledon-generated auxin is required for shade-induced hypocotyl growth in *Brassica rapa*. Plant Physiol 165:1285–1301. https://doi.org/10.1104/pp.114.241844
29. Michaud O, Fiorucci AS, Xenarios I et al (2017) Local auxin production underlies a spatially restricted neighbor-detection response in *Arabidopsis*. Proc Natl Acad Sci USA 114:7444–7449. https://doi.org/10.1073/pnas.1702276114
30. Iglesias MJ, Sellaro R, Zurbriggen MD et al (2018) Multiple links between shade avoidance and auxin networks. J Exp Bot 69:213–228. https://doi.org/10.1093/jxb/erx295
31. Ruzza V et al (2014) Auxin coordinates shoot and root development during shade avoidance response. In: Zažímalová E, Petrášek J, Benková E (eds) Auxin and its role in plant development. Springer, Vienna, pp 389–412. https://doi.org/10.1007/978-3-7091-1526-8_17
32. Konstantinova N, Korbei B, Luschnig C (2021) Auxin and root gravitropism: addressing basic cellular processes by exploiting a defined growth response. Int J Mol Sci 22:2749. https://doi.org/10.3390/ijms22052749
33. Kawamura E, Horiguchi G, Tsukaya H (2010) Mechanisms of leaf tooth formation in Arabidopsis. Plant J 62:429–441. https://doi.org/10.1111/j.1365-313X.2010.04156.x
34. Navarro-Cartagena S, Micol JL (2023) Is auxin enough? Cytokinins and margin patterning in simple leaves. Trends Plant Sci 28:54–73. https://doi.org/10.1016/j.tplants.2022.08.019

Chapter 8

Gene Expression Analysis to Investigate the Response of Salicylic Acid in Plant Immunity

Binoop Mohan, Zhengzhi Tan, Yiqing Wang, and Shahid Mukhtar

Abstract

Salicylic acid (SA) functions as a key signaling molecule in plant defense, orchestrating responses to diverse biotic and abiotic stresses. A clear understanding of SA-mediated immunity requires systematic analysis of SA-responsive transcriptional programs at both targeted and genome-wide levels. This chapter describes methodological strategies for examining SA-responsive gene expression through quantitative polymerase chain reaction (qPCR) as well as bulk transcriptome profiling using the Bio-Rad SEQuoia Complete Stranded RNA Kit. The workflow covers experimental design, sample preparation, data acquisition, and result interpretation within the framework of plant immune responses. Integrating targeted qPCR assays with SEQuoia-enabled transcriptome analysis enables robust characterization of regulatory networks underlying SA-induced immunity and supports reproducible comparisons across experiments. This combined approach provides insights into the complexity of plant defense signaling and its transcriptional control.

Key words Salicylic acid, Plant immunity, Gene expression, qPCR, Transcriptome analysis, Bio-Rad SEQuoia

1 Introduction

Plant immunity relies on a sophisticated, multilayered defense system, with salicylic acid (SA) acting as a crucial signaling molecule that regulates both local defense and systemic acquired resistance (SAR) to pathogens [1, 2]. SA serves as a critical phytohormone in plant defense responses, particularly against biotrophic and hemibiotrophic pathogens [3]. In Arabidopsis, the biosynthesis of SA primarily occurs through the isochorismate pathway, where *ICS1/SID2* encodes isochorismate synthase, the rate-limiting enzyme catalyzing the conversion of chorismate to isochorismate [4, 5]. This pathway requires *PBS3* (AvrPphB Susceptible 3), which facilitates isochorismate conjugation, and EPS1 (Enhanced Pseudomonas Susceptibility 1), which is essential for SA accumulation during pathogen challenge [6, 7]. SA signaling operates

Shahid Mukhtar (ed.), *Plant Hormones: Methods and Protocols*, Methods in Molecular Biology, vol. 3026,
https://doi.org/10.1007/978-1-0716-5214-5_8,

through a tiered receptor system involving Nonexpressor of Pathogenesis-Related genes including NPR1 proteins [8]. NPR1 functions as the master transcriptional co-regulator, which translocates to the nucleus upon SA perception to activate defense gene expression [9, 10]. The paralogous receptors *NPR3* and *NPR4* regulate *NPR1* degradation in an SA concentration–dependent manner, thereby fine-tuning immune responses [11–13]. At high SA levels *NPR3* promotes *NPR1* turnover, whereas at low SA levels *NPR4* maintains *NPR1* stability, creating a dynamic regulatory circuit that allows graded immune activation [14, 15].

Advances in transcriptomics, especially high-throughput RNA sequencing (RNA-seq), have transformed our ability to study gene regulation in plants by providing comprehensive, genome-wide views of how model and crop species respond to both biotic and abiotic stresses [16–23]. RNA-seq has been particularly powerful for dissecting plant immune signaling [24], since it enables sensitive and quantitative profiling of salicylic acid (SA) responsive gene expression, revealing the timing, magnitude, and pathway specificity of SA-mediated defense activation [25]. These transcriptomic datasets can then be leveraged for co-expression network analyses to identify regulatory hubs, modules, and candidate genes that coordinate SA-mediated defense responses and broader stress adaptation mechanisms across plant systems [26–28]. Bulk RNA-seq captures the average transcriptomic landscape of entire tissues, providing an efficient and cost-effective method to investigate defense pathways, including genes encoding pathogenesis-related (PR) proteins, antimicrobial compounds, and secondary metabolites [29]. Quantitative PCR (qPCR) remains the gold standard for sensitive and accurate validation of targeted gene expression changes, although it is limited in throughput and scope [30, 31]. qPCR complements bulk RNA-seq by offering discovery-driven, transcriptome-wide insights, but it requires high-quality RNA, careful library preparation and appropriate bioinformatic analysis. In plant tissues, biological variation, tissue heterogeneity, secondary metabolites and RNA degradation can reduce library quality and reproducibility [32, 33]. To overcome these challenges, the Bio-Rad SEQuoia Complete Stranded RNA Library Prep Kit provides a streamlined bulk RNA-seq workflow optimized for plant tissues. This kit tolerates a wider range of RNA qualities, reduces input requirements, and improves library complexity, which is particularly useful when working with Arabidopsis leaves after pathogen infection or SA treatment. Integrating qPCR for targeted gene validation with SEQuoia-enabled bulk RNA-seq for broad transcriptomic analysis leverages the strengths of both methods to deliver precise, reproducible and biologically meaningful insights into SA-mediated defense. In this chapter, we present step-by-step experimental procedures, from plant growth and SA treatment to pathogen inoculation, RNA extraction, cDNA synthesis, qPCR, library preparation and differential-expression analysis, to facilitate reproducible SA-response studies in Arabidopsis.

2 Materials

2.1 Chemicals and Reagents

1. *Arabidopsis thaliana* seeds Ecotype: Columbia-0 (Col-0).
2. Standard Potting Soil.
3. 1020 Trays heavy duty with holes/no holes.
4. Traditional Inserts.
5. Salicylic acid: Weigh salicylic acid (138.12 g/mol), dissolve it in a small volume of ethanol, dilute with distilled water, add Silwet L-77 surfactant to a final concentration of 0.02% (v/v), and adjust the solution pH to 6.5.
6. Ethanol, Absolute (200 Proof), Molecular Biology Grade.
7. Bioworld SILWET L-77 10ML.
8. Liquid Nitrogen.
9. TRIzol Reagent.
10. National Target All-Plastic Disposable Syringes.
11. GeneJET Plant RNA Purification Kit.
12. Thermo Fisher Scientific Plant RNA Isolation Aid.
13. SEQuoia Complete Stranded RNA Library Prep Kit, 24 reactions.
14. Bio-Rad SEQuoia Dual Indexed Primers Set.
15. 2X Universal SYBR® Green Fast qPCR Mix.
16. ABScript II RT Master Mix for qPCR.
17. ABScript II RT Master Mix for qPCR with DNase I.
18. ABScript II cDNA First-Strand Synthesis Kit.
19. Thermo Scientific Pierce Microcentrifuge Tubes, 1.5 mL.
20. UltraPure™ DNase/RNase-Free Distilled Water.

2.2 Instruments

1. Conviron Arabidopsis Chamber.
2. FastPrep-24 5G bead beating grinder and lysis system.
3. Applied Biosystems™ QuantStudio™ 5 Real-Time PCR System.
4. Thermo Scientific Sorvall Legend Microcentrifuges.
5. Digital Vortex Mixers.
6. NanoDrop One/OneC Microvolume UV-Vis Spectrophotometer.

2.3 Software

1. QuantStudio Design & Analysis or equivalent for qPCR.
2. FastQC.
3. Trimmomatic.
4. STAR (or HISAT2).
5. featureCounts (or HTSeq).
6. DESeq2, edgeR.
7. GO-enrichment tool such as clusterProfiler.

3 Methods

3.1 Seed Setting, Transplanting, and Sample Collection

1. The tubes containing Col-0 seeds were transferred to a round pot of 3-inch size and wrapped with aluminum foil and placed in a dark container.
2. The seeds were stratified at 4 °C in a cool room for 72 h to synchronize germination.
3. The pots are transferred to the growth chamber after removing the aluminum wrap and let to grow for 2–3 days.
4. The trays were covered with transparent plastic domes for the first few days to maintain high humidity during germination.
5. Grow plants under the following growth conditions: Temperature: 22 °C, Photoperiod 12 h light/12 h darkness, Light intensity: 120 μmol m^{-2} s^{-1} and humidity at 55%.
6. Remove the plastic domes after germination.
7. When seedlings develop two true leaves, transfer individual plants to a 6 × 12 growing tray.
8. Water plants every 3 days, or as needed, to maintain consistent soil moisture.
9. Allow plants to grow in these trays for up to 4 weeks.
10. Spray-treat the plants.

3.2 RNA Extraction and Quality Assessment

1. Harvest fresh Arabidopsis leaf tissue (~100 mg). Immediately freeze in liquid nitrogen.
2. Grind to a fine powder using a homogenizer.
3. Transfer powdered tissue to an RNase-free microcentrifuge tube.
4. Add 1 mL of TRIzol reagent per 100 mg tissue. Vortex thoroughly and incubate at room temperature for 5 min.
5. Add 200 μL of chloroform per 1 mL TRIzol. Cap the tube tightly, shake vigorously for 15 s. Incubate at room temperature for 3 min.

6. Centrifuge at 12,000× *g* for 20 min at 4 °C. Carefully transfer the aqueous phase to a new RNase-free tube.
7. Add 500 μL of chilled isopropanol to the aqueous phase. Mix gently and incubate at room temperature for 10 min. Centrifuge at 12,000× *g* for 10 min at 4 °C. Discard supernatant and retain the RNA pellet.
8. Wash pellet with 1 mL of 75% ethanol. Vortex briefly and centrifuge at 7500× *g* for 5 min at 4 °C. Remove ethanol and air-dry the pellet.
9. Add DNase I and buffer as per manufacturer's instructions. Incubate at 37 °C for 30 min. Clean up RNA using RNA magnetic beads.
10. Dissolve RNA in 30 μL RNase-free water. Incubate at 55 °C for 10 min.
11. Check RNA quality by running it on a Bioanlyzer and check for RIN value >7, if RIN value is lower try to optimize the RNA extraction protocol.

3.3 Identifying Salicylic Acid-Specific Genes and Designing qPCR Primers

1. Target genes of interest ICS1, PBS3, NPR1, NPR2, NPR3, and reference genes UBQ10 were identified from the Arabidopsis genome database TAIR.
2. Gene sequences were retrieved in FASTA format from NCBI.
3. Use primer design tools to design gene-specific primers. Primers were designed with the following specifications, Length: 18–22 base pairs, melting temperature (Tm): 58–62 °C, GC content: 40–60% and amplicon size: 80–150 base pairs (Table 1) and ordered through IDT service.
4. Design the primers to span exon-exon junctions where it is possible to prevent amplification of genomic DNA.
5. Reconstitute the primers in nuclease-free water to a stock concentration of 100 μM. Prepare the working solutions of 10 μM by dilution.

3.4 First-Strand cDNA Synthesis and qPCR Reaction Mix

1. Thaw all reagents on ice and gently mix by flicking or pipetting. Avoid vortexing the enzyme.
2. Prepare the reaction mix in a sterile, nuclease-free tube (Table 2).
3. For primer annealing, incubate RNA and primer at 65 °C for 5 min, then chill on ice for 2 min.
4. Perform reverse transcription by incubating the complete reaction mix at 42 °C for 30–60 min.
5. Inactivate the enzyme by heating at 70 °C for 10 min to terminate the reaction.

Table 1
Primers used to validate SA-responsive gene expression

Gene name	AGI code	Forward primer	Reverse primer
ICS1/ SID2	At1g74710	ATGGCTTCACTTCAATTTTCTTC TC	TTATTGTGAGAACCCTTA TCCCCC
PBS3	At5g13320	ATGAAGCATTTGTGTGTCACAC TTC	GTTGTAGTTTCCTTTGCCC TTTTG
EPS1	At1g09440	ATGGCTGGACTTAAGATC TGGCAAG	TCACCTGCCGATGCACACTTC
NPR1	At1g64280	ATGGACACCACCATTGATGGA TTC	TCACCGACGACGATGAGAG TTTAC
NPR3	At5g45110	ATGGCTACTTTGACTGAGCCATC	TCATGTTGTGTTGCAGGTCATC TC
NPR4	At4g19660	ATGGCTGCAACTGCAATAGAGCC	TCATGTTGGATTCTCTAAGGC TTC
UBQ10	At4g05320	ATGCAGATCTTTGTTAAGACTC TCACC	TTAGAAACCACCACGAAGCACG

Table 2
First-strand cDNA synthesis reaction mix

Component	Volume
RNA template	1 μg
Random primer	1 μL
dNTP mix (10 mM)	1 μL
5X RT buffer	4 μL
RNase inhibitor	0.5 μL
ABScript II RTase	1 μL
Nuclease-free water	Calculate based on RNA volume

6. Store synthesized cDNA at −20 °C for short-term or − 80 °C for long-term use.
7. Thaw the 2X SYBR Green mix as mentioned in Table 3 at room temperature and protect from light.
8. Briefly centrifuge before use.
9. Prepare a master mix with all the ingredients mentioned in Table 3 except for the template. Aliquot into each well.
10. Add template or NTC (No Template Control) separately. Mix gently and centrifuge briefly.

Table 3
q-PCR master mix

Component	Volume
Template cDNA	2 μL
Forward primer (10 μM)	0.4 μL
Reverse primer (10 μM)	0.4 μL
2X SYBR Green mix	10 μL
Nuclease-free water	7.2 μL

11. Seal with optically clear film. Centrifuge to collect contents at the bottom.
12. Follow the thermal cycling conditions Initial denaturation 95 ° C for 2 min and hold the cycle followed by denaturation at 95 ° C for 5 s and annealing/extension at 60 °C for 30–34 s for 40 cycles.
13. Determine the cycle threshold (Ct) values automatically using the instrument software at the point where fluorescence exceeds the threshold level.
14. Calculate mean Ct values from technical triplicates for each sample.
15. Calculate ΔCt values by subtracting the geometric mean Ct of the reference genes from the target gene Ct for each sample: ΔCt = Ct(target)—Ct (reference mean).
16. Calculate ΔΔCt values by subtracting the ΔCt of the calibrator from the ΔCt of each test sample: ΔΔCt = ΔCt(sample)—ΔCt (calibrator).
17. Determine the relative fold change in gene expression using the formula: 2^(-ΔΔCt).

3.5 Biorad SEQuoia RNA Library Preparation

1. Preheat the thermal cycler to 94 °C (or 70 °C for degraded RNA).
2. Prepare 1 μg to 1 ng of total RNA or 100 pg to 1 μg of pre-depleted RNA and adjust the sample volume to 16 μL with RNase-free water.
3. Add 2 μL of Fragmentation Mix (Reagent A), then vortex briefly and centrifuge to collect contents.
4. Incubate at the designated temperature for 1–10 min, depending on RNA integrity and desired fragment size, then place the reaction on ice for 2–5 min to stop fragmentation.

5. Set the thermal cycler to 37 °C for end repair. Add 2 μL of End Repair Enzyme (Reagent B) to each tube, then vortex and briefly centrifuge.
6. Incubate at 37 °C for 20 min, then at 70 °C for 4 min, and hold at 4 °C. Place the tubes on ice for 2–5 min.
7. Preheat the cycler to 16 °C for poly(A) tailing.
8. Prepare a master mix using 23.75 μL of Poly(A) Buffer (Reagent C) and 1.25 μL of Poly(A) Polymerase (Reagent D).
9. Add 25 μL of the master mix to each reaction tube, then vortex gently and centrifuge.
10. Incubate at 16 °C for 5 min, then at 70 °C for 2 min, and hold at 4 °C. Place the tubes on ice for 2–5 min.
11. Set the cycler to 34 °C for continuous synthesis.
12. Add 5 μL of SEQzyme Mix (Reagent E), then vortex and centrifuge.
13. Incubate at 34 °C for 60 min, then hold at 4 °C.
14. Perform cDNA purification using purification beads. Bring the beads to room temperature, vortex, and add 20 μL of beads to each tube.
15. Mix by pipetting and incubate at room temperature for 20 min, then place the tubes on a magnetic rack for 5–10 min.
16. Transfer the supernatant to a new tube.
17. Add 112 μL of SPRIselect Reagent (1.6×) to the purified cDNA, mix, and incubate at room temperature for 5–8 min.
18. Centrifuge, place on the magnet, and wash the beads twice with 200 μL of 80% ethanol.
19. Air-dry the beads for 1–4 min (do not over dry). Elute with 24 μL of 10 mM Tris–HCl, pH 8.0. After incubation and magnetic separation, collect 22.5 μL of the supernatant.
20. Assemble the library amplification reaction using 2.5 μL of dual-indexed primers, 25 μL of Amplification Mix (Reagent F), and 22.5 μL of cDNA library. Mix and centrifuge.
21. Run PCR using the following thermal profile: 95 °C for 180 s; 98 °C for 20 s, 61 °C for 15 s, 72 °C for 20 s (6–18 cycles); 72 ° C for 60 s; hold at 4 °C.
22. For post-amplification cleanup, add 60 μL of SPRIselect Reagent (1.2×), mix, and incubate at room temperature for 5–8 min.
23. Wash the beads twice with 200 μL of 80% ethanol and air-dry for 1–4 min.

24. Elute the amplified library with 20 μL of Tris–HCl buffer. After incubation and magnetic separation, collect 18 μL of the supernatant.
25. Store the final libraries at 4 °C or −20 °C and assess size distribution using a Bioanalyzer.

3.6 Bioinformatics Analyses and Data Presentation

1. Run FastQC on raw FASTQ files to evaluate base quality, GC distribution, adapter content, and overrepresented sequences.
2. Summarize all FastQC reports using MultiQC to generate an integrated visual summary of quality metrics across samples (*see* **Note 1**).
3. Trim adapter sequences and low-quality bases using Trimmomatic (or Cutadapt) with a minimum Phred score of 25 and read length cutoff of 50 bp.
4. Align the cleaned reads to the *Arabidopsis thaliana* reference genome (TAIR10) using STAR (preferred for splice awareness and speed) or HISAT2.
5. Use a comprehensive GTF annotation file that includes protein-coding genes, long noncoding RNAs (lncRNAs), and small/novel transcripts so that all RNA types detectable by the SEQuoia Complete Stranded RNA Library Prep Kit are quantified.
6. Use featureCounts (Subread) or HTSeq-count to generate raw read counts per gene (*see* **Note 2**).
7. Store all count matrices in a structured format (e.g., tab-delimited text or Excel) with clear sample identifiers for downstream analysis.
8. Import count matrices into DESeq2 (R) and define a design formula reflecting experimental conditions such as treatment (SA vs. mock).
9. Perform normalization using the median-of-ratios method to correct for library size differences. Estimate dispersion parameters and identify differentially expressed genes (DEGs) using the Wald test or likelihood ratio test (LRT) where appropriate.
10. Apply an adjusted p-value (FDR < 0.05) and $\log_2$ fold-change cutoffs ($|\log_2 FC| \geq 1$) to identify significantly SA-responsive genes.
11. Use volcano plots to visualize the magnitude and significance of DEGs, where the *x*-axis represents $\log_2$ fold change and the *y*-axis shows $-\log_{10}$(p-value).
12. Highlight canonical SA-responsive genes (e.g., PR1, PR2, PR5, ICS1, PBS3, NPR1) as positive controls to confirm treatment efficacy (*see* **Note 3**).

13. Generate Principal Component Analysis (PCA) plots to assess overall variance and clustering between biological replicates and treatments.
14. Create heatmaps of the top 50–100 DEGs using pheatmap or ComplexHeatmap (R packages) to visualize expression patterns across samples. Hierarchical clustering can be applied to reveal relationships between treatment groups and replicate consistency.
15. Additionally, use MA plots to illustrate global gene expression shifts and identify systematic biases or normalization issues.
16. Conduct Gene Ontology (GO) and KEGG pathway enrichment analysis for upregulated and downregulated gene sets using clusterProfiler or g:Profiler.
17. Emphasize enrichment in defense-related biological processes (e.g., systemic acquired resistance, cell wall modification, SA biosynthesis) and signaling pathways related to immune response and redox homeostasis.
18. Because the SEQuoia Kit captures both coding and noncoding RNAs, include differential expression profiles of small RNAs and lncRNAs when supported by annotation. Cross-reference differentially expressed noncoding RNAs with nearby defense-related coding genes to infer potential regulatory relationships.
19. Cross-validate RNA-seq-derived expression patterns using qPCR for key SA-responsive genes.
20. Provide sample metadata, raw read counts, normalized expression tables, and complete DEG lists as supplementary data to comply with MIAME-/MINSEQE guidelines. Include visualization outputs (PCA, volcano plots, heatmaps, MA plots) in the main or supplementary figures to facilitate data interpretation and reproducibility.

4 Notes

1. Samples failing base quality or adapter thresholds should be trimmed and re-evaluated before alignment.
2. For stranded libraries, ensure both sense and antisense reads are counted according to the library's strandedness.
3. Where small or noncoding RNAs are detected, report their differential expression separately and, when possible, map them to known defense-related regulatory RNAs.

Acknowledgments

This work was supported by NSF award IOS-2038872, PGRP-2318746, and OIA-2418230 to S.M.

References

1. Vlot AC, Dempsey DMA, Klessig DF (2009) Salicylic acid, a multifaceted hormone to combat disease. Annu Rev Phytopathol 47(1): 177–206
2. Boatwright JL, Pajerowska-Mukhtar K (2013) Salicylic acid: an old hormone up to new tricks. Mol Plant Pathol 14(6):623–634. https://doi.org/10.1111/mpp.12035
3. Fang X, Xie Y, Yuan Y et al (2025) The role of salicylic acid in plant defense responses against biotic stresses. Plant Hormones 1(1)
4. Li N, Han X, Feng D et al (2019) Signaling crosstalk between salicylic acid and ethylene/jasmonate in plant defense: do we understand what they are whispering? Int J Mol Sci 20(3): 671
5. Wildermuth MC, Dewdney J, Wu G et al (2001) Isochorismate synthase is required to synthesize salicylic acid for plant defence. Nature 414(6863):562–565
6. Torrens-Spence MP, Bobokalonova A, Carballo V et al (2019) PBS3 and EPS1 complete salicylic acid biosynthesis from isochorismate in Arabidopsis. Mol Plant 12(12):1577–1586
7. Rekhter D, Lüdke D, Ding Y et al (2019) Isochorismate-derived biosynthesis of the plant stress hormone salicylic acid. Science 365(6452):498–502
8. Kumar P, Pandey S, Pati PK (2025) Interaction between pathogenesis-related (PR) proteins and phytohormone signaling pathways in conferring disease tolerance in plants. Physiol Plant 177(2):e70174
9. Pajerowska-Mukhtar KM, Emerine DK, Mukhtar MS (2013) Tell me more: roles of NPRs in plant immunity. Trends Plant Sci 18(7):402–411. https://doi.org/10.1016/j.tplants.2013.04.004
10. Sun Y, Detchemendy TW, Pajerowska-Mukhtar KM et al (2018) NPR1 in JazzSet with pathogen effectors. Trends Plant Sci 23(6):469–472. https://doi.org/10.1016/j.tplants.2018.04.007
11. Spoel SH, Dong X (2024) Salicylic acid in plant immunity and beyond. Plant Cell 36(5): 1451–1464
12. van Butselaar T, Van den Ackerveken G (2020) Salicylic acid steers the growth–immunity tradeoff. Trends Plant Sci 25(6):566–576
13. Gao Q-M, Zhu S, Kachroo P et al (2015) Signal regulators of systemic acquired resistance. Front Plant Sci 6:228
14. Liu Y, Sun T, Sun Y et al (2020) Diverse roles of the salicylic acid receptors NPR1 and NPR3/NPR4 in plant immunity. Plant Cell 32(12):4002–4016
15. Ding Y, Sun T, Ao K et al (2018) Opposite roles of salicylic acid receptors NPR1 and NPR3/NPR4 in transcriptional regulation of plant immunity. Cell 173(6):1454–1467.e1415
16. Hussain A, Asif N, Pirzada AR et al (2022) Genome wide study of cysteine rich receptor like proteins in Gossypium sp. Sci Rep 12(1): 4885. https://doi.org/10.1038/s41598-022-08943-1
17. Liu J, Fakhar AZ, Pajerowska-Mukhtar KM et al (2022) A TIReless battle: TIR domains in plant-pathogen interactions. Trends Plant Sci 27(5):426–429. https://doi.org/10.1016/j.tplants.2022.01.011
18. Liu J, Majeed A, Kumar N et al. (2024) Single nuclei multiomics reveals the drought-driven gene regulatory atlas in Arabidopsis. bioRxiv. https://doi.org/10.1101/2024.01.11.575118
19. Liu J, Majeed A, Mukhtar MS (2022) Rooting through single-cell sequencing in phloem pole cells. Commun Biol 5(1):1194. https://doi.org/10.1038/s42003-022-04180-4
20. Naqvi RZ, Zaidi SS, Akhtar KP et al (2017) Transcriptomics reveals multiple resistance mechanisms against cotton leaf curl disease in a naturally immune cotton species, Gossypium arboreum. Sci Rep 7(1):15880. https://doi.org/10.1038/s41598-017-15963-9
21. Naqvi RZ, Zaidi SS, Mukhtar MS et al (2019) Transcriptomic analysis of cultivated cotton Gossypium hirsutum provides insights into host responses upon whitefly-mediated transmission of cotton leaf curl disease. PLoS One 14(2):e0210011. https://doi.org/10.1371/journal.pone.0210011

22. Thingujam D, Gouli S, Cooray SP et al (2025) Climate-resilient crops: integrating AI, multi-omics, and advanced phenotyping to address global agricultural and societal challenges. Plants (Basel) 14(17). https://doi.org/10.3390/plants14172699
23. Hussain A, Farooq M, Naqvi RZ et al (2020) Genome-wide identification and classification of resistance genes predicted several decoy domains in Gossypium sp. Plant Gene 24: 100250
24. Thingujam D, Tan Z, Wang Y et al (2025) PRIMER cells: immune hotspots in plants. Trends Plant Sci 30(10):1075–1078. https://doi.org/10.1016/j.tplants.2025.04.006
25. Hussain A, Liu J, Mohan B et al (2022) A genome-wide comparative evolutionary analysis of zinc finger-BED transcription factor genes in land plants. Sci Rep 12(1):12328
26. Kumar N, Mishra BK, Liu J et al (2023) Network biology analyses and dynamic modeling of gene regulatory networks under drought stress reveal major transcriptional regulators in Arabidopsis. Int J Mol Sci 24(8). https://doi.org/10.3390/ijms24087349
27. Kumar N, Mukhtar MS (2023) Integrated systems biology pipeline to compare co-expression networks in plants and elucidate differential regulators. Plants (Basel) 12(20). https://doi.org/10.3390/plants12203618
28. Kumar N, Mukhtar MS (2023) Ranking plant network nodes based on their centrality measures. Entropy (Basel) 25(4). https://doi.org/10.3390/e25040676
29. Dempsey DMA, Klessig DF (2012) SOS–too many signals for systemic acquired resistance? Trends Plant Sci 17(9):538–545
30. Mukhtar MS (2005) Defining WRKY factors involved in plant defense: characterization of Arabidopsis thaliana WRKY27, a gene affecting Ralstonia solanacearum disease development. Universität zu Köln
31. Nolan T, Hands RE, Bustin SA (2006) Quantification of mRNA using real-time RT-PCR. Nat Protoc 1(3):1559–1582
32. Meyers BC, Galbraith DW, Nelson T et al (2004) Methods for transcriptional profiling in plants. Be fruitful and replicate. Plant Physiol 135(2):637–652
33. Lightfoot S, Salowsky R, Buhlmann C (2005) RNA integrity number: towards standardization of RNA quality assessment for better reproducibility and reliability of gene expression experiments. Breast Cancer Res 7(Suppl 2):P7. 05

Chapter 9

Practical RNA-Seq with Spike-Ins: A Bench-to-Bioinformatics Guide

T. Blake Horton, Kanjana Laosuntisuk, and Colleen J. Doherty

Abstract

RNA-Sequencing (RNA-Seq) is a powerful, high-throughput technology for genome-wide transcriptional analysis, widely used to investigate gene expression changes in plants. A critical step in RNA-Seq data analysis is normalization, which enables accurate comparisons of gene expression between samples by reducing the effects of technical and biological factors that can confound expression measurements. Traditional normalization approaches, such as those implemented in DESeq2's median-of-ratios and edgeR's Trimmed Mean of M-values, operate under the assumption that the total transcript levels or the expression of most genes are stable across samples. However, this assumption is often violated, particularly in plant studies, where treatments, genotypes, or even time-of-sampling differences can affect overall transcription efficiency or RNA stability, producing unequal transcript abundance (Coate and Doyle, Genome Biol Evol 2:534–546, 2010; Laosuntisuk et al. Plant J 118:1241–1257, 2024; Forsythe et al. Proc Natl Acad Sci USA 119:e2204187119, 2022; Wang et al. New Phytol 230:1985–2000, 2021). Such violations can result in inappropriate scaling of expression levels, leading to incorrect identification of differentially expressed genes (DEGs). Exogenous RNA spike-ins offer a robust solution for controlling variations in total mRNA levels and technical biases. This chapter presents practical protocols for integrating exogenous RNA spike-ins at the bench and in downstream computational analysis. We show how to calculate spike-in amounts, implement spike-in-based normalization strategies, and contrast these with distribution-based methods. The goal of this chapter is to lower the barriers to adopting RNA spike-ins for plant transcriptional analysis, thereby improving the accuracy, specificity, and sensitivity of DEG calling in RNA-Seq experiments to ensure more reliable and interpretable biological conclusions.

Key words RNA-sequencing, Normalization, Synthetic RNA spike-ins, Differential expression, DESeq2, edgeR, Nu*Delta normalization, Plant RNA-Seq, Transcriptional abundance

1 Introduction

1.1 Why Normalization Matters

RNA Sequencing (RNA-Seq) enables genome-wide measurement of transcript abundance. It is widely used and has revolutionized molecular biology, offering a comprehensive view of the transcriptome. Critically, RNA-Seq enables the evaluation of gene expression changes in response to experimental treatments, such as

Shahid Mukhtar (ed.), *Plant Hormones: Methods and Protocols*, Methods in Molecular Biology, vol. 3026,
https://doi.org/10.1007/978-1-0716-5214-5_9,

exogenous hormone applications or environmental changes, even in non-model species. However, RNA-Seq is susceptible to biases that can obscure actual differences in gene expression. Within-sample factors, such as GC content, gene length, and contamination, can affect gene detection. Between-sample factors, especially variation in sequencing depth, impact comparability, making between-sample normalization essential.

Techniques like 3' RNA-Seq help mitigate gene length bias by quantifying only a single read per transcript and have been demonstrated in several plant species [5–12]. Still, sequencing depth, composition differences, and other systematic effects must be addressed through normalization approaches to ensure an accurate assessment of transcriptional changes.

1.2 Limits of Distribution-Based Normalization

Most standard normalization methods, including DESeq2's median-of-ratios [13] and edgeR's Trimmed Mean of M-values (TMM) [14], assume that the majority of genes have stable expression between samples. While often effective, this fundamental assumption can be violated. Global transcription levels change in many plant biology contexts, for example, when treatments alter global transcription rates, RNA stability, or tissue composition. In such cases, traditional distribution-based scaling methods can then inappropriately force expression levels to match, biasing fold changes and increasing the number of false positives/negatives [15]. For example, if a large group of genes is drastically upregulated (e.g., photosynthetic genes in plants during the day), distribution-based scaling can artificially repress other genes in that sample [2].

1.3 Why Exogenous RNA Spike-ins Help

Exogenous spike-in RNAs of known sequence and concentration provide a simple and effective solution. First developed for microarray and RT-qPCR analysis [16–22], spike-ins serve as external controls to normalize for variations in total RNA levels and technical biases across samples. While commonly used in mammalian and yeast systems, their application in plant RNA-Seq studies has been relatively limited.

Spike-in normalization, which accurately corrects for changes in global transcription and the proportional shares of the RNA pool, has been shown to result in improved accuracy, specificity, and sensitivity in identifying differentially expressed genes (DEGs) [2, 15, 23, 24]. RNA spike-ins can help distinguish true biological differences from technical variations introduced during sample preparation. Critically, the use of RNA spike-ins can reveal global changes in transcription, such as transcriptional amplification, that are masked by other normalization methods. Spike-ins can also be used to compare and optimize sequencing workflows [25]. This

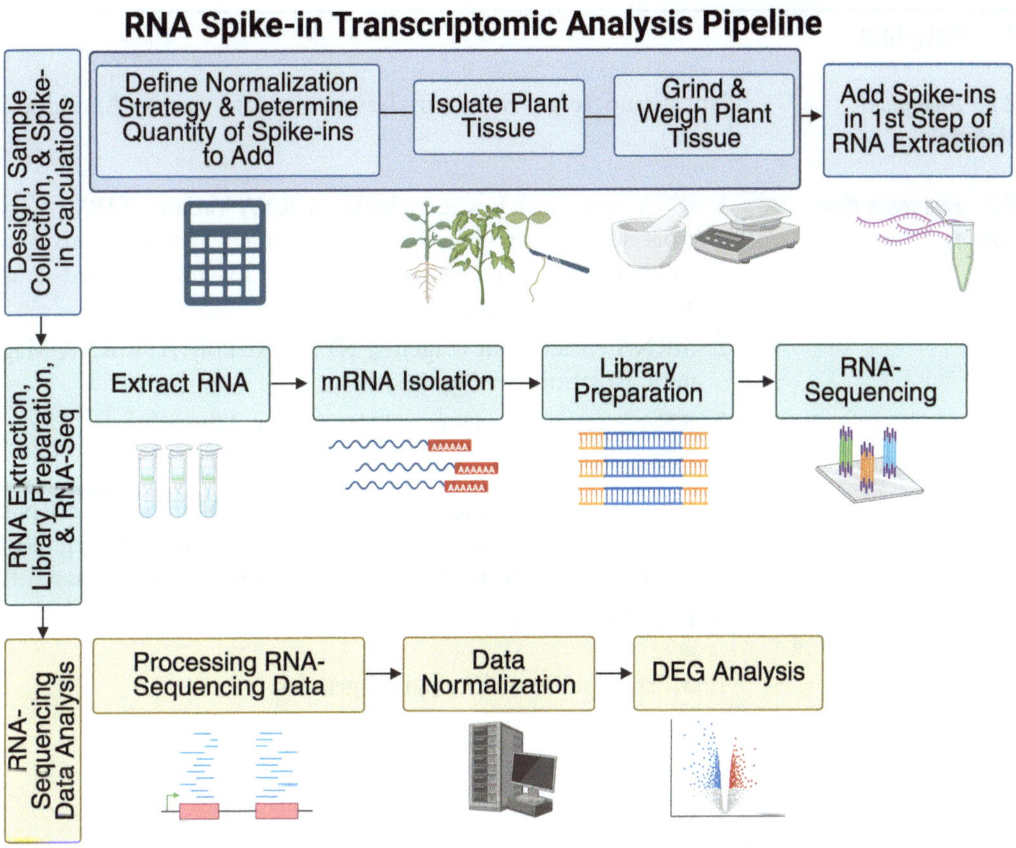

Fig. 1 Workflow for RNA-seq with exogenous RNA spike-ins. Overview of the end-to-end process used to control for global transcriptional variation with external RNA spike-ins. Rows depict: (1) Design, sample collection & spike-in calculations (set target spike-in fraction, plan dosing, add spike-ins prior to extraction), (2) RNA extraction, library preparation & RNA-Seq, and (3) RNA-sequencing data analysis. Subsequent figures expand each row. (Created with BioRender)

chapter details practical protocols for incorporating spike-ins into plant RNA-Seq experiments and implementing spike-in–based normalization in R (Fig. 1).

1.4 Methods Covered

This chapter describes practical protocols for incorporating spike-ins into plant RNA-Seq workflows, provides worked examples of spike-in calculations, and demonstrates three spike-in-based normalization methods in R:

- Nu*Delta Spike-in Normalization [26].
- Spike-in derived size factors using DESeq2 [27].
- Remove Unwanted Variation (RUV) using spike-ins [28].
- Median of Ratios Normalization (DESeq2) [29] (for comparison).

2 Materials

2.1 Biological Materials

Plant tissue (e.g., Sorghum leaves, Tomato roots, Arabidopsis shoots).

2.2 Reagents (See Note 1)

1. RNA Spike-ins: Lexogen Spike-in RNA Variant (SIRV) Controls set 3 (Lexogen, Vienna, Austria). Synthetic mixes or spike-ins from heterologous species [23] can be used as an alternative (*see* **Note 2**).
2. mRNA enrichment reagents: NEBNext Poly(A) mRNA Magnetic Isolation (*see* **Note 3**).
3. RNA extraction reagents: RNeasy Plant Mini Kit.
4. DNase I.
5. Library preparation kit: For 3′ RNA-Seq, QuantSeq 3′ mRNA-Seq Library Prep Kit for Illumina or for mRNA-Seq, NEB NEXT Ultra-directional RNA-Seq Library Preparation.
6. Liquid nitrogen.

2.3 Equipment

1. Mortar/pestle or mechanical grinder.
2. Micropipettes.
3. Vortex.
4. Centrifuge.
5. Microfuge.
6. Nanodrop or equivalent.
7. Agilent TapeStation.
8. Illumina NovaSeq 6000 or other sequencing platform.

2.4 Software (Examples)

1. *Quality Control:* FastQC (v0.12.1) https://www.bioinformatics.babraham.ac.uk/projects/fastqc/.
2. *Trimming/Adaptor Removal:* BBMap/BBDuk (v39.06) https://sourceforge.net/projects/bbmap/ [30].
3. *Alignment:* STAR aligner (v2.7.11a) https://github.com/alexdobin/STAR [31].
4. *Sorting:* SAMtools (v1.20) https://www.htslib.org/ [32].
5. *Feature Summarization:* Subread (v2.0.2) https://subread.sourceforge.net/featureCounts.html [33]/ HTSeq-count (v0.11.2) https://htseq.readthedocs.io/en/release_0.11.1/index.html.
6. *Statistical Analysis:* R ($\geq$ 4.0.0) with packages: `tidyverse`, `EDASeq`, `edgeR`, `DESeq2`, `RUVSeq`, `apeglm`.
7. *Workflow:* RStudio (Version 2024.04.2 + 764).

3 Methods

3.1 Determining Spike-in Amounts

The choice of normalization basis (e.g., per plant, per tissue weight, per cell number, or per total DNA content) significantly impacts the identified DEGs (Fig. 2). Normalization approaches should be carefully considered in relation to the experimental question (*see* **Note 4** and three examples demonstrating calculations for determining spike-in volumes in **Notes 5** and **6**) (Fig. 3).

1. Choose the normalization basis (e.g., per tissue, mass, seedling, or DNA content).
2. Select the target fraction of spike-in reads, F_{SIRV} typically 0.5–2% of reads.
3. Record fraction of target RNA species (e.g., mRNA) in total RNA, F_{RNA} (Typically ~0.03 for mRNA studies).
4. Record the total RNA input per library, M_{RNA} (ng).
5. Record the working spike-in concentration, C_{SIRV} (*see* **Note 7** for how to prepare the spike-in dilutions and **Note 8** concerning the final dilution amount).

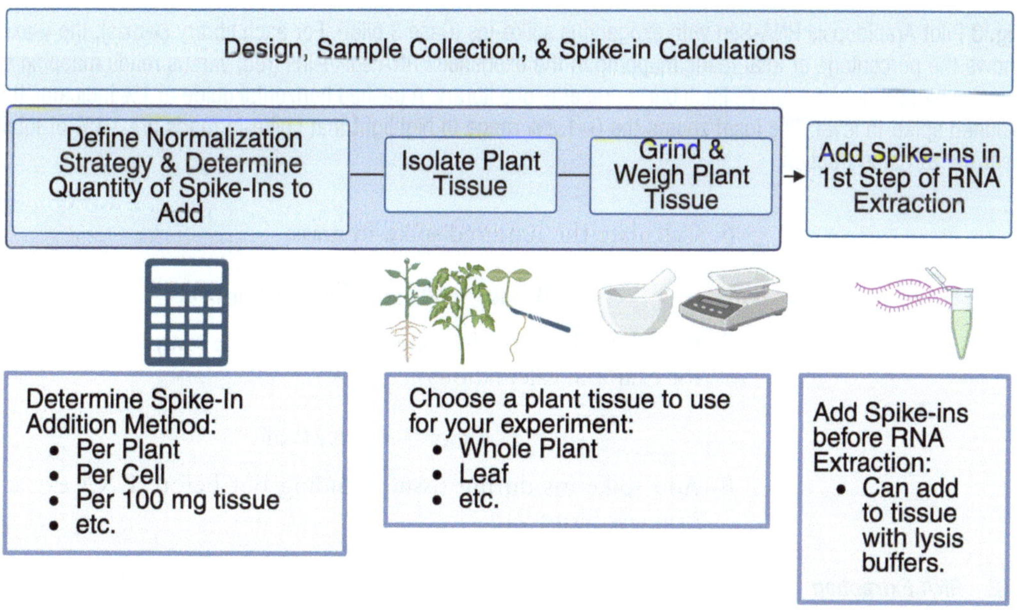

Fig. 2 Experimental design and calculations to optimize spike-in dosing for normalization. Schematic of the planning steps used to select a target spike-in read fraction, estimate RNA yield/complexity, and compute dilution and per-sample dosing. The boxed steps are iterated first using estimated inputs; a short pilot run then verifies detectability and coverage, after which spike-in amount/dilution is adjusted as needed to reach the target fraction for the full experiment. (Created with BioRender)

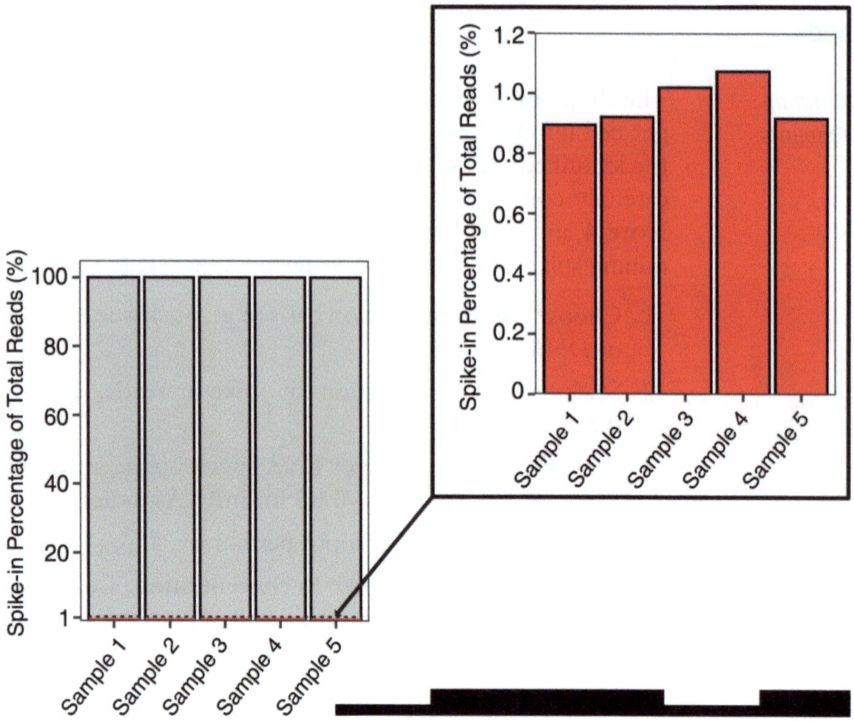

[redacted]

Fig. 3 Pilot Arabidopsis RNA-Seq with exogenous spike-ins (Case 3 pilot). For each library (*x*-axis), the *y*-axis shows the percentage of total reads mapping to the exogenous RNA spike-ins (red) versus reads mapping to the Arabidopsis genome (grey). Each bar represents one library. A dashed horizontal guide at 1% indicates the planned spike-in level. The inset zooms the 0–1.2% range to highlight that spike-in reads are ~1% of total, near the targeted fraction

6. Calculate the required spike-in mass:

$$M_{\mathrm{SIRV}} = (F_{\mathrm{SIRV}})(F_{\mathrm{RNA}})(M_{\mathrm{RNA}})$$

7. Calculate the spike-in volume to add to the sample (*see* **Note 6** for example calculations).

$$V_{\mathrm{SIRV}} = M_{\mathrm{SIRV}}/C_{\mathrm{SIRV}}$$

8. Add spike-ins during tissue grinding but before RNA extraction (*see* **Note 9**).

3.2 RNA Extraction and Library Preparation

Plant RNA extraction with spike-ins follows standard workflows with one critical modification: the calculated spike-in volume (V_{SIRV}) must be added early in the process to track RNA losses during extraction (*see* **Note 9**) (Fig. 4).

1. Flash-freeze tissue in liquid nitrogen and store at −80 °C until processing.

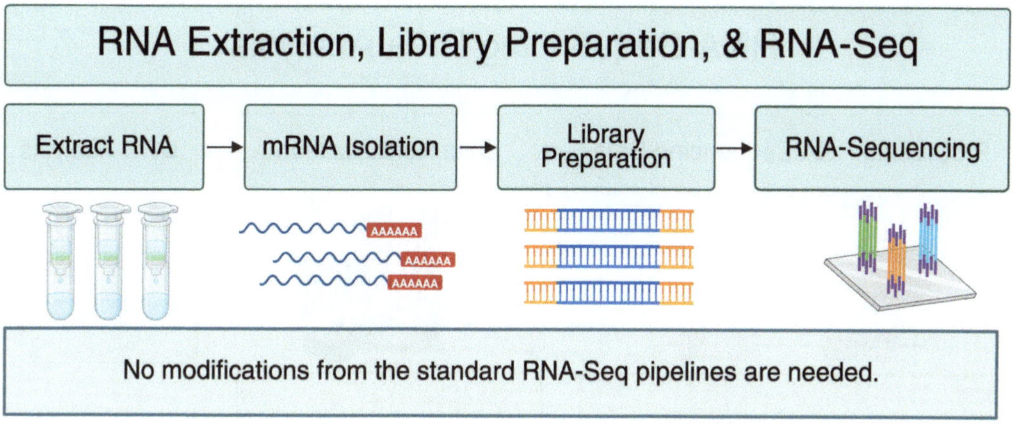

Fig. 4 Workflow schematic for RNA extraction, library preparation, and sequencing with spike-ins. Diagram of the standard wet-lab pipeline from RNA extraction, library preparation, and sequencing, showing that when exogenous RNA spike-ins are added upstream, before RNA extraction, they co-purify and are processed identically to sample RNA. No protocol steps, kit settings, or instrument parameters require modification due to the presence of RNA spike-ins. (Created with BioRender)

2. Grind to a fine powder using a mortar and pestle, ball mill, or equivalent homogenizer.
3. Add calculated V_{SIRV} directly to the ground tissue before or at the point of lysis (*see* **Notes** 7 and **10**).
4. Extract total RNA using Qiagen RNeasy Plant Mini Kit (or equivalent), including DNase I treatment to remove genomic DNA.
5. Quantify RNA and assess purity using Agilent TapeStation or Nanodrop plus an RNA gel.
6. Enrichment of target RNA. For mRNA analysis, enrich mRNA (poly(A) selection) or deplete rRNA. For 3′ RNA-Seq protocols, enrichment is not required (*see* **Note 3**).
7. Prepare libraries using standard kits or protocols (e.g., NEB-Next Ultra-Directional RNA-Seq Library Prep for full-length mRNA; Lexogen QuantSeq for 3′-Seq).
8. Quality Control (QC) of libraries with Agilent TapeStation to assess size distribution and concentration.
9. Store libraries individually at −20 °C and pool equimolarly prior to sequencing (*see* **Note 11**).
10. Sequence libraries (e.g., Illumina NovaSeq 6000).

3.3 Data Preprocessing (Fig. 5)

Preprocessing is identical to standard RNA-Seq until the alignment step. The key difference is aligning to both the experimental organism and the spike-ins. Creating a combined reference containing both the organism and the spike-in sequences (*see* Subheading 3.3.1). Generate BAM files and count tables for endogenous and spike-in features, then separate them for normalization.

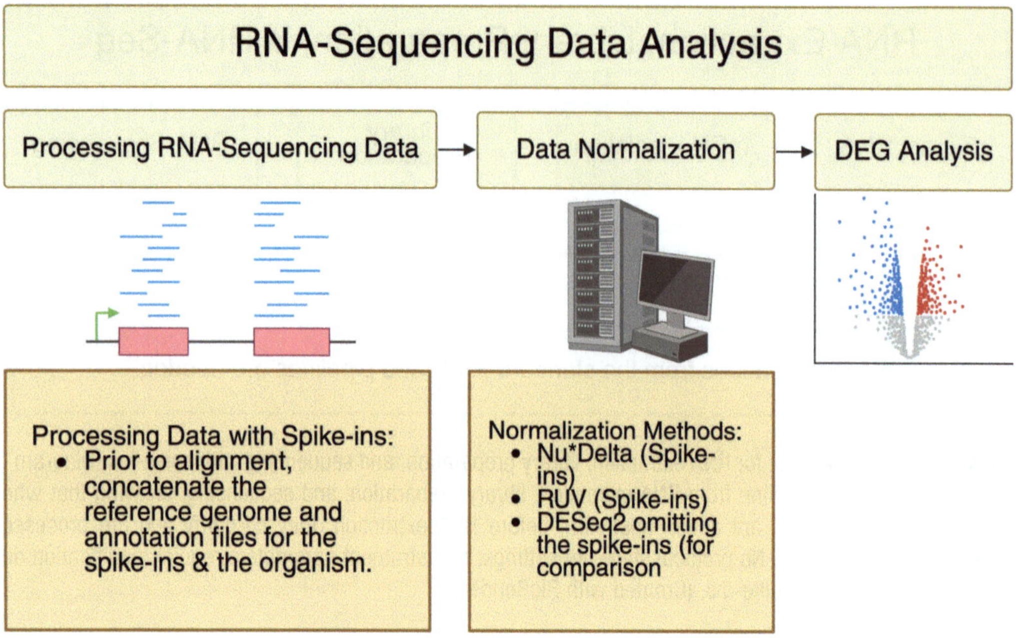

Fig. 5 RNA-sequencing data analysis. Diagram of the workflow for RNA-Seq analysis using RNA spike-ins. The alignment needs to be made to both the organism and the spike-in sequences. We outline the use of a concatenated reference (organism + spike-in FASTA/GTF). The concatenated reference ensures consistent mapping of both spike-in and endogenous reads. Normalization options illustrated include: Nu*Delta spike-in calibration (captures global transcriptional shifts), RUVg using spike-ins as negative controls (addresses unwanted technical variation), and DESeq2 median-of-ratios without spike-ins. Spike-ins can also be omitted from scaling; running a parallel analysis without them provides a useful comparison. (Created with BioRender)

3.3.1 Combining References Example

1- Linux command line example for generating the combined genome assembly.

```
cat Organism_reference.fasta Spikein_reference.fasta > Orga-
nism_Spikein_reference.fasta
```

2- Linux command line example for generating the combined annotation.

```
cat Organism_annotation.gtf Spikein_annotation.gtf > Orga-
nism_Spikein_annotation.gtf
```

3.3.2 Workflow

1. QC raw reads; *FastQC*.
2. Trim adapters: *BBDuk*.
3. Align reads to a concatenated FASTA (plant and spike-ins): *STAR*.
4. Sort/index BAM files: *SAMtools*.

5. Generate count tables: *HTSeq-count or Subread.*
6. Split spike-in features into a separate count table for downstream normalization.

3.4 Normalization Approaches

Spike-ins can be leveraged for normalization in multiple ways. We describe four representative approaches (Fig. 5) (*see* **Notes 12** and **13**).

3.4.1 DESeq2 Median-of-Ratios [29] (See **Note 14**)

1. Remove the spike-ins from the count table.

```
#For a count table imported from featureCounts named "BT_features.RData", ensure that gene IDs are a separate column for filtering.

# Load Combined Table:
BT <- readRDS("RData/BT_features.RData")

# Remove ERCC Spike-ins and name the dataframe "BTxGene":
BTxGene <- BT[!grepl("^ERCC", BTxGene$gene),]

#Verify that all ERCCs have been removed:
dim(BTxGene)

# Remove SIRV Spike-ins:
BTxGene <- BTxGene[!grepl("^SIRV", BTxGene$gene),]

#Verify that all SIRVs have been removed:
dim(BTxGene)

# export gene count table:
saveRDS(BTxGene, file = BTxgene.RData)
```

2. Normalize and identify DEGs with DESeq2.

```
library(DESeq2)
library(edgeR)

# Load gene read count tables
BTxgene <- readRDS("BTxgene.RData")

# Define experimental design
conditions <-  factor(rep(c("Condition1","Condition2","Condition3","Condition4"), each=4),
  levels=c("Condition1","Condition2","Condition3","Condition4"))
design <- model.matrix(~ 0 + conditions)
```

```
# Filter lowly expressed genes
y <- DGEList(BTxgene, group=conditions)
keep.exprs <- filterByExpr(y, design)
genefiltered <- y[keep.exprs,, keep.lib.sizes=FALSE]$counts

# Build DESeq2 dataset
dds <- DESeqDataSetFromMatrix(countData = genefiltered,
 colData = data.frame(conditions),
 design = ~ conditions)

# Normalize with median-of-ratios
dds <- DESeq(dds)
normalized_counts_deseq <- counts(dds, normalized=TRUE)
```

*3.4.2 Nu*Delta Normalization [26] (See **Note 15**)*

1. Quantify spike-ins and normalize using NuDelta.

```
# Select spike-ins from the count table:
BT - readRDS(RData/BT_features.RData)
# This step was done above,
ERCC - BT[grepl(^ERCC, BT$gene),]
SIRV - BT[grepl(^SIRV, BT$gene),]
# combine ERCC and SIRV spikes
spikes - bind_rows(ERCC, SIRV)
# export spikein count table
saveRDS(spikes, file = RData/BTxspike_new.RData)

library(tidyverse)
library(edgeR)
library(DESeq2)

# Calculating Nu Calibration Constant
# Load Spike-in and gene read count tables
BTxspike <- readRDS("RData/BTxspike_new.RData")
BTxgene <- readRDS("RData/BTxgene.RData")

# Define experimental design
conditions <-  factor(rep(c("Condition1","Condition2","Condi-
tion3","Condition4"), each=4),
  levels=c("Condition1","Condition2","Condition3","Condi-
tion4"))
design <- model.matrix(~ 0 + conditions)

# Filter genes
y <- DGEList(BTxgene, group=conditions)
keep.exprs <- filterByExpr(y, design)
genefiltered <- y[keep.exprs,, keep.lib.sizes=FALSE]$counts
```

```
# Filter spike-ins
spike_filtered <- BTxspike[apply(BTxspike, 1, function(x) any
(x>0)),]

# Load spike-in concentrations (attomol values)from Lexogen
spike_conc <- read.delim("input/spike_conc_BTx.txt", header=-
TRUE) %>%
 column_to_rownames(var="spikeID")
```

(*See* **Notes 16** and **17** for calculating the Spike-in Concentration Table.)

```
# Select the Spike-ins detected.
spikeconc <- spike_conc[rownames(spike_filtered),]

# Define the library sizes for the Spike-ins and the mRNA.
LibSize_spike <- colSums(spike_filtered)
LibSize_RNA <- colSums(genefiltered)

## Compute Nu
Nvec <- spikeconc[,"attomol"]
names(Nvec) <- rownames(spikeconc)

# Compute the empirical fraction of the total Spike-in counts
across all libraries accounted for by each Spike-in molecule i,
defined as fi.
f.vec <- rowSums(spike_filtered) / sum(spike_filtered)

#Determine the Spike-in with the largest proportion, to be
defined as the reference Spike-in f.ref
ref <- which.max(f.vec)

#Define the attomoles per cell of f.ref (should probably be fi
and repeated not JUST for MAX), referred to as n1.
f.ref <- f.vec[ref]; n.ref <- Nvec[ref]

#Calculate the correction factor nu.
nu <- (f.ref/n.ref) * LibSize_spike

## Compute Delta
#Take the log of the RNA matrix, then the mean of the abundance
of each transcript i over libraries in each condition.
matRNAplus1 <- genefiltered + 1
meanLogZplus1 <- t(apply(log(matRNAplus1), 1, function
(x) tapply(x, conditions, mean)))
MeanMatLogZplus1 <- t(apply(meanLogZplus1, 1, function(x) rep
(x, each=4)))
```

```
#Calculate Beta, the difference between the mean of log
(nominal abundance of transcript i in library j) and the mean
of log(transformation of nominal abundance of transcript i
among libraries in condition 1).

aux <- colMeans(log(matRNAplus1) - MeanMatLogZplus1)

#calculate delta
delta <- exp(aux)
```

2. Proceed with DEG analysis using DESeq2 and NuDelta:

```
# Combine Nu*Delta
nudelta <- nu * delta

#Take the geometric mean of NuDelta.
geomean <- exp(mean(log(nudelta)))

#Calculate the NuDelta normalization factor.
nudelta_cal <- nudelta / geomean

# Apply to DESeq2
dds <- DESeqDataSetFromMatrix(countData = genefiltered,
 colData = data.frame(conditions),
 design = ~ conditions)
sizeFactors(dds) <- nudelta_cal
dds <- DESeq(dds)
nudelta_norm_counts <- counts(dds, normalized=TRUE)
```

3.4.3 DESeq2 with Spike-in Size Factors [27] (See **Note 18**)

Code

```
library(tidyverse)
library(edgeR)
library(DESeq2)

# Load Spike-in and gene read count tables
BTxspike <- readRDS("RData/BTxspike_new.RData")
BTxgene <- readRDS("RData/BTxgene.RData")

#Define conditions for gene counts, make a design matrix, and
DEGList object for gene counts.

conditions <- as.factor(rep(c("Condition1", "Condition2",
"Condition3", "Condition4"), each=4))
conditions <- factor(conditions, levels = c("Condition1",
```

```
"Condition2", "Condition3", "Condition4"))
design <- model.matrix(~ 0 + conditions)
y <- DGEList(BTxgene, group = conditions)

# Filter lowly expressed spike-ins and genes; Keep reads > 0 in
at least 1 sample.
spike_filtered <- as.matrix(BTxspike[apply(BTxspike, 1, func-
tion(x) length(x[x>0])>=1),])
spikes <- rownames(spike_filtered)

keep.exprs <- filterByExpr(y, design)
genefiltered <- y[keep.exprs,, keep.lib.sizes=FALSE]$counts
genefilteredList <- rownames(genefiltered)

# Make a Spike-in DESeqDataSet
spike_dds <- DESeqDataSetFromMatrix(countData = spike_fil-
tered,
 colData = data.frame(conditions), design = ~ conditions)

# Estimate the size factors for the spike-ins
spike_dds <- estimateSizeFactors(spike_dds)
spike_sf <- sizeFactors(spike_dds)

# Make a gene DESeqDataSet
gene_dds <- DESeqDataSetFromMatrix(countData = genefiltered,
 colData = data.frame(conditions), design = ~ conditions)

# Use these Spike-ins for the size factors
sizeFactors(gene_dds) <- spike_sf

# Apply spike-in size factors
sizeFactors(gene_dds) <- spike_sf
gene_dds <- DESeq(gene_dds)
norm_counts_spikeSF <- counts(gene_dds, normalized=TRUE)
```

*3.4.4 Remove Unwanted Variation g (RUVg) with Spike-ins [28] (See **Note 19**)*

1. Use Spike-ins to determine W.

```
library(tidyverse)
library(RUVSeq)
library(DESeq2)
library(edgeR)

# Load Spike-in and gene read count tables
BTxspike <- readRDS("RData/BTxspike_new.RData")
BTxgene <- readRDS("RData/BTxgene.RData")

#Define conditions for gene counts, make a design matrix, and
```

```
DEGList object for gene counts.
conditions <- as.factor(rep(c("Condition1", "Condition2",
"Condition3", "Condition4"), each=4))
conditions <- factor(conditions, levels = c("Condition1",
"Condition2", "Condition3", "Condition4"))
design <- model.matrix(~ 0 + conditions)
y <- DGEList(BTxgene, group = conditions)

# Filter lowly expressed spike-ins and genes; Keep reads > 0 in
at least 1 sample.
spike_filtered <- as.matrix(BTxspike[apply(BTxspike, 1, func-
tion(x) length(x[x>0])>=1),])
spikes <- rownames(spike_filtered)

keep.exprs <- filterByExpr(y, design)
genefiltered <- y[keep.exprs,, keep.lib.sizes=FALSE]$counts
genefilteredList <- rownames(genefiltered)

# Combine genes and Spike-ins.
combined <- rbind(genefiltered, spike_filtered)

# Make a DESeq Make a list and object before RUV.
RUVg_set <- newSeqExpressionSet(as.matrix(combined),
 phenoData = data.frame(conditions, row.names=colnames(com-
bined)))

# Normalize and estimate unwanted variation with spike-ins
RUVg_set <- betweenLaneNormalization(RUVg_set, which="me-
dian")
RUVg_set2 <- RUVg(RUVg_set, spikes, k=1)

# Remove Spike-ins from the dataset before DESeq2.
RUVg_set3 <- RUVg_set2[rownames(RUVg_set2) %in% rownames(gen-
efiltered),]
```

2. Run DESeq2 with W_1 factor (*see* **Note 20**).

```
dds_ruv <- DESeqDataSetFromMatrix(countData = counts(RUVg_-
set3),
 colData = pData(RUVg_set3),
 design = ~ W_1 + conditions)
dds_ruv <- DESeq(dds_ruv)
norm_counts_ruv <- counts(dds_ruv, normalized=TRUE)
```

4 Notes

1. *Reagents, equipment, and software:* We have provided example sources for these items; however, these reagents are also available from several commercial suppliers. A variety of software options are available to complete the tasks outlined. The ones listed are used in the example protocols, but the user can replace any steps with their preferred analytical tools.
2. *Sources of exogenous RNA Spike-ins*: Spike-ins can be synthesized [23], purchased from commercial vendors, or drawn from heterologous species [34]. The workflow outlined here is intentionally simple and robust, requiring reliable detection of only a single exogenous RNA. For more in-depth analysis, many kits offer multiplexed mixes that span a gradient of concentrations, GC content, lengths, and isoform composition. These enable more advanced, model-based analyses that explicitly leverage multiple spike-ins and their distinct properties. Spike-ins that embed unique molecular identifiers (UMIs) enable molecule-level deduplication and error correction, provide an internal readout of capture efficiency, and are especially valuable for benchmarking single-cell RNA-Seq workflows [35].

   Regardless of the RNA spike-in source, the spike-in chemistry must match your RNA extraction and library-prep protocol. For example, in 3′ RNA-Seq, the spike-ins should be efficiently captured by the poly(dT) pulldown (i.e., have a poly(A)-region). We illustrate using the Lexogen Set 3, which is broadly compatible across RNA-Seq applications. However, other kits or custom spike-ins may be more economical or better tailored to specific assay requirements and experimental goals.
3. *mRNA enrichment or rRNA depletion:* This step is not always required and may not be suitable for some experimental analysis. For example, for 3' RNA-Seq, this step is incorporated into the library preparation. rRNA depletion may be more suitable for certain studies (e.g., miRNA and other noncoding RNAs), and rRNA depletion can also be used in conjunction with RNA spike-ins.
4. *Choice of normalization reference*: A key consideration when applying spike-in normalization is deciding which reference will serve as the basis for normalization. For example, per-leaf normalization may be suitable when leaf size and cell number are consistent (e.g., time-of-day comparisons in plants). However, when comparing samples with varying cell sizes due to ploidy differences, mutations, or environmental treatments

(e.g., chilling stress, which can reduce leaf size), other approaches, such as normalizing per cell or per total DNA content, might be considered.

5. *Amount of spike-ins to add*: The target for how much spike-ins to add is to aim for ~1% of total reads (the typical range is 0.5–2%) with a minimum of 250,000 spike-in-aligned reads per library to obtain reliable scaling estimates [27]. Fractions that are too high waste sequencing depth; fractions that are too low risk high noise: signal ratios, resulting in noisy estimates. Because spike-ins are costly, most users will keep the fraction minimal; ~1% is a practical target that balances detectability and cost. Run a small pilot experiment to verify recovery and count variance in your system, since the optimal fraction depends on the total read depth and library complexity. The pipeline described here supports normalization with a single spike-in, which works well at low fractions. More advanced, model-based analyses that leverage multiple spike-ins across a range of abundances may require allocating a higher fraction to ensure adequate sampling of those RNAs. Three example use cases are presented in **Note 6** to demonstrate the calculations for determining spike-in volumes using the Lexogen Spike-in RNA Variant (SIRV) Controls with Isoforms and ERCCs (Set 3, Lexogen, USA). For these SIRVs, the stock 1:10 dilution is 3.03 ng/μL. For illustration, we present a 1:2000 dilution (or 15.15 pg/μL) in Case 1 and a 1:1000 dilution (or 30.3 pg/μL) in Cases 2 and 3. Aim for between 0.5% and 2% of total reads from spike-ins per library [36] and a minimum of 250,000 spike-in reads [26]. An ideal scenario is to perform a pilot study to ensure spike-in levels are detectable without wasting excessive sequencing space.
6. *Example use cases to aid in calculating spike-in volumes:* Three use cases are presented here, along with calculations. These can be used as a guide for determining starting spike-in values for a pilot experiment (Table 1).

Case 1: Sorghum Leaf Tissue 3′ RNA-Seq [2]

Context: Total RNA extracted from leaf tissue; RNA spike-ins were added after tissue grinding but before RNA extraction, 3′RNA-Seq using Lexogen's QuantSeq and Illumina short-read sequencing.

Known values: $F_{SIRV} = 0.01$, $F_{RNA} = 0.03$, $M_{RNA} = 300$ ng (total RNA input into library preparation), $C_{SIRV} = 15.15$ pg/uL (1:2000 dilution working concentration).

Calculation:

M_{SIRV}: $M_{SIRV} = (0.01)(0.03)(300\text{ ng}) = 0.09\text{ ng} = 90\text{ pg}$.

V_{SIRV}: $V_{SIRV} = (90\text{ pg}) / (15.15\text{ pg/μL}) = 5.94\text{ μL}$ (per leaf).

A total of 5.94 μL of SIRV (1:2000 dilution) was added per leaf for this experiment.

Table 1
Calculation examples of spike-in volumes (V_{SIRV}) for Sorghum, tomato, and Arabidopsis experiments, based on RNA input, spike-in dilution, and desired read fraction

Case	Tissue	Normalization basis	M_{RNA} (ng)	F_{SIRV}	C_{SIRV} (pg/μL)	M_{SIRV} (pg)	V_{SIRV} (μL)
Sorghum (Laosuntisuk et al. 2024)	The second fully expanded leaf with a visible collar	Per leaf	300	0.01	15.15 (1:2000)	90	5.94
Tomato	Root	Per 100 mg tissue	1000	0.01	30.3 (1:1000)	300	9.9
Arabidopsis	Shoot of young seedlings	Per plant	1000	0.003333	30.3 (1:1000)	33.33	3.3

Case 2 Tomato Root Tissue

Context: Normalization using a per-weight basis for mRNA sequencing. Spike-ins were added to ground root tissue from tomato plants on a per-weight basis before the lysis buffer was added.

Known values: $F_{SIRV} = 0.01$, $F_{RNA} = 0.03$, $M_{RNA} = 1000$ ng (total RNA input into library preparation), $C_{SIRV} = 30.3$ pg/uL (1: 1000 dilution working concentration).

Calculations:

M_{SIRV}: $M_{SIRV} = (0.01)(0.03)(1000 \text{ ng}) = 0.3 \text{ ng} = 300 \text{ pg}$.

V_{SIRV}: $V_{SIRV} = (300 \text{ pg}) / (30.3 \text{ pg/μL}) = 9.9 \text{ μL}$ (per 100 mg of root tissue).

A total of 9.9 μL of SIRV (1:1000 dilution) was added per 100 mg of root tissue before the tissue was finely ground, and lysis buffer was added.

Case 3 Arabidopsis Seedlings

Context: Normalization using a per-seedling basis for mRNA sequencing. The shoot tissue was collected from 4 to 6 Arabidopsis seedlings grown on plates, and the number of seedlings per sample was recorded (Fig. 3). Spike-ins were added based on the number of seedlings in each set, and SIRVs were added to the ground plant tissue alongside the lysis buffer at the beginning of the RNA extraction.

Known values: $F_{SIRV} = 0.003333$, $F_{RNA} = 0.03$, $M_{RNA} = 1000$ ng (total RNA input into library preparation), $C_{SIRV} = 30.3$ pg/uL (1:1000 dilution working concentration).

Calculations:

M_{SIRV}: $M_{SIRV} = (0.003333)(0.03)(1000 \text{ ng}) = 0.09999 \text{ ng} = 99.99 \text{ pg}$.

V_{SIRV}: $V_{SIRV} = (99.99 \text{ pg}) / (30.3 \text{ pg/}\mu\text{L}) = 3.3\ \mu\text{L}$ (per seedling).

Based on these calculations, 3.3 μL of a 1:1000 dilution of SIRV set 3 was added per plant.

7. *Preparing the spike-in dilutions.* Prepare spike-ins using iterative dilutions from the RNA spike-in stock (e.g., 1:10 then 1: 100). To protect the RNA, avoid freezing highly dilute solutions; make dilutions beyond 1:10 fresh to avoid long storage. Dilute solutions should be stored for minutes, not hours. Freeze the initial (1:10) dilution in single-use aliquots to prevent repeated freeze-thaw cycles. For consistency across samples, prepare one common intermediate dilution for the entire experiment in a single session from the same stock, aliquot it, and then prepare the final working dilution at the time of RNA extraction for each library, discarding any leftovers. Work RNase-free, use low-bind tubes and tips, keep dilutions on ice, mix gently, and record lot numbers and dilution factors.
8. *Determining the spike-in dilution factor:* The optimal dilution depends on the input amount and the library type. Lexogen recommends a dilution of at least 1:100 for SIRVs Set 3. In our plant experiments, dilutions of 1:1000 to 1:2000 have worked well. You can prepare dilutions to 1:10,000 if needed. Aim for ~1% of total reads mapping to the RNA spike-ins, but a practical window is ~0.5% to 2%. Estimate a starting dilution based on expected RNA yields, and then run a small pilot using this estimated value. Measure the spike-in read fraction and adjust by either changing the dilution or the added volume of RNA spike-ins to achieve a value of approximately 1%. Once set, maintain a constant dilution throughout the batch for comparability.
9. *When to add spike-ins*: To capture both extraction losses and global differences in RNA output, add spike-ins at the earliest step, adding directly to tissue or in the RNA lysis buffer at a fixed amount per unit of normalization. It is ideal to add RNA Spike-ins before DNase I Treatment, Ribosomal RNA Depletion, or poly(A) selection to account for losses and potential sources of bias during these steps. Adding spike-ins later (during library prep) can help monitor technical variation in library construction, but it cannot correct for sample-to-sample differences in total RNA abundance. If desired, a second, library-prep-only spike-in can be used as a process control, but rely on pre-RNA-extraction spike-ins for quantitative normalization.
10. *Sample consistency*: Spike-in volume should be pipetted precisely; small deviations can propagate through normalization.

11. *Pooling strategy—keep libraries separate until the last possible moment, confirm concentrations, pool once, and load*: For best results, prepare, quantify, and normalize each library individually, then pool immediately before sequencing rather than storing a pooled mix. Keeping libraries separate avoids several pitfalls: small concentration drifts from evaporation or adsorption to plastics, uneven degradation across samples, and loss of the ability to rebalance under- or over-represented libraries after QC [37]. Store individual libraries at −20 °C (minimize freeze–thaw), re-quantify if they have been stored for more than a few days, and perform a final normalization just prior to pooling. Use unique dual indexes (UDIs) and ensure thorough cleanup to remove free adapters. These steps reduce index misassignment (e.g., index hopping) but do not change the recommendation to avoid long-term storage of pooled libraries.
12. *You do not have to normalize using the spike-ins*: A major benefit of adding spike-ins is optionality; you can ignore them and analyze your data with standard methods (e.g., DESeq2 median-of-ratios, TMM) and, in parallel, run a spike-in–scaled analysis. Comparing the two pipelines is informative. If the results are similar, spike-ins corroborate that global RNA abundance is stable. However, if they diverge (e.g., MA plots shift, many genes move in the same direction), that flags a global shift where spike-in scaling may be more appropriate. Even when not used for scaling, spike-ins are valuable QC: check their read fraction, consistency across samples, mapping/strand behavior, and detection across the spike-in dynamic range to spot pipetting issues, extraction loss, or batch effects. Even if your samples do not have global shifts in transcription, the presence of spike-ins can be leveraged to identify and correct for unwanted technical variation across batches, serving as "control features" as described in the section on remove unwanted variation (RUV) analysis with spike-ins. If spike-ins are not utilized, simply report that spike-ins were present, their read fraction, and why you chose (or declined) to use them for normalization. In short: adding spike-ins keeps your options open.
13. *Can be used with many methods*: We illustrate spike-in-based normalization with DESeq2, but the workflow is method-agnostic. The workflow is compatible with edgeR (TMM), set `samples$norm.factors` to spike-in factors, disable TMM (run `calcNormFactors(..., method = "none")`. The effective library size used by edgeR will be `lib.size * norm.factors`. For LIMMA-VOOM: compute spike-in-derived scaling factors and use them as the sample-level normalization. Pass counts via an edgeR `DGEList` with `norm.`

`factors` set to the spike-in factors (skipping TMM), then run `voom(y, design,...)`. Voom will carry those factors forward as offsets. Critically, avoid "double normalizing." If you adopt spike-in scaling as the primary normalization, keep TMM/median-ratio disabled.

14. *DESeq2 description, pros, cons, unique aspects, and general guidance*: Overview of the DESeq2 approach.

    Description: DESeq2 scales libraries by estimating size factors from the median of ratios relative to a geometric mean reference across samples. This approach assumes that most genes do not vary across samples (*see* **Notes 10** and **11**).

    Pros: Does not require spike-ins, widely used, robust to variations in sample processing and library preparation, with extensive tutorials and software support.

    Cons: Assumes most genes do not change between samples, therefore biases fold-change estimates when global transcriptional shifts occur.

    Unique aspects: Entirely distribution-based; normalization depends only on observed counts and does not require external standards.

    When to use: In experiments where changes are expected to affect only a subset of genes and not total transcriptional output.

15. *Nu*Delta description, pros, cons, unique aspects, and general guidance*: Overview of the Nu*Delta approach.

    Description: This approach uses spike-ins to calculate a calibration constant (Nu) reflecting total RNA abundance, and a correction factor (Delta) accounting for sample-specific variation between replicates. Together, Nu*Delta separates global scaling from technical variation, preserving biologically meaningful shifts in transcription.

    Pros: Captures global transcriptional changes (e.g., transcriptional amplification; diel cycles). Robust even when highly expressed genes dominate the transcriptome changes between treatments.

    Cons: Requires accurate spike-in quantification and metadata. Sensitive to spike-in pipetting or measurement noise (*see* **Note 7**).

    Unique aspects: Separates normalization into absolute RNA scaling (Nu) and sample-specific correction (Delta).

    When to use: Recommended when treatments alter total RNA content (e.g., growth rates, stress, hormone treatment, circadian experimental conditions).

16. *Calculating the spike-in concentration table*: Examples for calculating attomoles per spike-in for Nu*Delta calibration (Table 2; *see* **Note 21**).

Table 2
Worked example for calculating attomoles per spike-in for Nu*Delta calibration

SpikeID	Stock concentration (amol)	Dilution factor	Diluted concentration (amol/μL)	Amount of spike added (μL)	Mass of spike (amol)
SIRV1	60	100	0.6	3.3	1.98
SIRV2	60	100	0.6	3.3	1.98
SIRV3	60	100	0.6	3.3	1.98
SIRV4	60	100	0.6	3.3	1.98
SIRV5	60	100	0.6	3.3	1.98
SIRV6	60	100	0.6	3.3	1.98
SIRV7	60	100	0.6	3.3	1.98
ERCC-00130	1500	100	0.6	3.3	49.5
ERCC-00004	375	100	0.6	3.3	12.375
ERCC-00136	93.75	100	0.6	3.3	3.09375

How to convert a stock spike-in concentration to a working concentration and then to attomoles added per library. Using ERCC-00130 (stock 1500 amol/μL) diluted 1:100, apply $C_1V_1 = C_2V_2$ to obtain the working concentration: $C_2 = (1500$ amol/μL × 1 μL) / 100 μL = 15 amol/μL. If 3.3 μL of this working solution is added per sample, the amount added is $M = C \times V = 15$ amol/μL × 3.3 μL = 49.5 amol. Repeat this calculation for each spike-in (using its own dilution chain and addition volume) to populate the "attomoles per library" table used in Nu*Delta. Record all units and dilution steps (e.g., 1:10 → 1:10 = 1:100) to avoid rounding or unit-conversion errors

17. *Nu*Delta caution*: Requires accurate spike-in concentration metadata (attomol per molecule). Errors here directly distort scaling.

18. *DESeq2 with spike-in size factors description, pros, cons, unique aspects, and general guidance*: Overview of the DESeq2 with spike-ins as size factors approach.

    Description: Here, spike-in counts are used directly to calculate size factors, which are then applied to endogenous gene counts in DESeq2. This replaces distribution-based scaling with an external standard.

    Pros: Simple and easy to implement, integrates readily with DESeq2.

    Cons: Assumes spike-ins behave like endogenous RNA across all steps. Cannot distinguish between technical vs. biological shifts.

    Unique aspects: External reference-driven normalization without modeling global transcription.

When to use: Appropriate for experiments with stable total RNA, but suspected technical variation.

19. *Remove unwanted variance g (RUVg) description, pros, cons, unique aspects, and general guidance*: Overview of the RUVg factors approach.

    Description: RUVg treats spike-ins as negative controls to estimate hidden unwanted variation factors (W). These factors are then included in the DESeq2 design matrix to remove technical noise.

    Pros: Very effective at correcting batch and lane effects. Well-documented RUVSeq package, simple to implement.

    Cons: If global biological shifts occur, they may be mistakenly classified as technical variation and removed.

    Unique aspects: Focuses on variance partitioning rather than scaling.

    When to use: Best for experiments where technical noise is suspected to dominate, not appropriate when global changes in transcription are likely or are the biological question.

20. *RUVg limitations*: Use with caution when global RNA output is expected to change, as true biological shifts may be misinterpreted as technical noise.

21. How to convert a stock spike-in concentration to a working concentration and then to attomoles added per library. Using ERCC-00130 (stock 1500 amol/μL) diluted 1:100, apply $C_1 V_1 = C_2 V_2$ to obtain the working concentration: $C_2 = (1500\ \text{amol}/\mu\text{L} \times 1\ \mu\text{L}) / 100\ \mu\text{L} = 15\ \text{amol}/\mu\text{L}$. If 3.3 μL of this working solution is added per sample, the amount added is $M = C \times V = 15\ \text{amol}/\mu\text{L} \times 3.3\ \mu\text{L} = 49.5$ amol. Repeat this calculation for each spike-in (using its own dilution chain and addition volume) to populate the "attomoles per library" table used in Nu*Delta. Record all units and dilution steps (e.g., 1:10 → 1:10 = 1:100) to avoid rounding or unit-conversion errors.

Acknowledgments

We would like to acknowledge support for this project from the Chugach Heritage Foundation Graduate Scholarship and the Copper Mountain Foundation Graduate Scholarship (to T. Blake Horton). Funding for this work was provided by the US Department of Agriculture, HATCH Project #1017025, National Aeronautics and Space Administration (Space Biology #80NSSC18K1466), and the National Science Foundation (#2533400).

References

1. Coate JE, Doyle JJ (2010) Quantifying whole transcriptome size, a prerequisite for understanding transcriptome evolution across species: an example from a plant allopolyploid. Genome Biol Evol 2:534–546
2. Laosuntisuk K, Vennapusa A, Somayanda IM et al (2024) A normalization method that controls for total RNA abundance affects the identification of differentially expressed genes, revealing bias toward morning-expressed responses. Plant J 118:1241–1257
3. Forsythe ES, Grover CE, Miller ER et al (2022) Organellar transcripts dominate the cellular mRNA pool across plants of varying ploidy levels. Proc Natl Acad Sci USA 119: e2204187119
4. Wang G, DiTusa SF, Oh D-H et al (2021) Cross species multi-omics reveals cell wall sequestration and elevated global transcript abundance as mechanisms of boron tolerance in plants. New Phytol 230:1985–2000
5. Eveland AL, McCarty DR, Koch KE (2008) Transcript profiling by 3′-untranslated region sequencing resolves expression of gene families. Plant Physiol 146:32–44
6. Moll P, Ante M, Seitz A et al (2014) QuantSeq 3′ mRNA sequencing for RNA quantification. Nat Methods 11:i–iii
7. Kremling KAG, Chen S-Y, Su M-H et al (2018) Dysregulation of expression correlates with rare-allele burden and fitness loss in maize. Nature 555:520–523
8. Israeli A, Capua Y, Shwartz I et al (2019) Multiple auxin-response regulators enable stability and variability in leaf development. Curr Biol 29:1746–1759.e5
9. Kellenberger RT, Byers KJRP, De Brito Francisco RM et al (2019) Emergence of a floral colour polymorphism by pollinator-mediated overdominance. Nat Commun 10:63
10. Palmer NA, Basu S, Heng-Moss T et al (2019) Fall armyworm (Spodoptera frugiperda Smith) feeding elicits differential defense responses in upland and lowland switchgrass. PLoS One 14: e0218352
11. Silva KJP, Singh J, Bednarek R et al (2019) Differential gene regulatory pathways and co-expression networks associated with fire blight infection in apple (Malus × domestica). Hortic Res 6:35
12. Yu Y, Hu H, Doust AN et al (2020) Divergent gene expression networks underlie morphological diversity of abscission zones in grasses. New Phytol 225:1799–1815
13. Anders S, Huber W (2010) Differential expression analysis for sequence count data. Genome Biol 11:R106–R106
14. Robinson MD, Oshlack A (2010) A scaling normalization method for differential expression analysis of RNA-seq data. Genome Biol 11:R25
15. Lovén J, Orlando DA, Sigova AA et al (2012) Revisiting global gene expression analysis. Cell 151:476–482
16. Benes V, Muckenthaler M (2003) Standardization of protocols in cDNA microarray analysis. Trends Biochem Sci 28:244–249
17. Hartemink AJ, Gifford DK, Jaakkola TS et al (2001) Maximum-likelihood estimation of optimal scaling factors for expression array normalization. In: Bittner ML, Chen Y, Dorsel AN et al (eds) Microarrays. optical technologies and informatics. SPIE
18. Hill AA, Brown EL, Whitley MZ et al (2001) Evaluation of normalization procedures for oligonucleotide array data based on spiked cRNA controls. Genome Biol 2:RESEARCH0055
19. Bläsing OE, Gibon Y, Günther M et al (2005) Sugars and circadian regulation make major contributions to the global regulation of diurnal gene expression in Arabidopsis. Plant Cell 17:3257–3281
20. Czechowski T, Stitt M, Altmann T et al (2005) Genome-wide identification and testing of superior reference genes for transcript normalization in Arabidopsis. Plant Physiol 139:5–17
21. Piques M, Schulze WX, Höhne M et al (2009) Ribosome and transcript copy numbers, polysome occupancy and enzyme dynamics in Arabidopsis. Mol Syst Biol 5:314
22. Flis A, Fernández AP, Zielinski T et al (2015) Defining the robust behaviour of the plant clock gene circuit with absolute RNA timeseries and open infrastructure. Open Biol 5: 150042
23. Leshkowitz D, Feldmesser E, Friedlander G et al (2016) Using synthetic mouse spike-in transcripts to evaluate RNA-Seq analysis tools. PLoS One 11:e0153782
24. Lutzmayer S, Enugutti B, Nodine MD (2017) Novel small RNA spike-in oligonucleotides enable absolute normalization of small RNA-Seq data. Sci Rep 7:5913
25. Hofmann F, Schon MA, Nodine MD (2019) The embryonic transcriptome of Arabidopsis thaliana. Plant Reprod 32:77–91

26. Athanasiadou R, Neymotin B, Brandt N et al (2019) A complete statistical model for calibration of RNA-seq counts using external spike-ins and maximum likelihood theory. PLoS Comput Biol 15:e1006794
27. Brennecke P, Anders S, Kim JK et al (2013) Accounting for technical noise in single-cell RNA-seq experiments. Nat Methods 10: 1093–1095
28. Risso D, Ngai J, Speed TP et al (2014) Normalization of RNA-seq data using factor analysis of control genes or samples. Nat Biotechnol 32:896–902
29. Love MI, Huber W, and Anders S (2014) Moderated estimation of fold change and dispersion for RNA-seq data with DESeq2. https://doi.org/10.1101/002832
30. BBMap, https://sourceforge.net/projects/bbmap/
31. Dobin A, Davis CA, Schlesinger F et al (2013) STAR: ultrafast universal RNA-seq aligner. Bioinformatics 29:15–21
32. Danecek P, Bonfield JK, Liddle J et al (2021) Twelve years of SAMtools and BCFtools. Gigascience 10:giab008
33. Liao Y, Smyth GK, Shi W (2014) featureCounts: an efficient general purpose program for assigning sequence reads to genomic features. Bioinformatics 30:923–930
34. Schivre G, Wolff L, Mirasole FM et al (2025) Genome-scale transcriptome augmentation during Arabidopsis thaliana photomorphogenesis. Nat Commun. https://www.biorxiv.org/content/10.1101/2025.01.30.635720v1.abstract
35. Ziegenhain C, Hendriks G-J, Hagemann-Jensen M et al (2022) Molecular spikes: a gold standard for single-cell RNA counting. Nat Methods 19:560–566
36. Lexogen Spike-In RNA Variant Controls User's Guide, https://www.lexogen.com/wp-content/uploads/2021/11/050UG134V0210_SIRV-User-Guide_Set-2_Set-3_Set-4_2021-11-16.pdf
37. Costello M, Fleharty M, Abreu J et al (2018) Characterization and remediation of sample index swaps by non-redundant dual indexing on massively parallel sequencing platforms. BMC Genomics 19:332

Chapter 10

Single-Cell/Nucleus RNA-Sequencing for Phytohormone Signaling in Plants

Neslihan Yaşar, Dilara Sedef Karagöz, Halis Batuhan Ünal, Ceyhun Kayıhan, and Emre Aksoy

Abstract

A class of organic, naturally occurring compounds known as phytohormones affects physiological functions in plants at trace levels. Phytohormones play important roles in plant growth, development, nutrient transport, and survival by regulating plant responses to abiotic stress and form the basis of ability of plants for adapting to changing conditions. To carry out some specific and distinct functions, cells of complex organisms implement gene regulation dynamics according to their cell types, including during the synthesis and control of various plant hormones. Rather than the traditional approaches for phytohormone quantification, single-cell RNA sequencing (scRNA-seq) and/or single-nucleus RNA sequencing (snRNA-seq) technologies that enable high-resolution mapping of cellular heterogeneity at the transcriptomic level, provide a novel approach for the phytohormone signaling. Here, we detailed the methodology of single cell/nucleus technologies to investigate underlying gene regulatory networks of plant hormone biology.

Key words *Arabidopsis thaliana*, Nucleus isolation, Phytohormones, Protoplast isolation, Single-cell RNA sequencing (scRNA-seq), Single-nucleus RNA sequencing (snRNA-seq)

1 Introduction

Phytohormones, also known as plant growth regulators, are signaling molecules that play important roles in plant growth, development, and nutrient transport [1]. Near or far from their site of synthesis, these signaling molecules can influence various physiological processes such as cell division, differentiation, flowering, fruit development, and senescence [2]. Phytohormones regulate plant responses to abiotic stress and form the basis of plants' ability to adapt to changing conditions by activating defense mechanisms against environmental stress factors [3–5]. Although all phytohormones are involved in mitigating biotic and abiotic challenges, each phytohormone group has its own specific function. The hormones jasmonic acid (JA), strigolactones (SLs), and salicylic acid (SA) are

Shahid Mukhtar (ed.), *Plant Hormones: Methods and Protocols*, Methods in Molecular Biology, vol. 3026,
https://doi.org/10.1007/978-1-0716-5214-5_10,

generally involved in biotic stress management, while abscisic acid (ABA), cytokinin (CK), brassinosteroids (BR), ethylene (ET), gibberellin (GA), and auxin (AUX) are usually more vital in counteracting environmental stress conditions [6]. In order to study phytohormone biology, there are some traditional techniques for separating and purifying phytohormones. Among these techniques, chromatography and mass spectrometry are widely used [7, 8]. Moreover, analytical methods based on antigen-antibody interactions and spectrophotometric methods are also used [9–11]. However, these techniques require large amounts of plant tissue samples to analyze plant hormones and are based on different principles.

Alternatively, single-cell RNA sequencing (scRNA-seq) is a powerful technology that enables high-resolution investigation of cellular heterogeneity at the transcriptomic level [12]. This technique can be used to differentiate distinct cell types from each other in complex plant tissues and determine the gene expression profile of a single cell. In order to carry out some specific and distinct functions, cells of complex organisms implement gene regulation dynamics according to their cell types [13]. Despite the cellular heterogeneity of multicellular organisms, single-cell technologies enable us to understand the dynamic regulations of biological and metabolic processes including synthesis and control of various plant hormones [14]. In terms of phytohormone biology, scRNA-seq provides insight into how the same hormonal stimulus reveals distinct transcriptomic responses in various cells, and the distribution of hormone receptors on a cell-type basis [13, 15]. Moreover, this method can also be used to examine hormone responses that are time-dependent and caused by environmental stressors or developmental processes [16]. For example, the effects of hormones such as ABA on different cell populations can be monitored by trajectory analyses [17]. However, applications of scRNA-seq in plants have certain technical limitations. Since plant cells have rigid cell walls, isolation of living protoplasts is difficult in most tissues [18]. This may lead to changes in the gene expression profile or the exclusion of some cell types from the analysis. Single-nucleus RNA sequencing (snRNA-seq), overcomes this problem, enables RNA to be obtained directly from cell nuclei without the need for protoplast isolation [19]. Also, snRNA-seq is an ideal approach for tissues that are particularly difficult to isolate protoplasts using classical methods, such as roots or senescent leaves. Additionally, since this technique does not disrupt cell integrity, nucleus-based analysis allows gene expression patterns to be preserved, reflecting the physiological reality. Both approaches have been effectively used in model plants like rice, tomato, and *Arabidopsis thaliana*, yielding successful outcomes in the analysis of phytohormonal responses by cell type, the mapping of hormone receptor expression, and the modeling of ligand-receptor interactions

[20, 21]. These techniques can be used to investigate and understand the importance of the precise control of phytohormone mediated spatiotemporal, meaning time- and space-dependent, gene regulation that governs plant development and survival. For instance, Nolan et al. identified the cortex cells of *Arabidopsis thaliana* roots as the BR-dependent growth center via scRNA-seq [16]. In another study, snRNA-seq of peanut stem revealed that *AhWRKY70* has a tissue specific expression pattern that controls stem growth, and is substantially enriched in stem cortex and xylem cells [22]. Moreover, through altering the phytohormone signaling pathway and affecting the expression of 16 auxin and ethylene responsive genes, *AhWRKY70* overexpression in Arabidopsis resulted in faster stem growth. These and such studies may demonstrate how sc/snRNA-seq can precisely and spatiotemporally map phytohormone-mediated gene regulatory networks, thus revealing signaling elements specific to particular cell types in pathways ranging from BR and ABA to auxin and cytokinin. In spite of these advantages, there are significant compromises between sn- and scRNA-seq. Like all single-cell methods, scRNA-seq suffers from "dropouts," where low-abundance mRNAs (including hormone-induced transcripts) may go undetected [23]. Additionally, because it relies on enzymatic cell-wall digestion, it may introduce biases, as some cell types may be under-represented or transcriptionally altered by protoplasting. Since only nuclear RNA is analyzed, snRNA-seq typically captures fewer transcripts per cell, possibly eliminating cytoplasmic mRNAs or completely spliced transcripts [23].

In this work, we overviewed the importance of current single-cell/nucleus transcriptomics technology in phytohormone signaling and methodological details of these applications. Protocols detailed below comprise sample preparation, library construction, and quality control for both the sc- and snRNA-seq studies in model plant *Arabidopsis thaliana*. The protoplast isolation part of the protocol has been adapted from Ryu et al. [24]. The nucleus isolation part of the protocol has been adapted from Thibivilliers et al. [25].

2 Materials

2.1 Sample Preparation for scRNA-seq

1. Murashige & Skoog medium (MS medium).
2. Sucrose.
3. Plant agar.
4. Sodium hydroxide (NaOH).
5. Hydrochloric acid (HCl).
6. 50 mL centrifuge tubes.

7. 1.5 mL centrifuge tubes.
8. Pure ethanol (99.9%).
9. Sodium hypochlorite.
10. Distillated water (dH_2O).
11. Ultrapure water.
12. Scale.
13. Magnetic stirrer.
14. pH meter.
15. Autoclave.
16. Climate cabinet.

2.2 Protoplast Isolation

1. Mannitol.
2. MES monohydrate.
3. Potassium chlorite (KCl).
4. Calcium chloride ($CaCl_2$).
5. Bovine serum albumine.
6. Cellulase ONOZUKA R-10.
7. Pectolyase (P3026, Sigma-Aldrich).
8. Trypan blue.
9. Distilled water (dH_2O).
10. Thoma cell counting chamber.
11. Light microscope.
12. Forceps.
13. Scissor.
14. 40 μm tip strainer.
15. 70 μm cell strainer.
16. 35 mm petri dishes.
17. Parafilm.
18. Glass stick.
19. 15 mL centrifuge tubes.
20. 50 mL centrifuge tubes.
21. 1.5 mL centrifuge tubes.
22. 2.0 mL centrifuge tubes.
23. Vortex.
24. Shaker.
25. Centrifuge (with Swinging Bucket Rotor).
26. Light microscope.

2.3 Sample Preparation for snRNA-seq

1. MS medium.
2. Sucrose.
3. Plant agar.
4. Distilled water.
5. 120 mm petri dishes.
6. 70% EtOH.
7. 2.5% NaOCl.
8. Ultrapure water.
9. 0.01% agar solution.
10. Scale.
11. Magnetic stirrer.
12. pH meter.
13. Autoclave.
14. Climate cabinet.
15. 90 mm petri dishes.
16. Razor blade.
17. Scissors.
18. 40 μm cell strainer.
19. 30 μm cell strainer.
20. 50 mL falcon tubes.
21. 15 mL falcon tubes.
22. 1X Nuclei Isolation Buffer.
23. PBS+.
24. Protector RNase Inhibitor (Stock: 40 U/μL).
25. BSA (50 mg/mL).
26. 0.4% Trypan blue.
27. Centrifuge.
28. Platform shaker.
29. CountessTM II Automated Cell Counter (Invitrogen, cat. no. AMQAX1000).

2.4 Medium Preparation, Surface Sterilization, and Culturing of A. thaliana Seeds for sc/snRNA-seq

1. ½ MS Medium (for 1 L, pH: 5.7–5.8): 5 g sucrose, 2.2 g MS, 7 g plant agar.
2. 2.5% Sodium Hypochlorite (for 50 mL): 1.25 mL sodium hypochlorite, 48.75 mL ultrapure water.
3. 70% Ethanol (for 50 mL): 35 mL pure ethanol, 15 mL ultrapure water.
4. 0.01% Plant Agar (for 50 mL): 0.005 g plant agar, 50 mL dH_2O (should be autoclaved).

2.5 Protoplast Isolation for scRNA-seq

1. Solution-A (10 mL per sample, for 10 mL solution): 0.4 M mannitol [5 mL of 0.8 M mannitol], 20 mM MES (pH: 5.7), [1 mL of 0.2 M MES (pH:5.7)], 20 mM KCl [0.2 mL of 1 M KCl], 10 mM $CaCl_2$ [0.1 mL of 1 M $CaCl_2$], freshly prepared 0.1% BSA [0.1 mL of %10 BSA] and dH_2O [3.6 mL].
2. Enzyme Solution (4 mL per sample, for 10 mL solution): 5 mL of Solution-A, 25% Cellulase ONOZUKA R-10 [0.0625 g/5 mL], 0.1% Pectolyase [0.005 g/5 mL].

2.6 Nucleus Isolation for snRNA-seq

1. 1X NIB: 10 mL CelLyticTM PN Isolation/Extraction buffer 4X (Sigma-Alrich, cat. no: CELLYTPN1) is diluted with 30 mL of ultrapure water to prepare 1X NIB. This diluted solution is referred to as "NIB" (nuclei isolation buffer).
2. BSA (50 mg/mL): 50 mg of BSA (Sigma-Aldrich, A9418) was dissolved by vortexing in 1 mL of ultrapure water.
3. PBS+: 6.3375 mL of PBS is supplemented with 32.5 μL of Protector RNase Inhibitor 40 U/μL (Roche, cat. no. 03335399001) and 130 μL of BSA (stock concentration: 50 mg/mL).
4. 0.4% Trypan Blue: 0.4 mg of trypan blue is dissolved by heating in 80 μL PBS. Once the mixture has cooled to room temperature, the volume is adjusted to 100 μL with PBS.

3 Methods

3.1 ½ MS Medium Preparation, Surface Sterilization, and Culturing of *A. thaliana* Seeds for scRNA-seq

1. Dissolve 2.2 g MS and 5 g sucrose into 900 mL dH_2O.
2. Adjust the pH to 5.7–5.8.
3. Fill up the medium to 1 liter with dH_2O.
4. Add 7 g plant agar and sterilize the medium by autoclave.
5. Divide approximately 20 mL in each petri dish, after ½ MS medium has been sterilized.
6. Put 300 μL 70% ethanol into a 1.5 mL centrifuge tube containing seed and invert the tubes for 2 min.
7. Remove the ethanol.
8. Add 300 μL 2.5% sodium hypochlorite and invert the tubes for 10 min.
9. Remove the sodium hypochlorite.
10. Wash the seeds 3 times with 300 μL ultrapure water for 30 s. Ultrapure water should be removed in each step.
11. Add 300 μL 0.01% plant agar.
12. Draw a straight line across 1/3 of the medium and sow the seeds so that they are aligned with this line (*see* **Note 1**).

13. Wrap petri dishes in groups of five with cling film (*see* **Note 2**).
14. Store petri dishes upside down at 4 °C for 3 days.
15. Grow the seeds in a growth chamber at 24 ± 2 °C, 300 μmol. $m^{-2}.s^{-1}$, 70% humidity, and a 16/8 h photoperiod (light/dark).

3.2 Protoplast Isolation from Arabidopsis thaliana Root

1. Separate the roots from the shoots and lateral roots of *Arabidopsis thaliana* plants grown vertically under controlled conditions for 14 days.
2. Cut the roots into groups of five approximately 0.1 cm long using scissors (*see* **Note 3**).
3. Place a 70 μm cell strainer in a 35 mm petri dish.
4. Place the cut root samples in the strainer.
5. Add 4 mL of the enzyme solution and leave on the shaker at 90 rpm for 2–2.5 h.
6. Gently smash the samples with the help of a glass stick every 30 min.
7. Take samples from the petri dish with a pipette and collect them in a 15 mL Falcon tube using a 40 μm filter attached to the tip of the pipette (*see* **Note 4**).
8. Centrifuge at 500 *g* for 6 min at 22 °C.
9. Remove the supernatant gently without disturbing the pellet.
10. Dissolve the pellet with 500 μL 8% mannitol.
11. Take a 25 μL sample into 1.5 mL Eppendorf and add 2 μL 0.4% trypan blue (*see* **Note 5**).
12. Observe and count the root protoplasts under 60× objective of light microscope (*see* **Note 6**).

3.3 Library Construction for scRNA-seq

1. *Library Construction*: For the scRNA-seq library construction "10×Genomics Chromium Single Cell 3′ Reagents Kits (v3.1 Chemistry), User Guide, Revision D (Document Number: CG000204)" is recommended. Other platforms for scRNA-seq library construction are also compatible.
2. *Quality Control (QC) of Constructed Libraries*: For the quantification and quality control of the constructed cDNA and sequencing libraries, Agilent Bioanalyzer, Agilent TapeStation or Qubit are the recommended methods for accurate quantification.

3.4 Nuclei Isolation

1. Collect fresh roots from 14-day-old *Arabidopsis thaliana* plants by removing the stems with a razor blade (*see* **Note 7**).
2. Transfer the roots to a pre-cooled 90 mm petri dish using forceps.

3. Use scissors to cut the roots into smaller pieces.
4. Add 500 μL NIB to another clean, sterile and pre-cooled 90 mm petri dish.
5. Chop the roots with a razor blade by adding 1000 μL NIB for approximately 5 min.
6. Add extra 2 mL of NIB to keep the root tissues moist (additional NIB should only be added when necessary to aid in chopping and to maintain the sample's moisture).
7. Incubate the nuclear suspension for 21 min at +4 °C using gentle horizontal shaking (platform shaker).
8. At the end of incubation, add 1000 μL of NIB.
9. Place a 50 mL Falcon tube on ice and position a 40 μm filter on top.
10. Pre-wet the 40 μm filter with 500 μL of NIB before filtering the nucleus suspension.
11. Filter the suspension by using a pipette tip.
12. Add 1500 μL of NIB to the petri dish to collect the remaining samples in the petri dish.
13. Add extra 1000 μL to collect the remaining samples on the strainer.
14. Filter 5 mL of the 40 μm-filtered nuclear suspension through a 30 μm filter pre-wetted with 500 μL of NIB (*see* **Note 8**).
15. Centrifuge the filtered nucleus suspension for 10 min at 1000 × *g* and +4 °C to pellet the nuclei.
16. Remove the supernatant.
17. Repeat **step 15** (*see* **Note 9**).
18. Remove the supernatant.
19. Resuspend the pellet gently in 2.5 mL of PBS+ buffer supplemented with protective RNase inhibitor (0.2 U/μL, Roche) and ultrapure BSA (0.1%, Invitrogen).
20. Centrifuge the nucleus suspension for 10 min at 1000 × *g* and +4 °C to pellet the nuclei.
21. Remove 1.5 mL of the supernatant.
22. Resuspend the pellet in 1000 μL of supernatant.
23. Mix 18 μL of 0.4% trypan blue with 2 μL of nucleus suspension.
24. Load the mixture into the Countess™ II Automated Cell Counter (Invitrogen, cat. no. AMQAX1000) and observe the nuclei (*see* **Note 10**).

3.5 Library Construction for snRNA-seq

1. *Library Construction*: For the snRNA-seq library construction "10× Genomics Chromium Next GEM Single Cell 3′ Reagent Kits v3.1 (Dual Index) User Guide, Revision E (document number: CG000315)" is recommended. Other platforms for snRNA-seq library construction are also compatible.
2. *Quality Control (QC) of Constructed Libraries*: For the quantification and quality control of the constructed cDNA and sequencing libraries, Agilent Bioanalyzer, Agilent TapeStation, LabChip, Fragment Analyzer or Qubit are the recommended methods for accurate quantification.

4 Notes

1. If an 80 mm petri dish is used, a straight line is drawn at the 1/3 mark; if a 120 mm petri dish is used, lines are drawn at the 1/3 mark and at the center; if a square petri dish is used, lines are drawn at the top edge and at the center, and the seeds are sown.
2. When stretching the petri dishes in groups of five, care is taken to ensure that all areas where sowing is carried out are aligned, and one empty petri dish is added at the beginning and end to prevent water accumulation in the petri dishes.
3. Root lengths should be around 2 cm. Roots longer than 2 cm should be measured starting from the root tip and the excess should be cut off. In the study below 20 plant samples, the number of protoplasts obtained will be very small.
4. While passing samples through the 40 μm tip strainer, it is passed slowly and if the strainer becomes clogged, a new one can be used. After the samples are collected, 15 μL samples are taken to check the cell with trypan blue as an intermediate control step.
5. Sample and 0.4% trypan blue ratio is 10:0.8.
6. Samples are loaded onto a Thoma cell counting chamber at a volume of 1 μL. Live cells are counted under a microscope and calculated according to the equation cell/mL = A (number of cell) × DF (dilution factor) × 10,000.
7. The amount of root required for nuclei isolation and subsequent sequencing protocols is determined as 100–1000 mg [25], and sowing is carried out using 35 of 120 mm petri dishes.
8. To increase nuclear isolation efficiency at this stage, a 15 mL Falcon tube is utilized rather than a 50 mL one. Because the nuclei can sediment more compactly and there is less loss during supernatant removal, the smaller tube volume improves pelleting efficiency during centrifugation.

9. If the pellet could not be properly or fully precipitated and there is a cloudy structure, **step 15** can be repeated until a proper pellet structure is obtained.
10. Under normal conditions, trypan blue is used to stain dead cells with compromised cell membrane integrity. However, since nuclei were directly isolated in this protocol, membrane permeability has already been lost. Therefore, all nuclei are stained and detected as "dead" by the system. Therefore, nuclei that give a trypan blue-positive signal are suitable for analysis and should be considered viable.

Acknowledgments

These protocols were performed as part of two projects (121Z029 and 123Z251) supported by The Scientific and Technological Research Council of Türkiye (TÜBITAK).

References

1. Peleg Z, Blumwald E (2011) Hormone balance and abiotic stress tolerance in crop plants. Curr Opin Plant Biol 14(3):290–295
2. Davies PJ (ed) (2013) Plant hormones: physiology, biochemistry and molecular biology. Springer Science & Business Media
3. Skirycz A, Inzé D (2010) More from less: plant growth under limited water. Curr Opin Biotechnol 21(2):197–203
4. Depuydt S, Hardtke CS (2011) Hormone signalling crosstalk in plant growth regulation. Curr Biol 10-21(9):R365–R373
5. Eyidogan F, Oz MT, Yucel M, Oktem HA (2012) Signal transduction of phytohormones under abiotic stresses. In: Phytohormones and abiotic stress tolerance in plants, pp 1–48
6. Waadt R, Seller CA, Hsu PK, Takahashi Y, Munemasa S, Schroeder JI (2022) Plant hormone regulation of abiotic stress responses. Nat Rev Mol Cell Biol 23(10):680–694
7. Karady M, Hladík P, Cermanová K et al (2024) Profiling of 1-aminocyclopropane-1-carboxylic acid and selected phytohormones in *Arabidopsis* using liquid chromatography-tandem mass spectrometry. Plant Methods 20:41
8. Petřík I, Pěnčík A, Stýskala J, Tranová L, Amakorová P, Strnad M, Novák O (2024) Rapid profiling of cytokinins using supercritical fluid chromatography coupled with tandem mass spectrometry. Anal Chim Acta 1285: 342010
9. Jha Y, Yadav KA, Mohamed HI (2024) Plant growth-promoting bacteria and exogenous phytohormones alleviate the adverse effects of drought stress in pigeon pea plants. Plant Soil 505(1):163–183
10. Ullah A, Tariq A, Zeng F, Asghar MA, Sardans J, Peñuelas J (2024) Drought priming reduces Calligonum mongolicum sensitivity to recurrent droughts via coordinated regulation of osmolytes, antioxidants, and hormones. Plant Biol
11. Avalbaev A, Yuldashev R, Plotnikov A, Allagulova C (2025) A comparative analysis of the effect of 24-Epibrassinolide on the tolerance of wheat cultivars with different drought adaptation strategies under water deficit conditions. Plants 14(6):869
12. Jovic D, Liang X, Zeng H, Lin L, Xu F, Luo Y (2022) Single-cell RNA sequencing technologies and applications: a brief overview. Clin Transl Med 12(3):e694
13. Wang W, Garg V, Varshney RK, Liu H (2023) Single cell RNA-seq in phytohormone signaling: a promising future. Trends Plant Sci 28(11):1208–1210
14. Lin JL, Chen L, Wu WK, Guo XX, Yu CH, Xu M, Nie GB, Dun JL, Li Y, Xu B, Wang LJ, Chen XY, Gao W, Huang JQ (2023) Single-cell RNA sequencing reveals a hierarchical transcriptional regulatory network of terpenoid biosynthesis in cotton secretory glandular cells. Mol Plant 16(12):1990–2003

15. Zhang TQ, Xu ZG, Shang GD, Wang JW (2019) A single-cell RNA sequencing profiles the developmental landscape of Arabidopsis root. Mol Plant 12(5):648–660
16. Nolan TM, Vukašinović N, Hsu CW, Zhang J, Vanhoutte I, Shahan R, Taylor IW, Greenstreet L, Heitz M, Afanassiev A, Wang P, Szekely P, Brosnan A, Yin Y, Schiebinger G, Ohler U, Russinova E, Benfey PN (2023) Brassinosteroid gene regulatory networks at cellular resolution in the *Arabidopsis* root. Science 379(6639):eadf4721
17. Gao Z, Su Y, Jiao G, Lou Z, Chang L, Yu R, Xu C, Han X, Wang Z, Li J, Deng XW, He H (2025) Cell-type specific miRNA regulatory network responses to ABA stress revealed by time series transcriptional atlases in arabidopsis. Adv Sci (Weinh) 12(9):e2415083
18. Islam MT, Liu Y, Hassan MM, Abraham PE, Merlet J, Townsend A, Jacobson D, Buell CR, Tuskan GA, Yang X (2024) Advances in the application of single-cell transcriptomics in plant systems and synthetic biology. Biodes Res 6:0029
19. Sunaga-Franze DY, Muino JM, Braeuning C, Xu X, Zong M, Smaczniak C, Yan W, Fischer C, Vidal R, Kliem M, Kaufmann K, Sauer S (2021) Single-nucleus RNA sequencing of plant tissues using a nanowell-based system. Plant J 108(3):859–869
20. Xu C, Ma D, Ding Q, Zhou Y, Zheng HL (2022) PlantPhoneDB: a manually curated pan-plant database of ligand-receptor pairs infers cell–cell communication. Plant Biotechnol J 20(11):2123–2134
21. Sun Y, Sun J, Lin C, Zhang J, Yan H, Guan Z, Zhang C (2024) Single-cell transcriptomics applied in plants. Cells 13(18):1561
22. Wang X, Wang R, Huo X, Zhou Y, Umer MJ, Zheng Z, Jin W, Huang L, Li H, Yu Q, Li S, Varshney RK, Wang W, Xiao Y, Hong Y, Chen X, Lu Q, Liu H (2025) Integration of single-nuclei transcriptome and bulk RNA-seq to unravel the role of AhWRKY70 in regulating stem cell development in Arachis hypogaea L. Plant Biotechnol J 23(5):1814–1831
23. Grones C, Eekhout T, Shi D, Neumann M, Berg LS, Ke Y, Shahan R, Cox KL Jr, Gomez-Cano F, Nelissen H, Lohmann JU, Giacomello S, Martin OC, Cole B, Wang JW, Kaufmann K, Raissig MT, Palfalvi G, Greb T, Libault M, De Rybel B (2024) Best practices for the execution, analysis, and data storage of plant single-cell/nucleus transcriptomics. Plant Cell 36(4):812–828
24. Ryu KH, Huang L, Kang HM, Schiefelbein J (2019) Single-cell RNA sequencing resolves molecular relationships among individual plant cells. Plant Physiol 179:1444
25. Thibivilliers S, Anderson D, Libault M (2020) Isolation of plant root nuclei for single cell RNA sequencing. Curr Protocols Plant Biol 5(4):e20120

Chapter 11

Single-Cell Transcriptomics in Arabidopsis in Response to Salicylic Acid and Jasmonic Acid

Zhengzhi Tan, Ujjal J. Phukan, Yiqing Wang, Nicole A. Freeman, Sydney A. Greer, and Shahid Mukhtar

Abstract

Salicylic acid (SA) and jasmonic acid (JA) coordinate plant immunity through partially antagonistic signaling pathways that differ across cell types and over time. Single-cell and single-nucleus RNA sequencing now resolve these hormone responses at cellular resolution, allowing clear separation of early primary signaling waves from secondary and downstream transcriptional programs. In the context of plant–pathogen interactions, these approaches are particularly powerful for uncovering how distinct cell populations perceive SA and JA, reprogram defense networks, and balance growth-immunity trade-offs. This chapter provides end-to-end wet-lab protocols for nuclei-based snRNA-seq tailored to SA and JA treatments, including sample preparation, nuclei isolation, and library construction. In parallel, a bioinformatics analysis workflow is described, covering quality control, ambient RNA correction, doublet removal, dataset integration, cell type mapping, differential expression analysis, and pathway and module scoring. Together, these experimental and computational pipelines enable researchers to dissect hormone-driven immune responses at single-cell resolution and to generate cell-type-resolved hypotheses that can guide downstream genetic and physiological studies.

Key words Arabidopsis, Salicylic acid, Jasmonic acid, Single-nucleus RNA-seq, Protoplast, Nuclei isolation, Ambient RNA correction, Cell type mapping, Differential expression, Pathway scoring

1 Introduction

Salicylic acid (SA) and jasmonic acid (JA) are central regulators of plant immunity, orchestrating defense responses through distinct yet interconnected signaling pathways [1, 2]. SA primarily mediates resistance against biotrophic and hemi-biotrophic pathogens that grow on living host tissue [3–7]. SA is perceived by members of the NPR (Nonexpressor of Pathogenesis-Related Genes) protein family, particularly NPR1, NPR3, and NPR4, which function as bona fide SA receptors that fine-tune defense output. SA-dependent activation of the NPR1 signaling hub promotes its interaction with TGA transcription factors among other factors, leading to

Shahid Mukhtar (ed.), *Plant Hormones: Methods and Protocols*, Methods in Molecular Biology, vol. 3026,
https://doi.org/10.1007/978-1-0716-5214-5_11,

the induction of a suite of defense-related genes, including PR (pathogenesis-related) genes [3–7]. This leads to the establishment of systemic acquired resistance (SAR), a long-lasting, broad-spectrum immune state that primes distal tissues for enhanced defense.

In contrast, JA, often in cooperation with ethylene (ET), confers resistance to necrotrophic pathogens and herbivorous insects that kill host cells. JA perception is mediated by the COI1-JAZ co-receptor within the SCF^{COI1} ubiquitin ligase complex [1, 8, 9], where binding of the bioactive conjugate JA-isoleucine (JA-Ile) promotes JAZ repressor ubiquitination and degradation. This releases key basic helix-loop-helix (bHLH) transcription factors MYC2, MYC3, and MYC4, which drive wound and insect defense, and the Ethylene Response Factor 1 (ERF1) and Octadecanoid-Responsive Arabidopsis 59 (ORA59), which integrate JA and ET signaling during defense against necrotrophic pathogens [1, 8, 9]. The SA and JA pathways can act antagonistically, with SA frequently suppressing JA-responsive genes and, in some contexts, JA dampening SA signaling, thereby fine-tuning defense allocation and minimizing costly cross-activation [2, 7]. The balance between SA and JA is highly context-dependent, shaped by tissue identity, developmental stage, pathogen lifestyle, and the timing and strength of hormone perception [2, 7].

Over years bulk RNA sequencing has been instrumental in defining global transcriptional changes during SA and JA signaling. Downstream computational analyses such as co-expression network construction, and gene regulatory network inference help identify candidate regulators and effector nodes within these hormone-driven defense systems [10–15]. Co-expression networks group genes based on correlated expression patterns across treatments and cell types, highlighting modules associated with SA- or JA-responsive clusters. Integration of these networks with prior knowledge such as transcription factor binding motifs, chromatin accessibility, or protein-protein interactions allows prioritization of hub genes that may coordinate local or systemic immunity [14, 16, 17]. Network centrality metrics (degree, betweenness, and closeness) and module preservation analyses can further pinpoint core regulatory genes whose perturbation is most likely to alter the immune landscape.

Traditional bulk RNA sequencing provides averaged gene expression profiles from heterogeneous tissues, which mask the molecular diversity among individual cell types. In contrast, single-nucleus RNA sequencing (snRNA-seq) enables transcriptomic profiling at the level of individual nuclei and offers several important advantages [18–20]. It captures gene expression from single nuclei and therefore allows identification of distinct cell populations and rare cell types that are undetectable in bulk RNA-seq. In this chapter, we will provide a comprehensive laboratory and computational protocol for snRNA-seq in plants. We will describe optimized workflows for the isolation of high-quality nuclei from plant tissues suitable for downstream sequencing, library preparation and sequencing strategies compatible with high-throughput platforms such as the 10× Genomics Chromium system, and a complete bioinformatics pipeline that covers quality control, read alignment, transcript quantification, and downstream analyses such as clustering, cell type annotation, differential expression, and trajectory inference. By combining experimental and computational methods, this chapter is intended to serve as a practical guide for applying snRNA-seq to dissect transcriptional heterogeneity and cell type-specific regulatory networks in complex plant tissues.

2 Materials

2.1 Growth Media and Hormone Stocks

1. MS plates: 1× or 1/2 MS salts, sucrose 1–2% (w/v), agar 0.8% (w/v), adjust pH to 5.7.
2. Liquid MS: 1× MS salts, sucrose 1% (w/v), adjust pH to 5.7.
3. Salicylic acid stock: 100 mM in water.
4. Jasmonic acid or MeJA stock: 100 mM in ethanol.
5. Mock controls: Solvent-matched to each hormone stock and final working concentration.

2.2 Buffers and Solutions

All buffers should be prepared using RNase-free reagents and nuclease-free water. Keep on ice unless noted.

1. NIB (Nuclei Isolation Buffer): Tris–HCl 20 mM (pH 7.5), sucrose 250 mM, KCl 25 mM, $MgCl_2$ 5 mM, EGTA 2 mM, mannitol 250–400 mM (species-dependent), Ficoll 400 (2.5%) and Dextran T40 (1–5%) optional for dense tissues, DTT 1 mM, spermine 0.1 mM, spermidine 0.5 mM, Triton X-100 0.1–0.8% (v/v), BSA 0.1% (w/v), protease inhibitor 1×, RNase inhibitor 0.4 U/μL.
2. Nuclei wash buffer: Tris–HCl 20 mM (pH 7.5), sucrose 250 mM, KCl 25 mM, $MgCl_2$ 5 mM, BSA 0.1% (w/v), spermine 0.1 mM, spermidine 0.5 mM, protease inhibitor 0.1%, RNase inhibitor 0.4 U/μL.
3. PBS (RNase-free): 1× PBS, pH 7.4, used for washing and dilution.
4. NSB (Nuclei Suspension Buffer, Illumina T10 kit): Prepare working solution using kit components: 167 μL NSB + 40 μL 25% BSA + 40 U/μL RNase inhibitor + nuclease-free water to 1 mL total.
5. DAPI solution: 5 mg/mL stock; use at 5 μg/mL final concentration for nuclei staining.
6. 7-AAD: Add 25 μL per 1 mL nuclei suspension before FANS sorting.

2.3 General Reagents and Consumables

1. Nuclease-free water.
2. RNaseZap and 70% ethanol.
3. DTT (100 mM stock).
4. EGTA (100 mM stock).
5. Spermine (100 mM stock).
6. Spermidine (100 mM stock).
7. Ficoll 400 and Dextran T40.
8. BSA (25%).
9. Miracloth filters, razor blades, sterile Petri dishes.
10. Magnetic beads (IPB2 from Illumina kit).
11. Protease inhibitor cocktail.

2.4 Nuclei Isolation

1. Miracloth filters (1 layer, pre-wetted with NIB).
2. Cell strainers: 70 μm, 40 μm, 30 μm, and 20 μm (pre-wetted with NIB).
3. Razor blades: Sterile, single-use.
4. Petri dishes: Sterile 100 mm × 15 mm for tissue chopping.
5. RNase-free tubes and tips.
6. RNaseZap (Invitrogen AM9780) and 70% ethanol for cleaning pipettes and workspace.
7. Protease inhibitor cocktail (Sigma P9599 or Thermo 87785).
8. Laminar flow hood or RNA-dedicated biosafety cabinet.

2.5 Illumina T10 Kit Components

Include all reagents and their abbreviations as follows:

1. PAR: Partitioning Reagent.
2. CLB3: Chemical Lysis Buffer 3.
3. BBF: Breaking Buffer.
4. DPR: Departitioning Reagent.
5. PWB1: Wash Buffer.
6. CEB: cDNA Extraction Buffer.
7. IPB2: Illumina Purification Beads 2.
8. RTAM: Reverse Transcription Additive Mix.
9. RTEM: Reverse Transcription Enzyme Mix.
10. TSO1: Template Switch Oligo.
11. PMM4: 4× PCR Master Mix.
12. WTAP: Whole Transcriptome Amplification Primer.
13. LBP: Library Prep Buffer.
14. LPE: Library Prep Enzymes.

15. LPMA: Library Prep Mix A.
16. LAM: Library Adapter Mix.
17. UDI8: Unique Dual Index Mix Strip.
18. IDTE: DNA Elution Buffer (Tris–EDTA).

2.6 Equipment and Quality Control

1. Centrifuge: Refrigerated, capable of 1000 × *g*.
2. Dry bath or programmable heat block (66 °C, 42 °C, 85 °C cycles).
3. Magnetic stand (e.g., DynaMag-2 or equivalent).
4. 1 mL syringes and G22 blunt-end needles for PAR/DPR manipulations.
5. Microfluidic single-cell/nuclei partitioning system with 3′ chemistry (e.g., Illumina T10).
6. Illumina Single Cell 3′ RNA Prep Kit (T10) and Illumina Single Cell Library Prep Kit (for cDNA and library generation).
7. Fluorometer (Qubit 4) with High Sensitivity dsDNA kit.
8. Bioanalyzer or TapeStation (High Sensitivity D5000 ScreenTape).
9. Hemocytometer or automated counter for nuclei counting.
10. FANS (Fluorescence Activated Nuclei Sorting) for purification.
11. Cryostat (optional) for frozen tissue processing.
12. RNase-free consumables and low-bind microtubes.

2.7 Computing and Data Analysis

1. Workstation or server with sufficient memory (≥64 GB RAM) and GPU optional.
2. Software: fastqc, multiqc or trim_galore (quality control and adapter trimming), STAR, HISAT2, featureCounts or HTSeq-count (alignment), Cell Ranger, Alevin-fry, or kallisto | bus-tools (count matrix), CellBender, DoubletFinder, or Scrublet (Quality correction), Seurat or Scanpy (downstream clustering), DESeq2, edgeR, limma-voom, clusterProfiler, fgsea (Differential expression and pathway analyses).
3. Reference genome: Arabidopsis thaliana (Araport11) or other species-specific assemblies.
4. Optional: Illumina BaseSpace or bcl2fastq for basecalling and demultiplexing.

3 Methods

3.1 Experimental Design, Plant Growth, and Hormone Treatment

1. Use 7- to 10-day plate-grown seedlings for whole-seedling assays or 4- to 5-week rosettes for detached-leaf assays. Select one tissue type per experiment.
2. Include at least three biological replicates per condition and, if available, mutant controls such as *npr1* (for SA) or *coi1/dde2* (for JA) processed in parallel with wild type.
3. Define two sampling tiers per hormone: early (15–60 min) to capture signaling and immediate-early responses, and late (3–6 h) for downstream transcriptional programs.
4. Surface-sterilize seeds, stratify for 2–3 days at 4 °C dark, and germinate on MS plates containing 1–2% sucrose. Grow plants at 22 °C under a 12 h light/12 h dark photoperiod at 100–150 μmol/m^2/s.
5. For soil-grown plants, maintain identical conditions and collect fully expanded leaves from comparable rosette positions.
6. For plate assays, float intact seedlings or plates in liquid MS supplemented with the desired hormone or mock treatment (10–20 mL per plate) and agitate gently.
7. For detached leaves, excise selected leaves and float them abaxial side down on the treatment solution.
8. Start timing immediately upon first contact with the treatment.
9. At the end of exposure, blot samples gently and proceed directly to nuclei isolation, keeping all material and solutions cold (Fig. 1).
10. Randomize plate positions and processing order; if multiple days are required, balance treatments across days to minimize batch effects.
11. Maintain identical light, temperature, and agitation conditions throughout all treatments.

3.2 Nuclei Isolation from Plant Tissue

1. Chill all buffers, tubes, and equipment on ice before starting the procedure (*see* **Note 1**).
2. Place approximately 500 mg of fresh or frozen plant tissue in a sterile Petri dish.
3. Add 0.5 mL ice-cold NIB to the tissue.
4. Finely chop the tissue with a sterile razor blade for 4–5 min until a homogeneous slurry is obtained.
5. Add an additional 4.5 mL ice-cold NIB to the dish.
6. Incubate the slurry on ice for 5 min with gentle rocking or shaking.

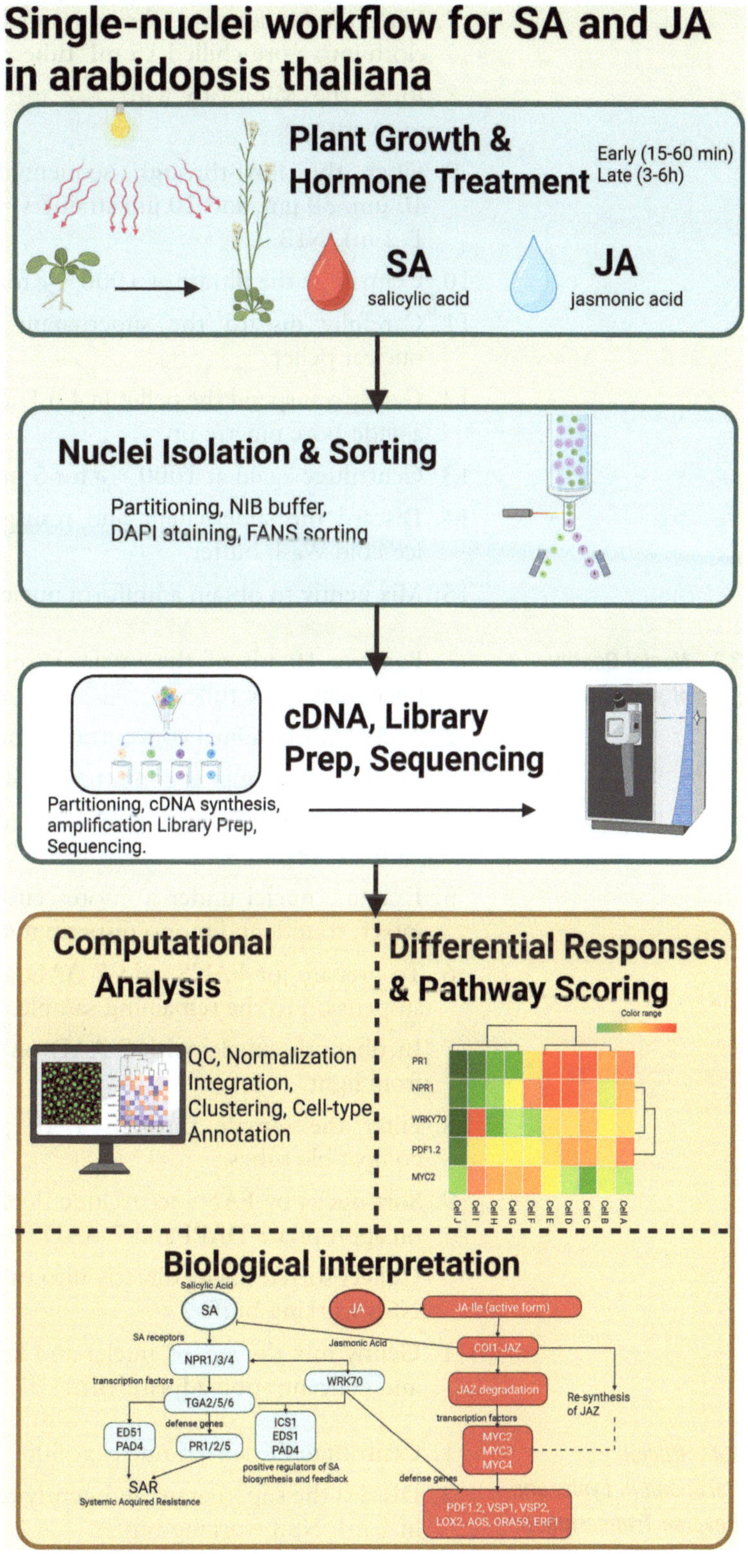

Fig. 1 Single-nuclei workflow for SA and JA in Arabidopsis. (Created in BioRender. Sharma, V. (2025) https:/BioRender.com/ascxuct)

7. Filter the homogenate through one layer of pre-wetted Miracloth into a pre-chilled 15 mL tube.
8. Rinse the Miracloth with 1–2 mL NIB to increase nuclei recovery.
9. Filter the flow-through sequentially through pre-wetted 40 μm, 30 μm, and 20 μm strainers, rinsing each strainer with 1–2 mL NIB.
10. Centrifuge the filtrate at 1000 × *g* for 5 min at 4 °C.
11. Carefully discard the supernatant without disturbing the nuclear pellet.
12. Gently resuspend the pellet in 4 mL ice-cold Wash buffer using a wide-bore pipette tip.
13. Centrifuge again at 1000 × *g* for 5 min at 4 °C.
14. Discard the supernatant and resuspend the pellet in 1 mL ice-cold Wash buffer.
15. Mix gently to obtain a uniform nuclei suspension (Fig. 1).

3.3 Nuclei Quality Control and FANS

1. Remove 10 μL of the nuclei suspension and transfer to a microcentrifuge tube.
2. Add DAPI to a final concentration of 5 μg/mL.
3. Incubate for 5 min at room temperature protected from light.
4. Place 5–10 μL of stained nuclei on a microscope slide and cover with a coverslip.
5. Examine nuclei under a fluorescence microscope to confirm intact, round, and mono disperse nuclei.
6. To prepare for FANS, add 7-AAD at 25 μL per 1 mL nuclei suspension to the remaining sample.
7. Incubate the nuclei with 7-AAD on ice for 30 min protected from light.
8. Filter the sample through a 40 μm strainer into FACS-compatible tubes.
9. Sort nuclei by FANS to exclude debris and aggregates, gating on appropriate DAPI and 7-AAD signals (*see* **Note 2**).
10. Collect sorted nuclei directly into tubes containing pre-chilled NSB working buffer.
11. Gently mix the sorted nuclei and keep on ice until counting and concentration adjustment.

3.4 Nuclei Partitioning, Lysis, and Reverse Transcription (Illumina T10)

1. Centrifuge the sorted nuclei at 500 × *g* for 5 min at 4 °C.
2. Discard the supernatant and gently resuspend the nuclei pellet in 1 mL NSB working buffer.
3. Centrifuge again at 500 × *g* for 5 min at 4 °C.

4. Discard the supernatant and resuspend the nuclei in an appropriate volume of NSB working buffer.
5. Count nuclei using a hemocytometer or automated counter.
6. Adjust the nuclei concentration to approximately 3400 nuclei/μL.
7. Aliquot 5 μL nuclei suspension (approximately 17,000 nuclei) for each T10 PIP tube (*see* **Note 3**).
8. Add 5 μL nuclei suspension directly into the PIP layer of the tube.
9. Add 40 U RNase inhibitor to each tube.
10. Add 320 μL Partitioning Reagent (PAR) along the wall of each tube.
11. Vortex horizontally for 15 s at 3000 rpm.
12. Vortex vertically for 2 min at 3000 rpm to form a stable emulsion.
13. Allow the emulsion to settle for 30 s at room temperature.
14. Remove the lower PAR phase carefully using a 1 mL syringe fitted with a G22 blunt-end needle, leaving the PIPs undisturbed.
15. In a separate tube, mix 90 μL Chemical Lysis Buffer 3 (CLB3) with 270 μL PAR.
16. Add the CLB3/PAR mixture to the PIP tube.
17. Invert the tube 10–15 times to mix thoroughly.
18. Place the tube on a dry bath and run Program B (66 °C for 45 min, 25 °C for 10 min, hold at 20 °C).
19. After lysis, remove the PAR phase using the same syringe and needle without disturbing the PIPs.
20. Add 300 μL Breaking Buffer (BBF) along the side of the tube.
21. Add 100 μL Departitioning Reagent (DPR).
22. Invert the tube 10–20 times to break the emulsion.
23. Centrifuge briefly (approximately 10 s) to separate phases.
24. Remove the lower pink phase with the syringe, leaving the PIPs in the upper aqueous phase.
25. Add 1 mL ice-cold PWB1 to the tube.
26. Vortex for 3 s and centrifuge for 1 min at 2000 × *g*.
27. Carefully remove the supernatant without disturbing the PIP pellet.
28. Repeat the wash with 1.2 mL PWB1 once and 1.0 mL PWB1 twice, each time vortexing and centrifuging as above.
29. Transfer the PIP pellet and residual buffer to a marked 0.5 mL tube.

30. Centrifuge for 1 min at 2000 × *g*, remove remaining supernatant, and place the tube on ice.
31. Prepare the RT Master Mix by combining RTAM, TSO1, and RTEM according to the kit instructions (100 μL total per reaction).
32. Add 100 μL RT Master Mix to each PIP tube.
33. Vortex briefly and centrifuge to collect contents.
34. Place tubes on a dry bath and run Program C (25 °C for 30 min, 42 °C for 90 min, 85 °C for 10 min, hold at 4 °C).
35. Store samples at 4 °C or proceed immediately to cDNA amplification.

3.5 cDNA Amplification and Cleanup

1. Prepare 0.5× PWB1 by mixing equal volumes of PWB1 and nuclease-free water (Fig. 1).
2. Centrifuge PIP tubes for 1 min at 2000 × *g*.
3. Remove and discard approximately 130 μL the supernatant without disturbing the PIPs.
4. Add 450 μL 0.5× PWB1 to the tube.
5. Vortex for 5 s and centrifuge for 1 min at 2000 × *g*.
6. Discard the supernatant carefully.
7. Repeat the wash step twice more using 350 μL 0.5× PWB1 each time.
8. Centrifuge a final time and remove all supernatant.
9. Prepare Whole Transcriptome Amplification (WTA) Master Mix according to kit instructions (35.2 μL total per reaction).
10. Add 35 μL WTA Master Mix to each tube containing PIPs.
11. Vortex briefly and centrifuge to collect the mixture.
12. Distribute the reaction mixture into two 68 μL aliquots in an 8-tube PCR strip.
13. Place tubes in a thermal cycler and run the recommended cDNA amplification program (95 °C 3 min; 5 cycles of 98 °C 15 s, 69 °C 10 min; 72 °C 5 min; hold 4 °C).
14. Briefly centrifuge the tubes after amplification.
15. Add 100 μL cDNA Extraction Buffer (CEB) to each reaction and transfer to a 1.5 mL tube.
16. Centrifuge at 2000 × *g* for 1 min.
17. Transfer 200 μL supernatant to a new 1.5 mL tube.
18. Add an additional 200 μL CEB to the pellet, vortex briefly, and repeat the spin.
19. Combine the second 200 μL supernatant with the first, yielding 400 μL total.

20. Add 0.8× volume (320 μL) IPB2 beads to the 400 μL cDNA solution.
21. Vortex thoroughly to mix and incubate at room temperature for 5 min.
22. Place the tube on a magnetic stand until the solution clears.
23. Remove and discard the supernatant without disturbing the beads.
24. Wash the beads with 1 mL freshly prepared 85% ethanol, incubate 30 s, and discard the ethanol.
25. Repeat the ethanol wash once more.
26. Air-dry the beads until a slight gloss remains and no visible liquid is present.
27. Remove the tube from the magnetic stand and add 42 μL IDTE buffer.
28. Resuspend the beads thoroughly and incubate at room temperature for 5 min.
29. Place the tube back on the magnetic stand and transfer 40 μL clear supernatant (purified cDNA) to a new tube.
30. Store cDNA at −80 °C or proceed to library preparation.

3.6 Library Preparation, Sequencing, and Data Processing

1. Thaw purified cDNA on ice and mix gently.
2. Transfer 40 μL cDNA to a PCR tube.
3. Prepare Library Prep Master Mix according to kit instructions (10 μL total per reaction).
4. Add 10 μL Library Prep Master Mix to the cDNA.
5. Vortex briefly and spin down.
6. Place the tube in a thermal cycler and run the fragmentation and end-repair program (30 °C for 6 min, 65 °C for 30 min, hold 4 °C).
7. Prepare the adapter ligation mix by combining LAM with nuclease-free water as specified by the kit.
8. Add the adapter mix and LPMA to each reaction according to the Illumina protocol.
9. Mix gently and incubate at 20 °C for 20 min.
10. Add 0.8× volume IPB2 beads to each ligation reaction.
11. Incubate for 5 min at room temperature and place the tubes on a magnetic stand.
12. Remove and discard the supernatant.
13. Wash the beads twice with 85% ethanol as described above.
14. Air-dry the beads briefly and elute the ligated DNA in 34 μL nuclease-free water.

15. Add 5 μL Unique Dual Index (UDI) mix and 12.5 μL PMM4 to each sample.
16. Mix and briefly centrifuge.
17. Run the library amplification program (98 °C 45 s; 15 cycles of 98 °C 15 s, 67 °C 30 s, 69 °C 45 s; 72 °C 1 min; hold 4 °C).
18. Add 0.8× volume IPB2 beads to each library, incubate 5 min, and place on a magnetic stand.
19. Remove supernatant and perform two 85% ethanol washes.
20. Elute libraries in 20 μL IDTE buffer.
21. Quantify 2 μL of each library using a Qubit High Sensitivity dsDNA kit.
22. Assess fragment size distribution using a TapeStation with High Sensitivity D5000 ScreenTape; confirm an average size of 370–550 bp.
23. Normalize and pool libraries to the desired molar concentration (*see* **Note 4**).
24. Load pooled libraries onto a NovaSeq 6000 or NovaSeq X Series instrument following the manufacturer's recommendations for 3′ single-cell RNA-seq.

3.7 Computational Processing, Integration, and Differential Responses

1. Generate a gene-by-cell UMI count matrix against the Araport11 reference using the appropriate single-cell/snRNA-seq pipeline (Fig. 1).
2. Filter out barcodes with very low total UMIs or detected genes, as well as those with high chloroplast or mitochondrial read fractions; start with global thresholds and refine per dataset (*see* **Note 5**).
3. Apply ambient RNA correction before dimensionality reduction and clustering.
4. Detect doublets using an appropriate tool and remove predicted multiplets, recording the fraction removed per sample.
5. Integrate biological replicates and treatment routes using a batch-aware integration method that preserves true treatment effects; visually inspect low-dimensional embeddings before and after integration.
6. Construct a shared nearest-neighbor graph on the integrated space and generate UMAP (or t-SNE) embeddings for visualization.
7. Assign cell types using canonical markers for epidermis, guard cells, mesophyll, vasculature, cortex, endodermis, and dividing populations, or by transferring labels from a compatible Arabidopsis cell atlas.

8. For each annotated cell type, fit a Treatment vs Mock differential expression model that includes replicate and batch terms; use a pseudobulk strategy as an orthogonal cross-check (*see* **Note 6**).
9. Score SA responses using PR-class genes and SA biosynthesis/signaling markers, and score JA responses using PDF-class genes and canonical JA pathway markers.
10. Compare early and late time points within each cell type to distinguish immediate-early transcriptional responses from downstream programs and pathway remodeling.

4 Notes

1. Excessive chopping, vortexing, or pipetting can shear nuclei and reduce recovery. For tissues with tougher cell walls (e.g., mature leaves), extend chopping time gradually rather than increasing force. Using wide-bore tips at all resuspension steps helps maintain intact nuclei.
2. DAPI and 7-AAD intensities vary across instruments and plant genotypes. Before large-scale sorting, run a small pilot to establish dataset-specific gates that reliably exclude debris, chloroplast fragments, and partially lysed nuclei.
3. Loading more than the recommended nuclei concentration can increase multiplet rates and reduce transcript capture uniformity. If tissue yields unusually high nuclei concentrations, dilute with NSB working buffer rather than forcing higher loads per tube.
4. Single-cell 3′ assays require balanced representation across index combinations to prevent sequence read skew on patterned flowcells. When pooling multiple treatments or time points, ensure all index pairs are represented at comparable molar concentrations.
5. Elevated plastid or mitochondrial RNA fractions often reflect tissue damage, excessive chopping, or prolonged handling at room temperature. Before filtering aggressively, confirm whether high organellar transcripts indicate a technical artifact or a biologically relevant hormone-induced stress signature.
6. Pseudobulk analyses provide a robust replicate-aware validation of cell-type-specific differential expression. Discrepancies between single-cell and pseudobulk results often point to integration artifacts, low cell counts, or inconsistent nuclei recovery.

Acknowledgments

This work was supported by NSF award IOS-2038872, PGRP-2318746, and OIA-2418230 to S.M.

References

1. Mu T, Luo S, Li L et al (2025) A review of the interaction mechanisms between jasmonic acid (JA) and various plant hormones, as well as the core regulatory role of MYC2. Plant Sci 353: 112407. https://doi.org/10.1016/j.plantsci.2025.112407
2. Boatwright JL, Pajerowska-Mukhtar K (2013) Salicylic acid: an old hormone up to new tricks. Mol Plant Pathol 14(6):623–634. https://doi.org/10.1111/mpp.12035
3. Sun Y, Detchemendy TW, Pajerowska-Mukhtar KM et al (2018) NPR1 in JazzSet with pathogen effectors. Trends Plant Sci 23(6):469–472. https://doi.org/10.1016/j.tplants.2018.04.007
4. Pajerowska-Mukhtar KM, Emerine DK, Mukhtar MS (2013) Tell me more: roles of NPRs in plant immunity. Trends Plant Sci 18(7):402–411. https://doi.org/10.1016/j.tplants.2013.04.004
5. Liu J, Fakhar AZ, Pajerowska-Mukhtar KM et al (2022) A TIReless battle: TIR domains in plant-pathogen interactions. Trends Plant Sci 27(5):426–429. https://doi.org/10.1016/j.tplants.2022.01.011
6. van Butselaar T, Van den Ackerveken G (2020) Salicylic acid steers the growth–immunity tradeoff. Trends Plant Sci 25(6):566–576
7. Spoel SH, Dong X (2024) Salicylic acid in plant immunity and beyond. Plant Cell 36(5): 1451–1464
8. Wasternack C, Song S (2017) Jasmonates: biosynthesis, metabolism, and signaling by proteins activating and repressing transcription. J Exp Bot 68(6):1303–1321. https://doi.org/10.1093/jxb/erw443
9. Wang J, Wu D, Wang Y et al (2019) Jasmonate action in plant defense against insects. J Exp Bot 70(13):3391–3400. https://doi.org/10.1093/jxb/erz174
10. Hussain A, Farooq M, Naqvi RZ et al (2020) Genome-wide identification and classification of resistance genes predicted several decoy domains in Gossypium sp. Plant Gene 24: 100250
11. Thingujam D, Gouli S, Cooray SP et al (2025) Climate-resilient crops: integrating AI, multi-omics, and advanced phenotyping to address global agricultural and societal challenges. Plants (Basel) 14(17). https://doi.org/10.3390/plants14172699
12. Naqvi RZ, Zaidi SS, Mukhtar MS et al (2019) Transcriptomic analysis of cultivated cotton Gossypium hirsutum provides insights into host responses upon whitefly-mediated transmission of cotton leaf curl disease. PLoS One 14(2):e0210011. https://doi.org/10.1371/journal.pone.0210011
13. Naqvi RZ, Zaidi SS, Akhtar KP et al (2017) Transcriptomics reveals multiple resistance mechanisms against cotton leaf curl disease in a naturally immune cotton species, Gossypium arboreum. Sci Rep 7(1):15880. https://doi.org/10.1038/s41598-017-15963-9
14. Kumar N, Mukhtar MS (2023) Integrated systems biology pipeline to compare co-expression networks in plants and elucidate differential regulators. Plants (Basel) 12(20). https://doi.org/10.3390/plants12203618
15. Hussain A, Asif N, Pirzada AR et al (2022) Genome wide study of cysteine rich receptor like proteins in Gossypium sp. Sci Rep 12(1): 4885. https://doi.org/10.1038/s41598-022-08943-1
16. Kumar N, Mukhtar MS (2023) Ranking plant network nodes based on their centrality measures. Entropy (Basel) 25(4). https://doi.org/10.3390/e25040676
17. Kumar N, Mishra BK, Liu J et al (2023) Network biology analyses and dynamic modeling of gene regulatory networks under drought stress reveal major transcriptional regulators in arabidopsis. Int J Mol Sci 24(8). https://doi.org/10.3390/ijms24087349
18. Thingujam D, Liu J, Majeed A et al (2024) Plant-microbiome dynamics through spatial metatranscriptomics and network biology. Trends Plant Sci 29(11):1176–1180. https://doi.org/10.1016/j.tplants.2024.07.007
19. Liu J, Majeed A, Mukhtar MS (2022) Rooting through single-cell sequencing in phloem pole cells. Commun Biol 5(1):1194. https://doi.org/10.1038/s42003-022-04180-4
20. Liu J, Majeed A, Kumar N et al (2024) Single nuclei multiomics reveals the drought-driven gene regulatory atlas in Arabidopsis. bioRxiv. https://doi.org/10.1101/2024.01.11.575118

Chapter 12

Stereo-Seq Transcriptomics in Arabidopsis Leaves in Response to Salicylic Acid

Yiqing Wang, Nicole A. Freeman, Ujjal J. Phukan, Zhengzhi Tan, Sydney A. Greer, and Shahid Mukhtar

Abstract

Spatially resolved transcriptomics captures gene expression in tissue context, enabling direct mapping of defense pathways across organs and cell layers. This chapter describes a complete Stereo-seq workflow for Arabidopsis leaves treated with salicylic acid (SA), from plant growth and hormone application through cryosectioning, fixation, bright-field imaging and optional histology, on-chip permeabilization, in situ reverse transcription and library construction on STOmics V1.3 1 × 1 cm chips, sequencing on DNBSEQ platforms, and analysis using STOmics Analysis Workflow (SAW) with downstream spatial statistics and pathway enrichment. Practical parameters are given for section thickness, fixation, imaging, permeabilization optimization, sequencing configuration, and quality control. The bioinformatics analyses include binning strategies, alignment to TAIR10, normalization, clustering, spatial differential expression, pathway testing, and integration with single-cell RNA-seq references for cell-type deconvolution.

Key words Arabidopsis, Salicylic acid, Stereo-seq, Spatial transcriptomics, STOmics V1.3, DNBSEQ, SAW, Spatial differential expression, Cell-type deconvolution

1 Introduction

Salicylic acid (SA) plays a central role in orchestrating plant immune responses, particularly against biotrophic and hemibiotrophic pathogens that rely on living host tissue [1–3]. In Arabidopsis, SA functions as a critical signaling molecule that activates a cascade of defense pathways leading to systemic acquired resistance (SAR), a long-lasting, broad-spectrum immune state [4]. SA accumulation is typically triggered after pathogen recognition, especially through effector-triggered immunity (ETI), which follows initial pattern-triggered immunity (PTI) [4, 5]. Once synthesized primarily through the isochorismate pathway via the enzyme Isochorismate Synthase 1 (ICS1), SA promotes the activation of key regulators such as Nonexpressor of Pathogenesis-Related genes 1 (NPR1) [6–

Shahid Mukhtar (ed.), *Plant Hormones: Methods and Protocols*, Methods in Molecular Biology, vol. 3026,
https://doi.org/10.1007/978-1-0716-5214-5_12,

10]. NPR1 acts as a transcriptional co-activator, interacting with TGA and WRKY transcription factors to induce PR genes, including PR1, a classical marker of SA-mediated defense [1, 4]. Through this signaling cascade, SA not only enhances local resistance but also mobilizes defense responses in distal tissues, effectively preparing the entire plant for potential pathogen attack [1, 4].

The bulk RNA-seq studies have established the association of SA signaling with defense gene activation [6, 9, 11, 12]. They have also enabled downstream co-expression analyses and network-based identification of regulatory hubs that coordinate these responses [13–15]. However, bulk measurements average signals across diverse cell types, masking cell-specific co-expression patterns and obscuring which hubs are truly active in particular tissues or spatial domains. Single-cell and spatial transcriptomics now reveal that SA signaling intensity and target gene expression vary among cell types, indicating that specific leaf or root regions contribute disproportionately to immunity [16–19]. This spatial regulation highlights the precision of SA-mediated defense, where localized accumulation and signaling enable plants to fine-tune immune responses without compromising growth or photosynthetic efficiency. Stereo-sequencing (Stereo-seq) is a high-resolution spatial transcriptomics technology developed by STOmics that captures genome-wide RNA expression directly from tissue sections [20, 21]. It uses patterned DNA nanoball arrays on specialized chips to record transcript positions with subcellular precision, preserving spatial context while maintaining high throughput. This method enables detailed mapping of gene activity across complex tissues, revealing cellular organization and functional heterogeneity that traditional RNA-seq cannot resolve.

This chapter provides a practical, end-to-end Stereo-seq workflow for *Arabidopsis thaliana* leaves treated with salicylic acid (SA). The protocol focuses on critical experimental steps that determine spatial transcriptome quality in small and fragile tissues, including proper orientation during OCT embedding, uniform 10 μm cryosectioning to minimize folds, brief warming for tissue adherence, immediate cold methanol fixation, and optimization of permeabilization time specific to leaf anatomy. Following chip preparation, sections are processed on STOmics Stereo-seq V1.3 chips, where tissue RNA is captured by spatially barcoded DNA nanoball arrays for subsequent reverse transcription and library preparation. For data analysis, we outline the complete Stereo-seq Analysis Workflow (SAW), including generation of gene-by-bin matrices, quality control metrics such as bin area, UMI counts, and mitochondrial and chloroplast RNA fractions, followed by clustering, spatial differential expression, and pathway enrichment analyses. Integration with a public single-cell RNA-seq atlas enables deconvolution of mixed spatial bins, allowing precise annotation of cell types such as epidermis, mesophyll, and vascular tissues under SA and mock conditions. This design supports temporal and spatial contrasts between SA and mock at matched time points and enables region-aware statistics across distinct leaf domains.

2 Materials

All reagents should be RNase-free, and all instruments and surfaces must be cleaned with RNaseZap followed by 70% ethanol. Work under a laminar flow hood and maintain separate stations for RNA and cDNA handling.

2.1 Reagents and Solutions

1. OCT compound (Optimal Cutting Temperature medium), pre-cooled on ice.
2. Methanol, pre-cooled to −20 °C.
3. RNase-free water.
4. 100% Ethanol.
5. 0.01 N HCl (pH 2.0 ± 0.1) for dissolving the Permeabilization Reagent (PR) enzyme.
6. Potassium Hydroxide, 8 M.
7. 0.1× SSC buffer.
8. 5× SSC buffer.
9. RI (RNase Inhibitor), 40 U/μL (from Stereo-seq kit).
10. TR buffer and TR enzyme (for tissue removal, from Stereo-seq kit).
11. Qubit™ ssDNA and dsDNA HS Assay Kits.
12. Agilent High Sensitivity RNA and DNA Kits.
13. AMPure® XP DNA cleanup beads.
14. TE buffer (10 mM Tris–HCl, 1 mM EDTA, pH 8.0).
15. Elute Additive and Neutralization Solution (from Stereo-seq kit).
16. cDNA Primer and 4× cDNA Primer Mix (from Stereo-seq kit).
17. RT Enzyme Mix, RT Oligo, and RT Buffer Mix (from Stereo-seq kit).
18. RT Plus (from Stereo-seq kit).
19. PR Enzyme (from Stereo-seq kit).
20. KMB and KME reagents (for multiple-displacement amplification, from Stereo-seq kit).
21. PCR Amplification Mix and PCR Barcode Primer Mix (1–16) (from Stereo-seq kit).
22. Glycerol (from Stereo-seq kit).

2.2 Equipment

1. Cryostat with adjustable specimen temperature (−20 °C chamber, −10 °C to −15 °C specimen disc).
2. PCR thermal cycler with flat heating block and adaptor for chip cassette.
3. Stereo-seq V1.3 Chip T (1 cm × 1 cm) and Chip P (4 × 1 cm^2).
4. Stereo-seq Accessory Kit (cassette, gasket, sealing tapes).
5. Stainless-steel or plastic embedding molds.
6. Pre-cooled forceps, razor blades, brushes.
7. Microscope glass coverslips, 18 mm × 18 mm and 24 mm × 60 mm.
8. Power dust remover (compressed air blower).
9. Centrifuge with temperature control (4 °C).
10. Magnetic Racks, <200 μL and 1.5–2 mL tubes.
11. Fluorescence microscope with FITC/TRITC channels and 10× objective.
12. Qubit 4 Fluorometer and Agilent TapeStation.

2.3 Biological Materials

1. Arabidopsis Col-0 seeds; potting mix; growth chamber.
2. SA working spray: 1 mM SA in water with 0.02% Silwet L-77.
3. Mock solution: Water with 0.02% Silwet L-77.

2.4 Software and Data

1. SAW (STOmics Analysis Workflow).
2. TAIR10 reference genome and GTF.
3. Python or R with Scanpy or Seurat.
4. Spatial statistics packages such as squidpy and SPARK-X.
5. Pathway tools such as g:Profiler or clusterProfiler.
6. Deconvolution tools such as cell2location.
7. Stereoscope, or Tangram.

3 Methods

3.1 Plant Growth and Hormone Treatment

1. Grow Col-0 to 4–5 weeks at 22 °C, 12 h light, 60% RH. Select fully expanded rosette leaves from the middle whorl.
2. Spray SA and harvest at 1 h, 6 h, and 24 h. Excise leaves at the petiole and proceed immediately to embedding. Keep a strict order of operations so each time point reflects true elapsed time (Fig. 1).
3. Replication: Minimum $n = 3$ plants per treatment per time point. Mix plants from different pots to avoid pot effects.

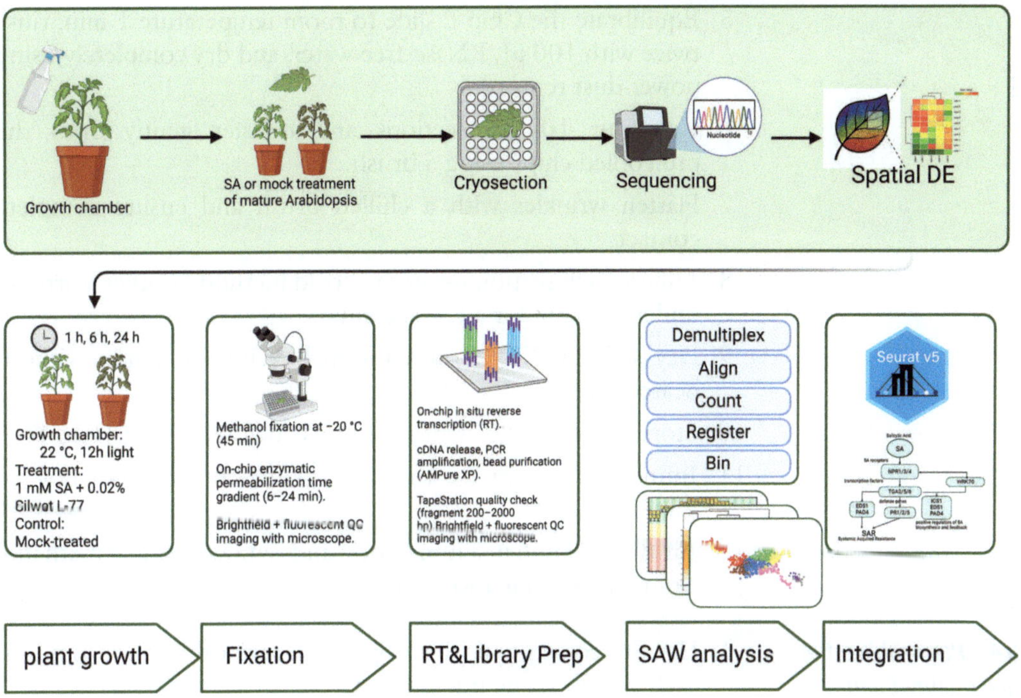

Fig. 1 Stereo-seq V1.3 workflow for SA in Arabidopsis. (Created in BioRender. Sharma, V. (2025) https:/ BioRender.com/zu7d0ma)

3.2 Tissue Embedding and Storage

1. Pre-cool OCT for ≥10 min on ice. Add to mold and continue cooling for ≥10 min.
2. Excise tissue <0.9 cm × 0.9 cm, blot off surface liquid, and embed within 30 min of harvest.
3. Coat the tissue evenly with OCT and orient it in the mold so that the sectioning face is downward.
4. Add additional chilled OCT to the mold to fully submerge the tissue, remove air bubbles using a syringe, and freeze horizontally on dry ice until opaque.
5. Wrap blocks in aluminum foil, seal in labeled bags, and store at −80 °C (Fig. 1).

3.3 Cryosectioning, Chip Mounting, and Tissue Fixation

1. Set cryostat chamber to −20 °C and specimen disc from −10 °C to −15 °C.
2. Equilibrate the OCT block inside the chamber for 30 min before sectioning.
3. Trim the block to expose the desired region (20 μm).
4. Record the Chip P serial number (SN) and avoid touching the active surface.

5. Equilibrate the Chip P slide to room temperature 1 min, rinse twice with 100 μL RNase-free water, and dry completely using power dust remover.
6. Cut four 10 μm sections and transfer gently onto the pre-cooled chips using a brush.
7. Flatten wrinkles with a chilled brush and ensure complete contact.
8. Adhere each section using the "cold method" (finger warming under the chip for a few seconds).
9. Dry slide at 37 °C on a thermal cycler for 8 min without heated lid.
10. Store slides at −80 °C for ≤4 weeks before fixation (Fig. 1).
11. Immediately immerse the slide in pre-cooled methanol for 45 min (−20 °C).
12. Air-dry 4–6 min in a fume hood and remove residual methanol from edges using a Kimwipe.

3.4 Permeabilization and Quality Control

1. Use Stereo-seq Chip P to optimize permeabilization time for each type of tissue understudy.
2. Prepare 10× PR stock by dissolving PR enzyme in 1 mL 0.01 N HCl; aliquot and store at −20 °C.
3. Dilute to 1× PR solution with 0.01 N HCl and pre-warm to 37 °C (≤30 min).
4. Apply 5 μL glycerol and a coverslip over the tissue, incubate 10 min.
5. Pre-warm gasket + cassette at 37 °C for 10 min, then remove coverslip and rinse slide in 0.1× SSC 3–5 s.
6. Incubate with 150 μL 1× PR solution at 37 °C for gradient time points (6, 12, 18, 24 min).
7. Wash with 200 μL wash buffer per chip (190 μL 0.1× SSC + 10 μL RI).
8. Add 100 μL RT QC mix (396 μL Buffer Mix + 44 μL Enzyme Mix per 4 chips), seal, and incubate 45 °C for 1 h.
9. Remove mix and wash twice with 400 μL 0.1× SSC per chip.

3.5 Tissue Removal and Imaging

1. Add 400 μL Tissue Removal Mix (1724.8 μL TR buffer + 35.2 μL TR enzyme per 4 chips) (Fig. 1).
2. Seal and incubate 55 °C for 1 h (up to 16 h if residual tissue persists).
3. Wash twice with 400 μL 0.1× SSC, rinse with 100 μL RNase-free water, and dry gently with power dust remover.

4. Image ssDNA signal under TRITC (532 nm) channel, 10× objective, identical settings across timepoints.
5. Select optimal permeabilization time showing maximal signal with minimal diffusion (*see* **Note 1**).

3.6 Reverse Transcription and cDNA Release

1. Perform tissue mounting, fixation, staining, and imaging on Stereo-seqStereo-seq ChipChromatin immunoprecipitation (ChIP) T according to Stereo-seqStereo-seq V1.3 transcriptomics workflow. Incubate the chip with Permeabilization Reagent for the optimal time period determined in Sect. 3.4.
2. Prepare RT mix (per chip: 166 μL RT Buffer Mix + 4 μL RT Plus + 10 μL RT Oligo + 20 μL RT Enzyme Mix).
3. Add 200 μL RT mix to chip, seal, and incubate 45 °C for 2 h (≤ 5 h).
4. Prepare cDNA Release Mix (195 μL 0.1 M KOH + 5 μL Elute Additive per chip).
5. Wash with 200 μL 0.1× SSC then add 180 μL cDNA Release Mix, seal and incubate 55 °C for 10 min.
6. Collect 180 μL solution in 1.5 mL tube, neutralize with 23 μL Neutralization Solution, and aliquot 66 μL × 3.

3.7 cDNA Amplification, Purification, Library Construction, and DNBSEQ Sequencing

1. Prepare cDNA PCR mix (76.5 μL 4× PCR Mix + 12 μL cDNA Primer + 13.5 μL water per reaction) (Fig. 1).
2. Add 34 μL PCR mix to 66 μL cDNA, amplify: 95 °C 5 min; 13 cycles (98 °C 20 s, 58 °C 20 s, 72 °C 3 min); 72 °C 5 min.
3. Purify with 0.8× AMPure XP beads; wash 2× 80% ethanol; elute 100 μL TE.
4. Analyze on Qubit and TapeStation: Fragment distribution 200–2000 bp, >100 ng.
5. First amplification: Mix 10 μL KMB + (35 − X) μL water + X μL cDNA (100 ng). Run 95 °C 5 min, 40 °C 3 min.
6. Add 5 μL KME and incubate 37 °C 10 min. Purify (0.8× beads), wash 2× 80% ethanol, and elute 15 μL TE.
7. Barcode PCR: 12.5 μL product + 25 μL PCR Amplification Mix + 12.5 μL Barcode Primer Mix. Run 95 °C 5 min; 13 cycles (98 °C 20 s, 58 °C 20 s, 72 °C 30 s); 72 °C 5 min.
8. Size-select: Add 0.55× beads → retain supernatant → add 7.5 μL beads → purify (2× wash, 80% ethanol). Elute 15 μL TE.
9. Secondary PCR: 100 ng selection product + 25 μL Amplification Mix + 12.5 μL Barcode Primer Mix + water to 50 μL; 5 cycles.
10. Final cleanup (0.8× beads). Elute 50 μL TE.

11. Analyze on Qubit and TapeStation: Main peak 200–600 bp, ≥ 200 ng.
12. Pool equimolarly and sequence PE75. A starting depth is 50–100 million read pairs per 1 × 1 cm chip; adjust for biology and desired bin size.

3.8 SAW Primary Processing

1. Export FASTQs and run folder metrics. Record Q30, cluster density, and yield (Fig. 1).
2. CID parsing and demultiplex: Run SAW to extract and error-correct CIDs.
3. Alignment: Map reads to TAIR10 and GTF. Enable intron-aware alignment if capturing pre-mRNA.
4. UMI collapse: Collapse with edit distance 1.
5. Counting: Produce a gene × CID matrix.
6. Registration: Register the tissue image using fiducials.
7. QC: Export per-CID metrics, including UMIs, genes, mitochondrial, chloroplast and ribosomal fractions, and distance to edges. Save SAW logs.

Example Scaffold

```
saw demux --in fastq/ --chip V1.3 --out demux/
saw align --genome TAIR10 --gtf TAIR10.gtf --in demux/ --out
aln/
saw count --in aln/ --umi-ed 1 --out counts/
saw register --image brightfield.tif --in counts/ --out reg/
saw bin --sizes 10,50 --in reg/ --out bins/
saw qc --in bins/ --out qc/
```

3.9 Secondary Analysis in Scanpy or Seurat

1. Filtering: Keep CIDs with ≥500 UMIs and ≥ 300 genes; adjust with depth.
2. Normalization: Library-size normalize and log1p or use SCTransform.
3. Batch effect assessment and correction: Visualize samples by batch (PCA, correlation heatmap). If batch separates samples, apply Harmony or regress out technical Feature selection and PCA: Identify highly variable genes and compute PCs.
4. Clustering: Build a kNN graph and run Leiden at several resolutions. Validate with marker genes.
5. Spatial differential expression: Use Moran's I to flag spatial structure and a spatial DE model such as SPARK-X for tests.
6. Pathway analysis: Run GO or pathway enrichment for SA and JA gene sets by cluster or region.

7. Contrasts: Build pseudo-bulk profiles per region or cluster and fit models for SA vs mock and JA vs mock at each time point, controlling for chip as a random effect.
8. Visualization: Plot spatial feature maps, cluster maps, and region-aggregated expression.

3.10 Integration with scRNA-seq References

1. Select an Arabidopsis leaf scRNA-seq reference with consistent gene IDs.
2. If needed, batch-correct reference and Stereo-seq embeddings with Harmony or Scanorama.
3. Run cell2location, stereoscope, or Tangram to estimate cell-type proportions per bin.
4. Compare cell-type shifts between SA and JA near veins, edges, and mesophyll. Validate with known markers.

3.11 Reporting, Replication, and Batch Control

1. Use ≥3 biological replicates per treatment and time point. Distribute conditions across chips and runs. Include a shared control per batch for integration.
2. Report chemistry (V1.3), section thickness, permeabilization time, RT and PCR cycles, sequencing mode and depth, SAW version, genome, filters, and all statistical settings.
3. Share a minimal, runnable script and an example dataset for reproducibility.

4 Note

1. The permeabilization time-course described in Subheadings 3.4 and 3.5 is a stand-alone QC protocol and is not followed by reverse transcription and cDNA release. For the actual Chip T Stereo-seq workflow, the optimal permeabilization time is selected from the imaging results of the Chip P assay based on maximized signal intensity and minimized diffusion (Subheading 3.4). The Chip T permeabilization, staining, and imaging workflow differs from the permeabilization test workflow described in Subheadings 3.4 and 3.5. Once an optimal permeabilization time is selected using Chip P, follow the Stereo-seq V1.3 transcriptomics workflow for Chip T up through tissue imaging before this protocol can be resumed at Subheading 3.6.

Acknowledgments

This work was supported by NSF award IOS-2038872, PGRP-2318746, and OIA-2418230 to S.M.

References

1. Boatwright JL, Pajerowska-Mukhtar K (2013) Salicylic acid: an old hormone up to new tricks. Mol Plant Pathol 14(6):623–634. https://doi.org/10.1111/mpp.12035
2. Thingujam D, Gouli S, Cooray SP et al (2025) Climate-resilient crops: integrating AI, multi-omics, and advanced phenotyping to address global agricultural and societal challenges. Plants (Basel) 14(17). https://doi.org/10.3390/plants14172699
3. van Butselaar T, Van den Ackerveken G (2020) Salicylic acid steers the growth–immunity tradeoff. Trends Plant Sci 25(6):566–576
4. Spoel SH, Dong X (2024) Salicylic acid in plant immunity and beyond. Plant Cell 36(5): 1451–1464
5. Liu J, Fakhar AZ, Pajerowska-Mukhtar KM et al (2022) A TIReless battle: TIR domains in plant-pathogen interactions. Trends Plant Sci 27(5):426–429. https://doi.org/10.1016/j.tplants.2022.01.011
6. Hussain A, Farooq M, Naqvi RZ et al (2020) Genome-wide identification and classification of resistance genes predicted several decoy domains in Gossypium sp. Plant Gene 24: 100250
7. Sun Y, Detchemendy TW, Pajerowska-Mukhtar KM et al (2018) NPR1 in JazzSet with pathogen effectors. Trends Plant Sci 23(6):469–472. https://doi.org/10.1016/j.tplants.2018.04.007
8. Pajerowska-Mukhtar KM, Emerine DK, Mukhtar MS (2013) Tell me more: roles of NPRs in plant immunity. Trends Plant Sci 18(7):402–411. https://doi.org/10.1016/j.tplants.2013.04.004
9. Hussain A, Asif N, Pirzada AR et al (2022) Genome wide study of cysteine rich receptor like proteins in Gossypium sp. Sci Rep 12(1): 4885. https://doi.org/10.1038/s41598-022-08943-1
10. Torrens-Spence MP, Bobokalonova A, Carballo V et al (2019) PBS3 and EPS1 complete salicylic acid biosynthesis from isochorismate in Arabidopsis. Mol Plant 12(12):1577–1586
11. Naqvi RZ, Zaidi SS, Mukhtar MS et al (2019) Transcriptomic analysis of cultivated cotton Gossypium hirsutum provides insights into host responses upon whitefly-mediated transmission of cotton leaf curl disease. PLoS One 14(2):e0210011. https://doi.org/10.1371/journal.pone.0210011
12. Naqvi RZ, Zaidi SS, Akhtar KP et al (2017) Transcriptomics reveals multiple resistance mechanisms against cotton leaf curl disease in a naturally immune cotton species, Gossypium arboreum. Sci Rep 7(1):15880. https://doi.org/10.1038/s41598-017-15963-9
13. Kumar N, Mukhtar MS (2023) Ranking plant network nodes based on their centrality measures. Entropy (Basel) 25(4). https://doi.org/10.3390/e25040676
14. Kumar N, Mukhtar MS (2023) Integrated systems biology pipeline to compare co-expression networks in plants and elucidate differential regulators. Plants (Basel) 12(20). https://doi.org/10.3390/plants12203618
15. Kumar N, Mishra BK, Liu J et al (2023) Network biology analyses and dynamic modeling of gene regulatory networks under drought stress reveal major transcriptional regulators in Arabidopsis. Int J Mol Sci 24(8). https://doi.org/10.3390/ijms24087349
16. Thingujam D, Liu J, Majeed A et al (2024) Plant-microbiome dynamics through spatial metatranscriptomics and network biology. Trends Plant Sci 29(11):1176–1180. https://doi.org/10.1016/j.tplants.2024.07.007
17. Thingujam D, Tan Z, Wang Y et al (2025) PRIMER cells: immune hotspots in plants. Trends Plant Sci 30(10):1075–1078. https://doi.org/10.1016/j.tplants.2025.04.006
18. Liu J, Majeed A, Mukhtar MS (2022) Rooting through single-cell sequencing in phloem pole cells. Commun Biol 5(1):1194. https://doi.org/10.1038/s42003-022-04180-4
19. Liu J, Majeed A, Kumar N et al (2024) Single nuclei multiomics reveals the drought-driven gene regulatory atlas in Arabidopsis. bioRxiv. https://doi.org/10.1101/2024.01.11.575118
20. Bawa G, Liu Z, Yu X et al (2024) Introducing single cell stereo-sequencing technology to transform the plant transcriptome landscape. Trends Plant Sci 29(2):249–265. https://doi.org/10.1016/j.tplants.2023.10.002
21. Xia K, Sun HX, Li J et al (2022) The single-cell stereo-seq reveals region-specific cell subtypes and transcriptome profiling in Arabidopsis leaves. Dev Cell 57(10):1299–1310 e1294. https://doi.org/10.1016/j.devcel.2022.04.011

Chapter 13

Automated Flow-Cytometric Readout of Reporter-Gene Activation in Transiently Transformed Protoplasts

Samuel Chisholm, Joseph S. Taylor, R. Clay Wright, and Bastiaan O. R. Bargmann

Abstract

Reporter-gene activation studies are essential for dissecting gene regulatory mechanisms, yet traditional whole-plant assays are often low-throughput and difficult to quantify. This chapter presents a streamlined method for analyzing reporter-gene activity using transiently transformed *Arabidopsis thaliana* mesophyll protoplasts. We describe the use of the pBeaconRFP vector for positive-fluorescent selection, enabling the isolation of successfully transformed cells via flow cytometry and fluorescence-activated cell sorting. Additionally, we introduce the pEvTV Gateway-compatible vector for flexible reporter-gene construct delivery and an automated pipeline for reproducible flow-cytometric data analysis. These methods facilitate rapid, robust, and scalable quantification of transcriptional responses, exemplified by the activation of the DR5 auxin-responsive reporter by gain-of-function Auxin Response Factor expression. The protocols are adaptable to other tissues and species, offering a versatile platform for high-throughput functional genomics.

Key words Reporter-gene activation, Protoplasts, Flow cytometry, Automated analysis

1 Introduction

Reporter-gene activation studies can give insight into the molecular basis of the regulation of gene expression. However, the examination of reporter-gene activity in whole plants can be labor intensive, time-consuming, challenging to quantify, and hence relatively low-throughput. In contrast, transient transformation of protoplasts coupled with flow-cytometric quantification of fluorescent reporter-gene activity can be relatively rapid, reproducibly quantified by automated analysis, and therefore better suited to higher-throughput studies.

We have developed the use of positive-fluorescent selection to isolate responses in successfully transformed protoplasts [1]. This system makes use of the pBeaconRFP transient transformation

Shahid Mukhtar (ed.), *Plant Hormones: Methods and Protocols*, Methods in Molecular Biology, vol. 3026, https://doi.org/10.1007/978-1-0716-5214-5_13,

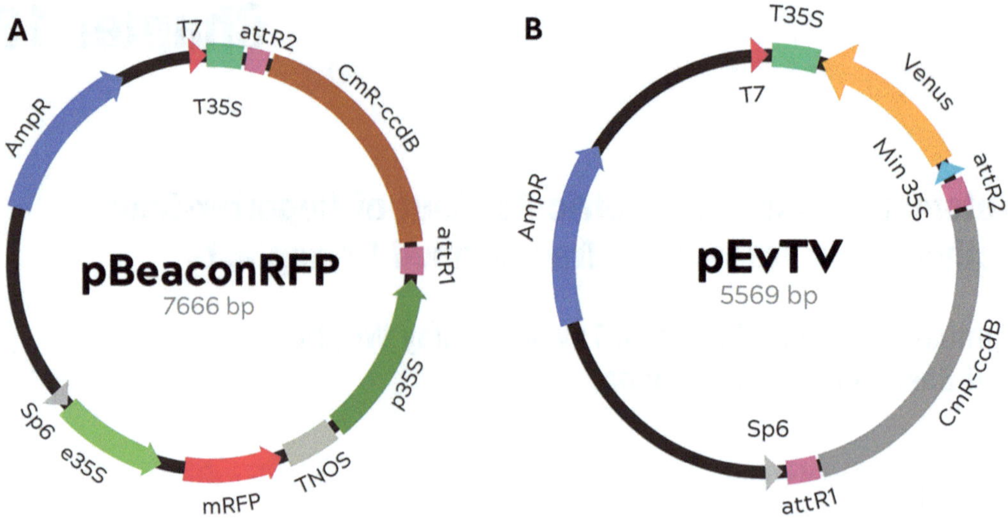

Fig. 1 Plasmids used for protoplast reporter-gene activation studies. (**a**) pBeaconRFP is a Gateway-compatible vector for effector expression, it also contains an mRFP positive selection marker to mark successfully transformed cells. (**b**) pEvTV is a Gateway-compatible vector for reporter-gene expression, a promoter-of-interest is inserted upstream of a minimal 35S promoter driving the expression of a Venus yellow-fluorescent protein. Both are high-copy plasmids with Ampicillin resistance markers

vector, where red fluorescent protein (RFP) is expressed alongside your gene-of-interest that can be inserted into the Gateway-compatible cloning site (Fig. 1a). We have also described the use of flow cytometry and fluorescence-activated cell sorting to identify and separate populations of fluorescently marked protoplasts [2]. In addition, we have compared integrated and transiently transformed fluorescent reporter systems [3] and have established the use of a Gateway-compatible vector (pEvTV; Fig. 1b) for transient transformation of reporter-gene constructs [4, 5]. Lastly, we have developed an automated analysis pipeline for reproducible interpretation of flow-cytometric data in reporter-gene activation studies [6].

In this chapter, we walk through the steps involved in the quantitative analysis of reporter-gene activation studies in transiently transformed protoplasts. This includes protoplast isolation and transformation, flow-cytometric measurement of the fluorescent properties of protoplasts, and the automated quantification and representation of flow-cytometry results. These procedures allow for the rapid, robust, and reproducible analysis of effectors and reporters in transient assays. The methods described here apply to mesophyll protoplasts obtained from two-week-old *Arabidopsis thaliana* seedlings but can easily be adapted to experiments using other tissues or species. We provide example data for measurement of the effects of gain-of-function Auxin Response Factor (ARF) transcription factor expression on the activation of the DR5 auxin-responsive reporter gene [7].

2 Materials

Prepare all solutions using ultrapure water (prepared by purifying deionized water, to attain a sensitivity of 18 MΩ/cm at 25 °C) and analytical grade reagents. Prepare and store all reagents at room temperature (unless indicated otherwise). Diligently follow all regulations when disposing of waste materials.

2.1 Plant Growth

1. Plant growth medium: 2.2 g/L MS salts + vitamins (Murashige and Skoog), 1% w/v sucrose, 1% w/v agar. Adjust to pH 5.8 with KOH. Autoclave at 120 °C to sterilize. Pour into 10 × 10 cm square plates. Store at 4 °C.
2. 70% v/v ethanol, 20% v/v bleach (1.21% w/v sodium hypochlorite), and sterile water.
3. Plant growth chamber, 100 μMol/m^{2}s PAR and 22 °C with 18 h-light/6 h-dark.

2.2 Generation of Protoplasts

1. Enzyme solution: 1.25% w/v Cellulase R-10 (Yakult, Japan), 0.3% w/v Macerozyme R-10 (Yakult, Japan), 0.4 M mannitol, 20 mM MES (2-(N-morpholino)ethanesulfonic acid), 20 mM potassium chloride, 0.1% w/v bovine serum albumin, 10 mM calcium chloride, 5 mM β-mercaptoethanol. Adjust to pH 5.7 with 1 M Tris–HCl pH 7.5. Heat the solution to 55 °C for 10 min (the solution should become clear), cool to room temperature. Store at −20 °C. Filter sterilize with 0.22 μm filters (*see* **Notes 1** and **2**).
2. Small scissors, rotary shaker, 250 mL flasks, 40 μm cell strainer, 15 mL conical tubes, large orifice P1000 pipette tips, swing-bucket centrifuge (100 G), hemacytometer, compound microscope with 10× objective.

2.3 Transformation of Protoplasts

1. Plasmids: pBeaconRFP, pEvTV (available upon request) (Fig. 1).
2. MMg solution: 0.4 M mannitol, 15 mM magnesium chloride hexahydrate, 4 mM MES. Adjust to pH 5.7 with 1 M potassium hydroxide solution. Make fresh. Filter sterilize with 0.22 μm filters.
3. MIDIprep kit. Store DNA at −20 °C (*see* **Note 3**).
4. PEG solution: 40% w/v polyethylene glycol 4000, 0.2 M mannitol, 0.1 M calcium chloride. Make fresh. Filter sterilize with 0.22 μm filters (*see* **Note 4**).
5. Incubation solution: (Enzyme solution without enzymes or β-mercaptoethanol) 0.4 M mannitol, 20 mM MES (2-(N-morpholino)ethanesulfonic acid), 20 mM potassium chloride, 0.1% w/v bovine serum albumin, 10 mM calcium chloride. Adjust

to pH 5.7 with 1 M Tris–HCl pH 7.5. Store at −20 °C. Filter sterilize with 0.22 μm filters.

6. 24-Well plates, epifluorescence microscope equipped with YFP and RFP (or equivalent) filters.

2.4 Flow Cytometry

1. Flow cytometer: Sony SH800, PBS buffer for sheath fluid, calibration beads (*see* **Note 5**).
2. Sample preparation: 40 μm cell strainer, 1.5 mL microfuge tubes.
3. Automated flow-cytometry analysis: R statistical computing software, R Studio Desktop, Automated Protoplast Analysis R markdown (.Rmd) file.

3 Methods

Carry out all procedures at room temperature unless otherwise specified. Use aseptic techniques for all procedures (*see* **Note 2**).

3.1 Plant Growth

1. Sterilize 100 μL of dry Arabidopsis seed in a 15 mL tube by a 5-min incubation with 10 mL 70% ethanol, followed by a 10-min incubation with 10 mL 20% bleach, and rinsing three times with 50 mL sterile water (*see* **Note 6**).
2. Plate seeds in four rows on a square plate with plant growth medium using a sterile transfer pipette (Fig. 2). Plates are sealed with micropore tape and placed horizontally in a growth chamber.

3.2 Generation of Protoplasts

1. Harvest the shoots of 14-day-old seedlings using a sterilized small pair of scissors and transfer to a 250 mL flask with 50 mL enzyme solution.
2. Shake the flask with the shoot tissue in enzyme solution at 50 rpm at room temperature for 3–4 h, then increase to 100 rpm for 10 min to expel protoplasts. The enzyme solution should become cloudy, indicating the release of protoplasts.
3. Filter the protoplasts by passing them over a 40 μm cell strainer into a fresh flask (*see* **Note 7**).
4. Transfer the filtered protoplast suspension to 15 mL conical tubes and centrifuge for 5 min at 100 × *g* with no brake. Remove the supernatant by aspirating, leaving some enzyme solution behind to resuspend pellet (*see* **Note 8**).
5. Resuspend pellets in the remaining enzyme solution, combine into one tube and bring up to 2 mL with Incubation solution.
6. In a new 15 mL tube, overlay protoplasts gently on top of 6 mL 0.6 M sucrose cushion and centrifuge 100 × *g* for 5 min. Gently transfer the green band of protoplasts at the interface into a new 15 mL conical tube.

Fig. 2 Plant growth of seedling-shoot source tissue for the generation of protoplasts. Sterilized Arabidopsis seeds are plated on square 10 × 10 cm plates with ½ MS + 1% sucrose. Shoots are harvested after 2 weeks

3.3 Transformation of Protoplasts

1. Wash the protoplasts with 15 mL MMg solution and centrifuge for 5 min at 100 × *g* with no brake. Remove the supernatant by aspirating.
2. Resuspend the protoplasts in a volume appropriate for quantification of protoplast density using a hemacytometer. Assess the protoplast density and adjust the volume of the MMg solution to achieve a density of 4×10^6 protoplasts per mL (*see* **Note 9**).
3. Prepare a 15 mL conical tube for each transformation (and one mock transformation with no DNA) by labeling them and adding 12.5 μg of the pBeaconRFP effector plasmid and 12.5 μg of the pEvTV reporter plasmid to the bottom of the tube (from 1 μg/μL plasmid stocks) (*see* **Note 10**).
4. Add 250 μL (1×10^6) protoplasts to each tube. Add 250 μL PEG solution to each tube and mix well by vortexing for 5 s (*see* **Notes 11** and **12**).
5. Wash the protoplasts by adding 15 mL Incubation solution to each tube and spinning them down for 5 min at 100 × *g* with no brake. Remove the supernatant by aspirating.
6. Resuspend the protoplasts in 1 mL Incubation solution and transfer them to a 24-well plate.
7. Incubate the protoplasts overnight at room temperature in the dark.
8. The next day, inspect the protoplasts with an epifluorescence microscope equipped with YFP and RFP (or equivalent) filters to check for successful transformation (Fig. 3) (*see* **Note 13**).

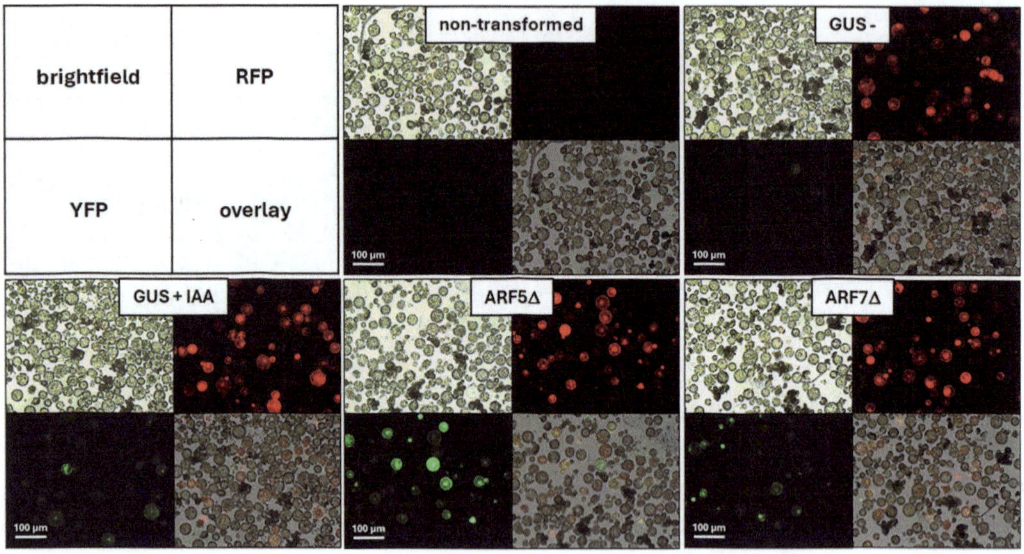

Fig. 3 Inspection of transformed protoplasts. Protoplasts were either mock transformed or transformed with the pEvTV_DR5 auxin-responsive reporter and three different effector constructs: pBeaconRFP_GUS (an inert control mock-treated or treated with 100 nM indole-3-acetic acid [IAA] overnight), pBeaconRFP_ARF5Δ (an irrepressible Auxin Response Factor), or pBeaconRFP_ARF7Δ (another irrepressible Auxin Response Factor). Protoplasts incubated in a 24-well plate were imaged using an inverted epifluorescence microscope (Olympus X81) equipped with RFP and YFP filters. The scale bars represent 100 μm

3.4 Flow Cytometry with Manual Analysis

Unless stated otherwise, flow-cytometry parameters are presented as area of event signals (e.g., FSC-A) in arbitrary units. Side scatter and back scatter parameters are functionally equivalent and only differ between cytometer manufacturers depending on the arrangement of the optics.

1. Transfer protoplast suspensions to 1.5 mL microfuge tubes by passing over a 40 μm cell strainer using a P1000 pipette with large orifice tips.
2. Set up the flow cytometer, run the start-up protocol and performance test, apply compensation setting for YFP vs. RFP.
3. Set up two dot plots for forward scatter vs. back scatter (one for All Events and one for RFP-positive Events), one dot plot for forward scatter area vs. forward scatter height (for discriminating Singlet Cells) two dot plots for YFP (488 nm excitation, 525/50 emission) vs. RFP (561 nm excitation, 617/30 emission) fluorescence emission (one for All Events and one for Singlet Cells). Adjust the photo-multiplier tube voltages so that the bulk of the detected events falls within the plots (Fig. 4a and b) (*see* **Notes 14** and **15**).
4. Use the mock-transformed protoplasts to set up a gate for RFP-positive events on the All Events YFP vs. RFP dot plot (Fig. 4b).

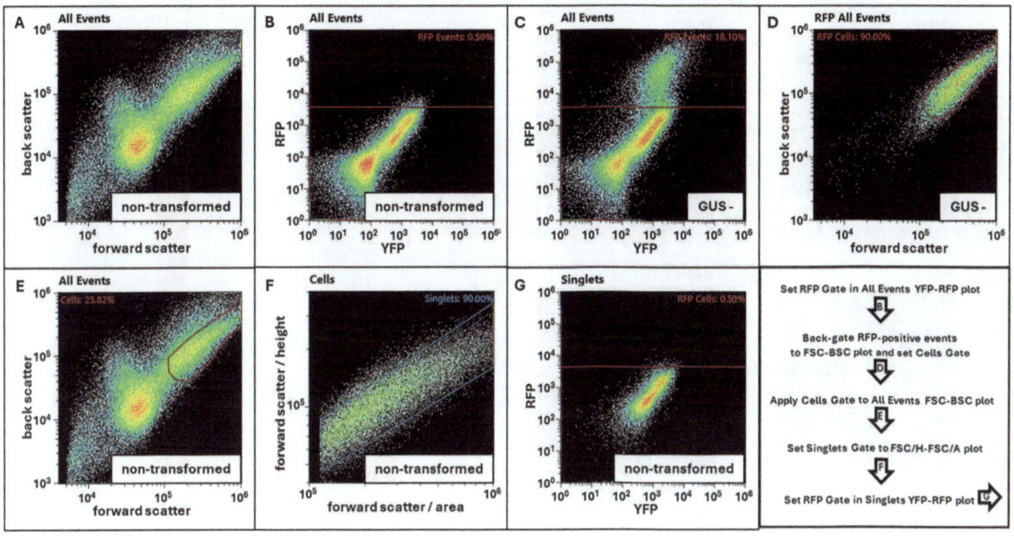

Fig. 4 Gating strategy to measure YFP intensity in RFP-positive cells. (**a**) A dot plot with forward scatter vs. back scatter showing All Events for a non-transformed control. (**b**) A dot plot with YFP vs. RFP showing All Events for a non-transformed control, a gate is drawn for RFP-positive events that excludes 99.5% of non-transformed events. (**c**) A dot plot with YFP vs. RFP showing All Events for a sample transformed with pEvTV_DR5 and pBeaconRFP_GUS and mock treated, RFP-positive events fall within the RFP gate. (**d**) A dot plot with forward scatter vs. back scatter showing back-gated RFP-positive events for a sample transformed with pEvTV_DR5 and pBeaconRFP_GUS and mock treated, a gate is drawn to capture 90% of RFP-positive events, these are considered live Cells. (**e**) A dot plot with forward scatter vs. back scatter showing All Events for a non-transformed control, the Cells gate enriches for live cells and discards dead cells and debris. (**f**) Singlets are identified by visualizing Cells on a dot plot with forward scatter area vs. forward scatter height and drawing a gate that encompasses 90% of the cells closest to the diagonal. (**g**) A dot plot with YFP vs. RFP showing Singlets for a non-transformed control, a gate is drawn that excludes 99.5% of non-transformed singlet cells and can identify RFP-positive singlet cells for quantification of YFP intensity in transformed samples

5. Run a sample with RFP-positive cells and back-gate the RFP-positive events to the second forward scatter vs. back scatter dot plot (RFP-positive Events; Fig. 4c and d).
6. Draw a gate encompassing the central 90% of the back-gated events, this gate defines on the All Events forward scatter vs. back scatter dot plots where cells are found (Fig. 4d).
7. Copy the gate that was just drawn onto the first All Events forward scatter vs. back scatter dot plot, this is the "Cells" gate (Fig. 4e).
8. Display the Cells gate from the All Events forward scatter vs. back scatter dot plot on the Cells forward scatter area vs. forward scatter height dot plot and draw a gate encompassing 90% of the cells nearest the diagonal, these are the Singlets (Fig. 4f).

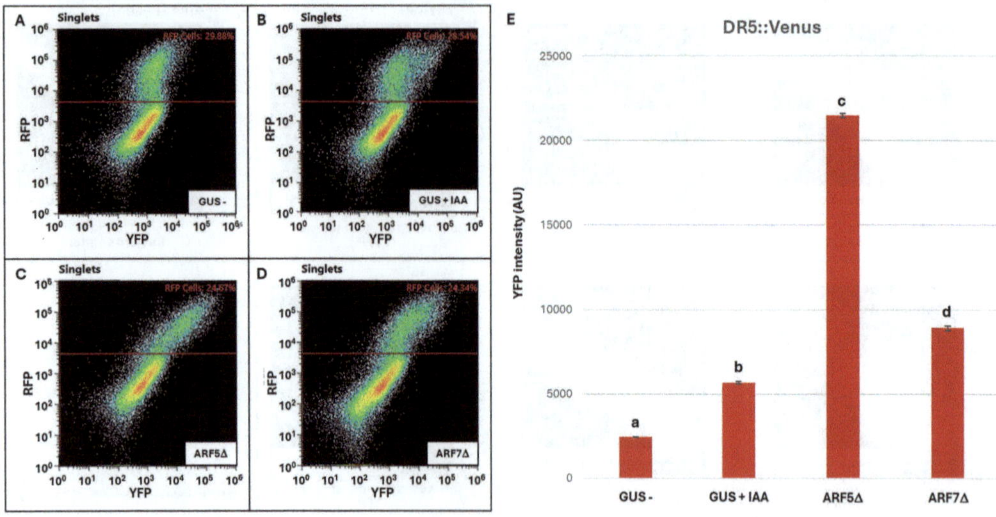

Fig. 5 Measurement of YFP reporter intensity in RFP-positive cells. Protoplasts were transformed with the pEvTV_DR5 auxin-responsive reporter and three different effector constructs: pBeaconRFP_GUS (an inert control mock-treated [**a**] or treated with 100 nM indole-3-acetic acid overnight [**b**]), pBeaconRFP_ARF5Δ (an irrepressible Auxin Response Factor, **c**), or pBeaconRFP_ARF7Δ (another irrepressible Auxin Response Factor, **d**). Results are shown in dot plots with YFP vs. RFP showing Singlet Cells for **a**–**d**. (**e**) The mean YFP intensity (arbitrary units) in RFP-positive Singlet Cells for three technical replicates is shown in a bar graph with standard deviation and statistical significance was assayed by one-way ANOVA followed by a Tukey post hoc test; letters indicate homogeneous subsets ($p < 0.05$)

9. Display the Singlets gate from the Cells forward scatter area vs. forward scatter height dot plot on the Singlet Cells YFP vs. RFP dot plot (Fig. 4g).
10. Using data from a non-transformed sample, draw a cut-off gate on the Singlet Cells YFP vs. RFP dot plot that excludes 99.5% of the cells, this gate captures the "RFP-positive Singlet Cells" that exceed the autofluorescence signal of protoplasts in the red channel (Fig. 4g). Note the mean YFP signal in the "RFP-positive Singlet Cells," this is the reporter-gene read-out for each sample (Fig. 5).
11. Collect data for >1000 RFP-positive Singlet Cells for each sample, but more is better.
12. Export the FCS files.

3.5 Automated Analysis of Transformed Protoplasts

1. Download the Automated Protoplast Analysis Template folder from GitHub (*see* **Notes 16** and **17**).
2. Move the folder containing FCS files exported in Subheading 3.4, **step 12**, into the Automated Protoplast Analysis Template folder. Rename this folder "FCS_files" and make sure only.fcs files are in this folder.

3. Download and install R and RStudio. Open the ProtoplastAutomationScript.Rmd inside of the Automated Protoplast Analysis Template folder.
4. If running for the first time, run the chunk on line 12 to install the required packages. This will install packages from the R and Bioconductor repositories.
5. Run the setup chunk on line 30 to load the required packages into your workspace. Make sure that the message reads as all "TRUE" to verify successful loading of all packages.
6. Enter the experiment name and date into lines 54 and 56 in chunk 1. Run chunk 1 on line 52 to generate a template.csv annotation file that will be exported into the Automated Protoplast Analysis Template folder. Open this file in a spreadsheet editor, like Excel, and fill in the template according to your experiment's details. Save the file.
7. Run chunk 2 on line 77 to import the raw FCS data as a flowSet and annotate the flowSet with experimental variables.
8. Run chunk 3 on line 105 if you want to add additional variable metadata. Remove or alter example metadata in the chunk to describe the variables in your experiment.
9. Run chunk 4 on line 121 to save the annotated flowSet. This code will export your annotated flowSet into the Automated Protoplast Analysis Template folder (*see* **Note 18**).
10. Run chunk 5 on line 131 to read the annotated flowSet back into R. This annotated flowSet can be stored or shared and is re-read here to ensure that it was saved properly.
11. Run the code on line 147 to get a list of sample IDs. Identify the non transformed control and baseline control (pBeaconRFP_GUS in this example) ID and enter these into lines 150 and 151, respectively. Run chunk 6 on line 145.
12. Run chunk 7 on line 156 to define the preset laser abbreviations used by your cytometer. If using a different cytometer than the Sony SH800S or ThermoFisher Attune NxT, you may have to set these abbreviations manually (*see* **Note 19**).
13. Run chunk 8 on line 196 to generate and save forward scatter (FSC) vs. back scatter (BSC) plots that allow you to assess cell size and complexity across all of your samples (Fig. 6a). This step corresponds with Subheading 3.4, **step 3**, in the manual protoplast analysis method. Plots are saved to the "plots/AllEvents_FSCvSSC" folder (*see* **Note 20**).
14. Run chunk 9 on line 294 to apply compensation and log transformation to the flowSet (*see* **Note 21**).

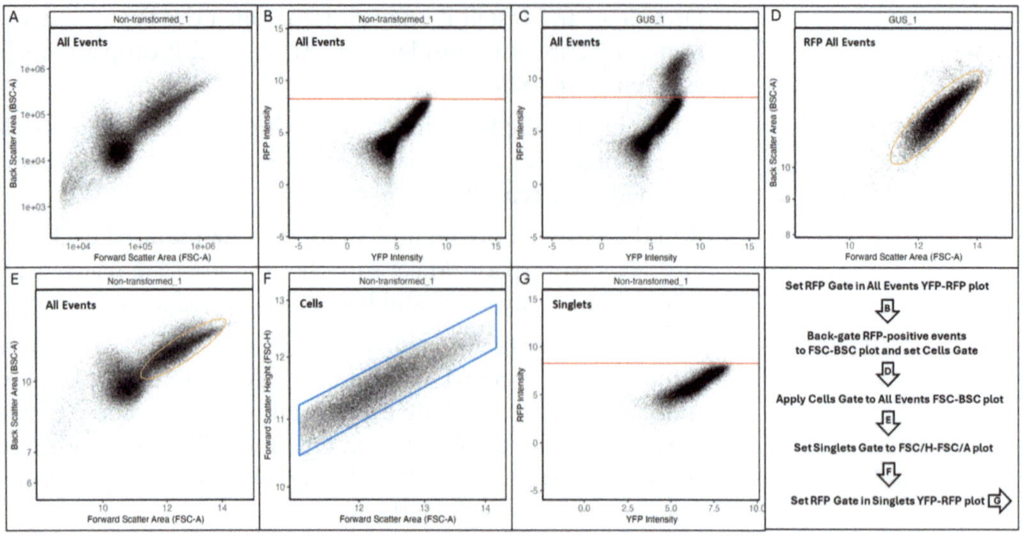

Fig. 6 Automated gating strategy to measure YFP intensity in RFP-positive cells. (**a**) A dot plot with forward vs. back scatter showing All Events for a non-transformed control. (**b**) A dot plot with YFP vs. RFP showing All Events for a non-transformed control, a gate is drawn for RFP-positive events that exclude 99.5% of non-transformed events. (**c**) A dot plot with YFP vs. RFP showing All Events for a sample transformed with pEvTV_DR5 and pBeaconRFP_GUS and mock treated, RFP-positive events fall within the RFP gate. (**d**) A dot plot with forward scatter vs. back scatter showing back-gated RFP-positive events for a sample transformed with pEvTV_DR5 and pBeaconRFP_GUS and mock treated, a gate is drawn to capture 90% of RFP-positive events, these are considered live Cells. (**e**) A dot plot with forward scatter vs. back scatter showing All Events for a non-transformed control, the Cells gate enriches for live cells and discards dead cells and debris. (**f**) Singlets are identified by visualizing Cells on a dot plot with forward scatter area vs. forward scatter height and drawing a gate that encompasses 90% of the cells closest to the diagonal. (**g**) A dot plot with YFP vs. RFP showing Singlets for a non-transformed control, a gate is drawn that excludes 99.5% of non-transformed singlet cells and can identify RFP-positive singlet cells for quantification of YFP intensity in transformed samples

15. Run chunk 10 on line 352 to generate and save YFP vs. RFP plots of all events. This step corresponds with Subheading 3.4, **step 3**, in the manual protoplast analysis method. Plots are saved to the "plots/AllEvents_YFPvRFP" folder.
16. Run chunk 11 on line 451 to set the RFP-positive gate and plot all events on a YFP vs. RFP scale with the RFP-positive gate overlaid (Fig. 6b and c). This step corresponds with Subheading 3.4, **step 4**, in the manual protoplast analysis method. Plots are saved to the "plots/AllEvents_YFPvRFP_RFPeventsGate" folder (*see* **Note 22**).
17. Run chunk 12 on line 557 to create a new flowSet that includes only RFP positive events and set the Cells gate based on this subset. This step corresponds with Subheading 3.4, **steps 5** and **6**, in the manual protoplast analysis method. This chunk plots the Cells gate over both the RFP events (Fig. 6d) and all events (Fig. 6e) to assess the uniformity of the fit of the gate

around different samples (*see* **Note 23**). Plots are saved to "plots/RFPevents_CellsGate_FSCvBSC" and "plots/AllEvents_CellsGate_FSCvBSC" folders.

18. Run chunk 13 on line 769 to create a new flowSet that includes only the events inside of the Cells gate. This chunk defines the Singlet Cells gate based on this subset. This step corresponds with Subheading 3.4, **step 8**, in the manual protoplast analysis method. This chunk plots the Cells flowSet on an FSC-A v. FSC-H scale and overlays the Singlet Cells gate (Fig. 6f) (*see* **Note 24**). Plots are saved to the "plots/Cells_singletsGate_FSCAvFSCH" folder.
19. Run chunk 14 on line 860 to create a new flowSet that includes only the events inside of the Singlet Cells gate. This chunk defines the RFP-positive Singlet Cells gate based on this subset. This chunk plots the Singlet Cells flowSet on a YFP v. RFP scale and overlays the RFP-positive Singlet Cells gate (Fig. 7a–d) (*see* **Note 25**). Plots are saved to the "plots/singletCells_RFPsingletsGate_YFPvRFP" folder.

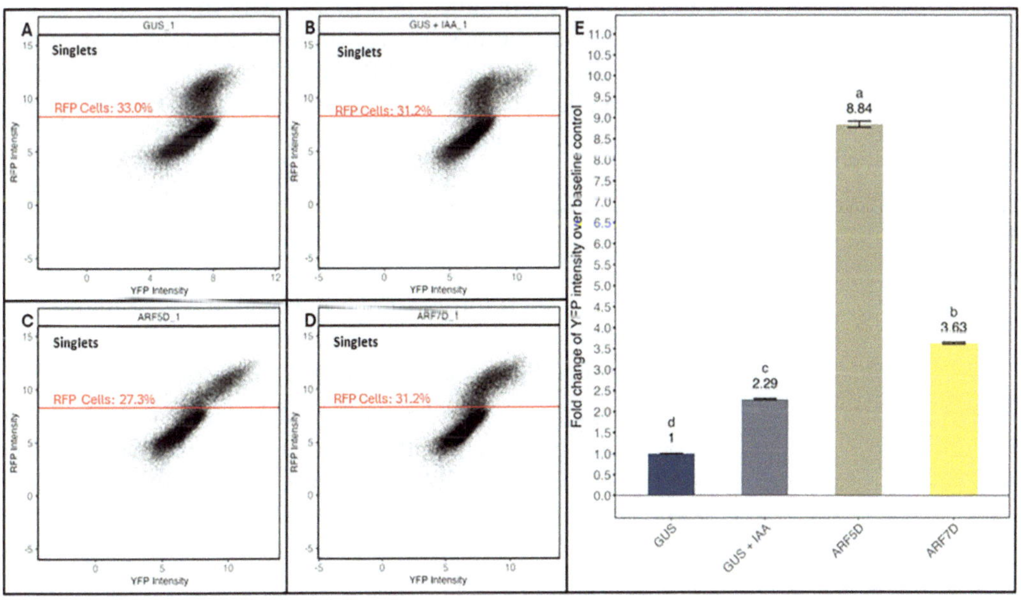

Fig. 7 Automated measurement of YFP reporter intensity in RFP-positive cells. Protoplasts were transformed with the pEvTV_DR5 auxin-responsive reporter and three different effector constructs: pBeaconRFP_GUS (an inert control mock-treated [**a**] or treated with 100 nM indole-3-acetic acid overnight [**b**]), pBeaconRFP_ARF5Δ (an irrepressible Auxin Response Factor, **c**), or pBeaconRFP_ARF7Δ (another irrepressible Auxin Response Factor, **d**). Results are shown in dot plots with YFP vs. RFP showing Singlet Cells for **a**–**d**. (**e**) The mean fold change of YFP intensity relative to the mock-treated baseline control (GUS) in RFP-positive Singlet Cells for three technical replicates is shown in a bar graph with standard deviation. Statistical significance was assayed by one-way ANOVA followed by Tukey's Honest Significant Difference post-hoc test; letters indicate homogeneous subsets ($p < 0.05$)

20. Run chunk 15 on line 952 to create a new flowSet that includes only the events inside of the RFP-positive Singlet Cells gate. This subset of the Singlet Cells flowSet includes only singlet cells which are RFP-positive and therefore have been transformed with at least the pBeaconRFP:effector construct. This chunk will also generate a table of transformation efficiencies based on the percentage of events that pass the RFP-positive Singlet Cells gate.
21. On line 1004, enter the order that the constructs should be plotted. Run chunk 16 on line 995. This chunk converts the RFP-positive Singlet Cells flowSet to a data frame, removes the non-transformed control sample for plotting, and reorders the constructs for plotting.
22. Run chunk 17 on line 1011 to plot the distribution of YFP signal intensity across samples and replicates in a ridgeline plot. The plot is saved to the "plots/TablesFigures" folder.
23. Run chunk 18 on line 1038 to generate summary statistics, run ANOVA, conduct tests for normality, and conduct post-hoc analysis.
24. Run chunk 19 on line 1045 to calculate fold change of YFP intensity over the baseline control and generate a bar plot of average fold change of YFP intensity over baseline control across replicates (Fig. 7e). The plot is saved to the "plots/TablesFigures" folder.

4 Notes

1. The use of β-mercaptoethanol is optional. It has been reported to improve yield and/or viability, but we have found no significant effect and generally leave it out.
2. These experiments can be conducted under non-sterile conditions, since the experiment only takes 18–24 h. This may speed up the process. However, for reproducibility, and especially if you are studying factors involved in plant-pathogenesis responses, it is recommended to perform the experiment using aseptic techniques. The enzyme solution may clog the 0.22 μm filters and require several extra filters to be used.
3. Resuspend the DNA pellet in a small volume of sterile de-ionized water (50 μL). Alternatively, use 10 mM Tris–HCl pH 8.0 (DNA pellets may dissolve better in slightly buffered water, note that EDTA inhibits transfections and it is in almost all kit elution buffers (TE)). It is ideal for the final concentration to be 1 μg/μL. Be sure to dilute plasmid DNA at least tenfold before measuring the concentration via UV-spectrometry, for accuracy. Other plasmid purification

kits have been successfully used (e.g., ZymoPURE Plasmid Midiprep or Maxiprep kits (Zymo Research), PureLink™ HiPure Expi Plasmid Megaprep kit (Invitrogen)); however, the quality of the DNA is important and other purification methods/kits may need to be tested on a small scale first.

4. PEG of differing average molecular weight have been used successfully in our hands (1500–8000). Due to its viscosity, filter sterilization of the PEG solution may take more time.
5. The Sony SH800 fluorescence-activated cell sorter equipped with a 100 μm nozzle chip is used in this example but any flow cytometer or fluorescence-activated cell sorter (FACS) capable of analyzing particles up to 40 μm in size and that can measure emission spectra for RFP and YFP as well as forward scatter and back(side) scatter can be used here.
6. 100 μL of seed is generally enough for one square plate and one plate of two-week-old seedling shoots generally yields 5×10^6 protoplasts, although this yield may vary. Scale up as needed.
7. Cell strainers can be washed, sterilized with 70% ethanol, and reused.
8. The use of 15 mL conical tubes ensures the formation of a compact pellet and prevents loss of protoplasts. Similarly, the use of centrifugation with no (or minimal) braking ensures maximum recovery of protoplasts.
9. Be gentle when resuspending the protoplasts. Large orifice pipette tips can be used to minimize shearing forces. A viability stain (e.g., fluorescein diacetate) can be used for a more secure count of live protoplasts.
10. For statistical analysis, we recommend a minimum of three independent transformations per tested effector.
11. Plasmid preps that contain too much salt will cause a precipitation of the DNA when mixed with the PEG solution and adversely affect transformation efficiency.
12. Traditionally, the protoplast/PEG/DNA solution was incubated at room temperature for 15 min, but we have found that brief vortexing at moderate speed and immediate washing can work better and saves time.
13. We generally see protoplast transformation efficiencies ranging between 20% and 30% using Arabidopsis seedling shoot protoplasts.
14. We have successfully used 488 nm excitation for both YFP and RFP detection. The excitation and emission frequencies listed here are not strictly the only ones that can be used.
15. A detailed methodology for setting up a flow-cytometer or FACS for analyzing protoplasts is available in video format [2].

16. An example folder can be downloaded from the project's GitHub repository (https://github.com/PlantSynBioLab/positive-fluorescence-selection). We recommend starting by working through this example. For each new protoplast flow cytometry experiment, create a new folder using the exact same structure as the example, and import your FCS files into it.
17. *Important:* Download a fresh copy of the example folder for every new experiment. Treat each folder as a self-contained unit of analysis. This approach ensures that your R Markdown working directory is automatically set to the correct location, making navigation simple and ensuring that your FCS files are found without additional configuration.
18. All the steps that precede the chunk on line 131 are only necessary for the first import of the FCS files into the automated analysis script. Subsequent analyses can begin from line 131 by reimporting the existing, annotated flowSet.
19. To determine the laser abbreviations used by a different cytometer, uncomment and run the code on lines 187–188. This should query the flowSet object to return the list of names used to describe each laser, as well as many other abbreviations. Enter the abbreviations for the RedDetector, GreenDetector, Forward Scatter Area (FSC_A), and Back Scatter Area (BSC_A) into cytometer preset 3, which is left blank for use as a template.
20. Plots are automatically saved by the ggsave() code beginning on line 242. The file type, image size, and resolution can be customized in the ensuing lines. All data visualizations will be automatically saved to folders predefined within the "plots" folder.
21. Compensation was not applied to the example experiment that accompanies this chapter, since the emission spectra of the two fluorescent proteins used have minimal overlap. The code needed to apply compensation is still included in the template R Markdown file. Running this chunk without applying custom compensation will apply a compensation of 1, which will not change the data. If using a different cytometer, uncomment and run line 298. Search through the output keywords to find the spillover matrix. Replace existing "SPILL" keyword on line 303 with the correct keyword for your cytometer if applying compensation.
22. The RFP events gate is defined by the highest 0.5% of RFP fluorescent non-transformed cells. At this threshold, there is a high likelihood that any cells that exceed this threshold are transformed, and therefore viable, cells. On the contrary, cells below this threshold are likely non-transformed or nonviable.

23. The Cells gate is defined by clustering 90% of the RFP events in a single sample, on a FSC vs. BSC scale, into an elliptical gate. The gate, defined by a single control sample, is used to define the viable cells for all other samples.
24. The Singlets gate was defined by drawing a rectangular region along the diagonal encompassing ~90% of events that exhibit a near 1:1 ratio between FSC-A and FSC-H. This relationship indicates that cell size scales proportionally with area, suggesting the cells are likely round, single protoplasts rather than doublets, or clumps.
25. The RFP-positive Singlet Cells gate is defined in the exact same way as in Subheading 3.4, **step 16**, using the non-transformed control and gating to exclude 99.5% of non-transformed Singlet cells. The difference here is that the events in the non-transformed control have been gated to only include singlet cells. This improves the accuracy of the RFP-positive threshold by excluding dead cells and debris from the non-transformed control sample which may have auto-fluorescent properties, which would erroneously alter the threshold.

Acknowledgments

Research in the Bargmann lab is supported by U.S. National Science Foundation (United States) IOS 2343702, NSF IOS 2314549, USDA National Institute of Food and Agriculture, HATCH project VA-160260, and Multistate S-009 project VA-136423. Research in the Wright lab is supported by U.-S. National Science Foundation (United States) IOS 2343702, United States Department of Agriculture National Institute of Food and Agriculture (USDA-NIFA) Hatch Project VA-160244, and the National Institute of General Medical Sciences of the National Institutes of Health under award number R35GM150856. The content is solely the responsibility of the authors and does not necessarily represent the official views of the National Institutes of Health. J.S.T. was supported by the George Washington Carver Program (Virginia Tech).

References

1. Bargmann BO, Birnbaum KD (2009) Positive fluorescent selection permits precise, rapid, and in-depth overexpression analysis in plant protoplasts. Plant Physiol 149:1231–1239
2. Bargmann BO, Birnbaum KD (2010) Fluorescence activated cell sorting of plant protoplasts. J Vis Exp
3. Taylor JS, Villaseñor EA, Rashkovsky J et al (2025) A comparative analysis of transfected and integrated auxin reporter systems reveals sensitivity advantages in protoplast transient expression assays. MicroPublication Biol
4. Beard KM, Boling AWH, Bargmann BOR (2021) Protoplast isolation, transient transformation, and flow-cytometric analysis of

reporter-gene activation in Cannabis sativa L. Ind Crops Prod 164:113360

5. Lieberman-Lazarovich M, Yahav C, Israeli A et al (2019) Deep conservation of cis-element variants regulating plant hormonal responses. Plant Cell 31:2559–2572
6. Taylor JS, Shoppell EA, Bargmann BOR et al (2025) Glow with the flow: reproducible analysis of transiently transformed protoplasts using dual fluorescent reporters in R. https://www.biorxiv.org/content/10.1101/2025.07.26.666912v1
7. Gonzalez JH, Taylor JS, Reed KM et al (2021) Temporal control of morphogenic factor expression determines efficacy in enhancing regeneration. Plants 10:2271

Chapter 14

Protocols for Chromatin Immunoprecipitation Sequencing and qPCR to Investigate Pytohormone-Responsive Transcriptional Regulation in Plants

Emre Aksoy, Sefa Ayten, and Ceyhun Kayıhan

Abstract

Plant hormones regulate a wide array of developmental and stress-response pathways by modulating gene expression through specific transcription factors (TFs). Understanding how these TFs interact with their genomic targets is essential for dissecting hormone signaling networks. Chromatin immunoprecipitation (ChIP), coupled with quantitative PCR (ChIP-qPCR) or next-generation sequencing (ChIP-seq), offers a powerful approach to study in vivo protein–DNA interactions. However, ChIP in plants poses unique technical challenges due to the presence of rigid cell walls, abundant secondary metabolites, and chromatin accessibility constraints. Here, we present a detailed and optimized protocol for performing ChIP-qPCR and ChIP-seq in *Arabidopsis thaliana* to investigate hormone-responsive gene regulation. The protocol includes steps for hormone treatment, tissue fixation, chromatin extraction, immunoprecipitation using specific antibodies, reverse crosslinking, DNA purification, and downstream quantification. This method enables high-resolution mapping of transcription factor occupancy at target loci and can be adapted for different hormones and plant species. Compared to indirect transcriptomic approaches, ChIP-based assays provide direct evidence of TF–DNA binding and allow identification of primary targets in signaling cascades. The protocol supports mechanistic studies of hormone action in plants and facilitates the construction of gene regulatory networks.

Key words Chromatin, Immunoprecipitation, Transcription factor binding, ChIP-seq, ChIP-qPCR, Plant hormone signaling, *Arabidopsis thaliana*

1 Introduction

Plant hormones are integral to virtually all aspects of plant life, including development, environmental adaptation, and stress response. These small molecules, such as abscisic acid (ABA), auxins, gibberellins, salicylic acid, jasmonates, and cytokinins, coordinate complex physiological and molecular responses by regulating gene expression through intricate signaling networks [1, 2]. Central to these pathways are hormone-responsive transcription factors

Shahid Mukhtar (ed.), *Plant Hormones: Methods and Protocols*, Methods in Molecular Biology, vol. 3026,
https://doi.org/10.1007/978-1-0716-5214-5_14,

(TFs), which bind specific cis-regulatory DNA elements to modulate transcriptional activity of target genes.

Dissecting hormone signaling pathways requires identification of these direct TF–DNA interactions. Chromatin immunoprecipitation (ChIP), coupled with quantitative PCR (ChIP-qPCR) or next-generation sequencing (ChIP-seq), provides a powerful means to identify genome-wide binding sites of transcription factors or histone modifications in vivo [3, 4]. These tools are particularly useful in hormone biology, where rapid and dynamic TF binding to chromatin plays a central regulatory role. While ChIP-qPCR is widely used to validate known or candidate binding sites due to its specificity and sensitivity, ChIP-seq allows for unbiased, genome-wide profiling of TF occupancy or histone marks in response to hormonal stimuli [5, 6]. In comparison to alternative approaches such as RNA-seq or promoter-reporter assays, ChIP offers a direct readout of DNA–protein interaction, distinguishing primary regulatory events from indirect transcriptional changes [7].

ChIP in plant tissues presents unique technical challenges due to rigid cell walls, secondary metabolites, and chromatin accessibility issues [8]. Optimization of crosslinking, chromatin extraction, antibody selection, and DNA recovery is essential for obtaining reliable results. Here, we provide a detailed, optimized protocol for performing ChIP followed by qPCR and/or sequencing in *Arabidopsis thaliana* treated with plant hormones such as ABA. This workflow enables researchers to map transcription factor binding sites and investigate hormone-responsive regulatory mechanisms with precision and reproducibility.

2 Materials

1. *Arabidopsis thaliana* seedlings: 10–14 days old on plates or soil-grown rosettes (*see* **Note 1** for plant growth and hormonal treatment).
2. 1× PBS (phosphate-buffered saline) (pH 7.4): Dissolve 8.0 g of NaCl, 0.2 g of KCl, 1.44 g of Na_2HPO_4, and 0.24 g of KH_2PO_4 in approximately 800 mL of distilled water. Mix until all components are fully dissolved, then adjust the pH to 7.4 using HCl or NaOH if necessary. After pH adjustment, bring the final volume up to 1 L with distilled water. The solution can be sterilized by autoclaving or filtered through a 0.22 μm membrane filter if sterility is required. Prepared PBS can be stored at room temperature or at 4 °C for long-term use.
3. 1% formaldehyde in PBS (for crosslinking): Dilute a 16% methanol-free formaldehyde stock solution using 1× phosphate-buffered saline (PBS, pH 7.4). For example, to make 50 mL of 1% formaldehyde, mix 3.125 mL of the 16%

stock with 46.875 mL of PBS. Gently invert the tube to mix thoroughly. The solution should be prepared fresh on the day of use and kept on ice until needed, as formaldehyde gradually degrades and loses its crosslinking efficiency. (*see* **Note 2** for safety procedures!)

4. 0.125 M glycine: Weigh 0.938 grams of glycine (molecular weight: 75.07 g/mol) and dissolve it in distilled or deionized water. Adjust the final volume to 100 mL. The solution does not require pH adjustment for use in quenching formaldehyde during crosslinking procedures. Once fully dissolved, the glycine solution can be used immediately or stored at 4 °C for short-term use.
5. Extraction Buffer 1 (EB1): 0.4 M sucrose, 10 mM Tris–HCl (pH 8.0), 10 mM $MgCl_2$, 5 mM β-mercaptoethanol (add fresh), 1 mM PMSF (add fresh), 1% PVP-40 (polyvinylpyrrolidone), 1× protease inhibitor cocktail (EDTA-free, add fresh) (Prepare fresh, keep cold (4 °C), and protect from light).
6. Extraction Buffer 2 (EB2): Same as EB1 but with 0.25 M sucrose.
7. Extraction Buffer 3 (EB3): Same as EB1 but with 1.7 M sucrose and no PVP-40 (optional).
8. Lysis buffers: 50 mM Tris–HCl, pH 8.0, 10 mM EDTA, 1% SDS, 1 mM PMSF (add fresh), 1× protease inhibitor cocktail (EDTA-free, add fresh) (Note: Prepare just before use. Pre-warm to room temperature only if SDS has solidified, then chill again on ice).
9. ChIP Dilution Buffer (pH 8.0): 20 mM Tris–HCl (pH 8.0), 150 mM NaCl, 2 mM EDTA, 1% Triton X-100, 1× protease inhibitor cocktail (add fresh before use) (Optional: Add 0.01% SDS if reduced stringency is desired for antibody binding) (Prepare fresh or sterile-filter and store at 4 °C).
10. Antibody specific to your protein of interests.
11. Protein A/G magnetic or agarose beads.
12. Elution Buffer (for ChIP Reverse Crosslinking): 100 mM Sodium bicarbonate, 1% SDS.
13. DNase-free RNase A (10 mg/mL) and Proteinase K (20 mg/mL).
14. Library prep kit for ChIP-seq.
15. qPCR primers for ChIP-qPCR.
16. qPCR master mix (e.g., SYBR Green or probe-based mix with Taq polymerase).
17. Nuclease-free water.

3 Methods

3.1 Crosslinking

1. Harvest ~2 g of fresh Arabidopsis tissue (seedlings or leaves) (*see* **Note 1**).
2. Immerse in 1% formaldehyde in PBS and apply vacuum infiltration at −60 kPa to −70 kPa (450–525 mmHg or −0.6 bar to −0.7 bar) for 10 min to ensure effective penetration of the fixative into the plant tissues. Perform this step in a vacuum desiccator or chamber and gently release the vacuum after 10 min to avoid tissue damage (*see* **Note 2**).
3. Quench the crosslinking reaction by transferring the tissue into 0.125 M glycine prepared in PBS and apply vacuum infiltration at −60 kPa to −70 kPa (450–525 mmHg) for 5 min. This ensures thorough penetration of glycine to neutralize unreacted formaldehyde. After 5 min, gently release the vacuum to avoid tissue rupture.
4. Rinse the tissue briefly with cold PBS to remove residual glycine.
5. Blot briefly on paper towel to remove excess liquid.
6. Flash-freeze tissue in liquid nitrogen and store at −80 °C or proceed immediately.

3.2 Chromatin Extraction (see Note 3 Before Starting This Step)

1. Place tissue in a chilled mortar with 5–10 mL of cold Extraction Buffer 1.
2. Grind thoroughly on ice with pestle until you obtain a homogenous slurry (at least 5–10 min) (*see* **Note 4**).
3. Filter the homogenate through two layers of Miracloth or a 40–70 μm nylon mesh into a pre-chilled 50 mL conical tube.
4. Gently squeeze or press to recover as much liquid as possible.
5. Centrifuge filtrate at 4000 × *g* for 20 min at 4 °C.
6. Carefully discard the supernatant without disturbing the pellet (*see* **Note 5**).
7. Gently resuspend the pellet in 10 mL of cold Extraction Buffer 2 using a wide-bore pipette tip or soft swirling.
8. Repeat centrifugation at 4000 × *g* for 20 min at 4 °C and then discard supernatant similar to **step 6**.
9. Carefully resuspend the final pellet in 500 μL to 1 mL of Extraction Buffer 3.
10. Transfer the resuspension to a new ultracentrifuge tube (if needed).
11. Gently overlay the suspension onto 4–5 mL of Extraction Buffer 3 that has been pre-chilled in the bottom of the tube (do not mix layers) (*see* **Note 6**).

12. Centrifuge at 16,000 × *g* for 1 h at 4 °C in a fixed-angle rotor.
13. After centrifugation, you should observe a small, tight, white nuclear pellet at the bottom. Carefully remove the supernatant without disturbing the pellet.
14. The nuclear pellet can now be directly lysed in SDS lysis buffer for sonication or snap-frozen in liquid nitrogen and stored at −80 °C if not used immediately (*see* **Note 7**).

3.3 Chromatin Shearing (Sonication)

1. Resuspend the nuclear pellet in lysis buffer at a ratio of ~1 mL per 2 g tissue starting material.
2. Pipette gently or flick tube to mix (*see* **Note 8**).
3. Incubate on ice for 10 min to lyse nuclei and allow chromatin solubilization (*see* **Note 9**).
4. Transfer lysate into pre-chilled microtubes compatible with your sonicator (*see* **Note 10**).
5. Keep samples on ice-water mix or in the sonicator's cooling system throughout (*see* **Note 11**).
6. After sonication, centrifuge at 12,000 × *g* for 10 min at 4 °C to pellet debris.
7. Transfer the clear supernatant to a new, pre-chilled tube. This contains your sheared chromatin.
8. (Optional) Take 10 μL of sonicated sample and reverse cross-link (by incubating at 65 °C overnight), treat with RNase A and Proteinase K, purify DNA, and run on a 1.5% agarose gel (*see* **Note 12**). Ideal size range: 200–600 bp smear for ChIP-seq; 200–1000 bp smear for ChIP-qPCR (Fig. 1).
9. Use chromatin lysate immediately for immunoprecipitation or aliquot and snap-freeze in liquid nitrogen, then store at −80 °C.

3.4 Immuno-precipitation

1. Take the cleared chromatin supernatant (from sonication) and dilute it 1:10 in ice-cold ChIP dilution buffer (For example, 100 μL of chromatin +900 μL of dilution buffer). This reduces SDS concentration from lysis buffer and provides optimal ionic strength for antibody binding.
2. For pre-clearing to reduce nonspecific binding, add 30–50 μL of pre-washed Protein A/G magnetic or agarose beads (no antibody) to the diluted chromatin.
3. Incubate at 4 °C for 1 h on a rotator or nutator set to a gentle speed of 10–15 rpm (or equivalent low setting) (*see* **Note 13**).
4. Pellet beads by magnetic stand or brief centrifugation (if agarose), then transfer the cleared supernatant to a new tube (*see* **Note 14**).

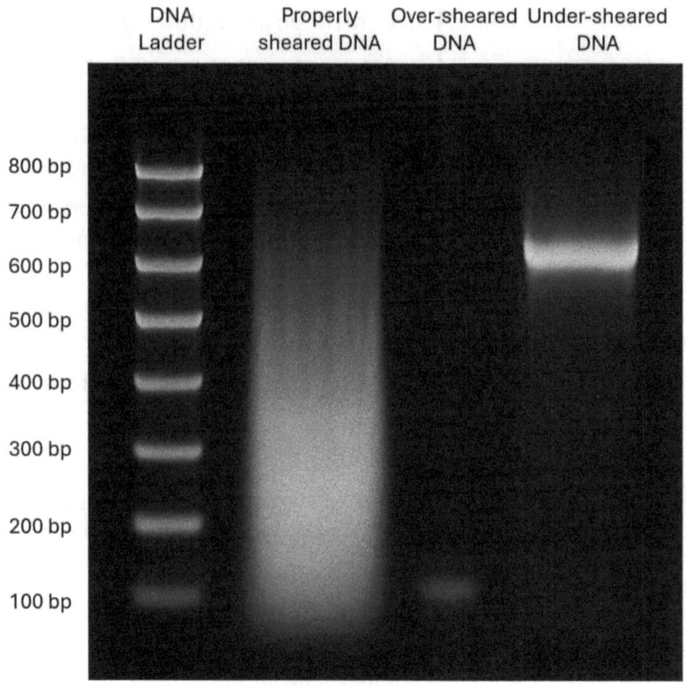

Fig. 1 Example gel image after sonication. Lane 1. 100 bp DNA ladder. Lane 2. Properly sheared DNA that shows a smear between ~200 and 600 bp, ideal for ChIP or NGS. Lane 3. Over-sheared DNA represented by a smear or band concentrated below ~200 bp, often <100 bp. Lane 4. Under-sheared DNA represented by high molecular weight band(s) above ~600 bp, possibly >1000 bp

5. Add the specific antibody to the pre-cleared chromatin at an amount of 1–5 μg of antibody per IP, depending on its affinity and ChIP quality (*see* **Notes 15** and **16**).
6. Incubate the mixture overnight (12–16 h) at 4 °C with gentle rotation.
7. On the next day, after the antibody incubation step, add 30–50 μL of Protein A/G beads to each sample to capture the antibody–chromatin complexes. Before adding them, the beads must be pre-blocked to minimize nonspecific interactions, especially important when working with complex plant extracts (*see* **Note 17** for pre-blocking of beads).
8. Incubate at 4 °C for 2–4 h with rotation (10–15 rpm) to allow the beads to bind the antibody–chromatin complexes (*see* **Note 18**).
9. After incubating the chromatin-antibody mixture with pre-blocked Protein A/G beads, the bead-bound immune complexes must be collected and washed thoroughly to remove nonspecific interactions before elution. Bead separation can be done using magnetic or agarose beads (*see* **Notes 19** and **20**).

10. For magnetic beads, pace the tube on a magnetic rack and let the beads migrate to the side (typically 1–3 min at room temperature). Once the solution becomes clear and beads are fully pulled to the magnet side, carefully remove the supernatant using a pipette without disturbing the bead pellet. If residual droplets remain, briefly spin (e.g., 500 × *g*, 10 s) and return to magnet to collect remaining liquid.
11. For agarose beads, centrifuge the tubes at 1000–2000 × *g* for 30 s to 1 min at 4 °C to pellet the beads. Using a pipette, gently remove the supernatant without touching the soft bead pellet. Optional: Tilt the tube slightly while removing supernatant to minimize bead disturbance.
12. After antibody-bound chromatin has been captured on Protein A/G beads, perform a series of washes to remove nonspecific binders. Each wash buffer serves a specific purpose in gradually increasing stringency (*see* **Note 21**).
13. After final TE wash, proceed to elution of chromatin, or store beads on ice if there is a short delay.

3.5 Reverse Crosslinking & Enzymatic Digestion

1. After the final TE wash, resuspend the beads (with bound chromatin) in elution buffer (*see* **Note 22**).
2. Add NaCl to a final concentration of 200 mM. For example, add 8 μL of 5 M NaCl to 192 μL of sample.
3. Incubate the mixture at 65 °C for 6–16 h (typically overnight) in a thermomixer or heating block with the lid on to prevent evaporation (*see* **Note 23**).
4. After the 65 °C incubation is complete, cool the sample briefly to 37 °C.
5. Add RNase A to a final concentration of ~50 μg/mL (e.g., add 1 μL of 10 mg/mL RNase A to 200 μL) (*see* **Note 24**).
6. Incubate at 37 °C for 30 min.
7. Add Proteinase K to a final concentration of ~100–200 μg/mL (e.g., add 2–4 μL of 20 mg/mL Proteinase K).
8. Incubate at 45 °C for 1 h.

3.6 DNA Purification

After digestion is complete, proceed to purify the DNA using one of the following methods.

3.6.1 Option A: Phenol–Chloroform Extraction (Classical Method) (Work in Fume Hood When Using Phenol)

1. Add equal volume of phenol:chloroform:isoamyl alcohol (25:24:1) to the sample.
2. Vortex thoroughly for 10–15 s.
3. Centrifuge at 12,000 × *g* for 5–10 min at room temperature.
4. Carefully transfer the upper aqueous phase to a new tube (avoid interface).

5. Add 2.5 volumes of cold 100% ethanol +1/10 volume of 3 M sodium acetate (pH 5.2) and 20 μg glycogen (optional).
6. Precipitate overnight at −20 °C or for 1 h at −80 °C.
7. Centrifuge at 14,000 × *g* for 30 min at 4 °C, wash pellet with 70% ethanol, air-dry, and resuspend in 30–50 μL TE buffer or nuclease-free water.

3.6.2 Option B: For Column-Based Purification (Preferred for Speed and Safety), Use a Commercial PCR Purification Kit

1. Mix the sample with the kit's binding buffer (usually 5:1 or 7:1 ratio).
2. Load onto the column.
3. Wash with provided buffer (typically containing ethanol).
4. Elute in 30–50 μL of TE buffer (pH 8.0) or nuclease-free water (*see* **Note 25**).
5. Measure concentration using Qubit or fluorometric methods. Nanodrop may be inaccurate due to low yield.
6. Store purified ChIP DNA at −20 °C (short-term) or −80 °C (long-term).

3.7 Library Preparation for NGS

1. Run 1 μL on Bioanalyzer (or TapeStation) to confirm fragment size (~200–600 bp expected after sonication) (*see* **Note 26**).
2. Use the NEBNext Ultra II DNA Library Prep Kit and combine ChIP DNA with End Prep Reaction Buffer and Enzyme Mix. Then, incubate in thermocycler (30 min at 20 °C, 30 min at 65 °C).
3. For adapter ligation, add diluted Illumina-compatible barcoded adapters to A-tailed DNA. Incubate at 20 °C for 15 min. Add USER enzyme if using NEBNext adapters, incubate at 37 °C for 15 min (*see* **Note 27**).
4. Use AMPure XP beads for (double-sided) size selection. First add 0.5× volume of beads → discard bound (large) fragments. Then, add 0.3× additional volume → bind desired ~300–500 bp fragments.
5. Wash twice with 80% ethanol and elute in 17–20 μL nuclease-free water (*see* **Note 28**).
6. For library amplification, use 10–15 cycles depending on input DNA amount as follows: <1 ng: 15 cycles; 1–5 ng: 12 cycles; 5 ng: 10 cycles.
7. Use the following thermocycler conditions for the library amplification in a PCR reaction of Library template + primers + high-fidelity polymerase: 98 °C, 30 s; 10–15 cycles: 98 °C for 10 s, 65 °C for 75 s; Final extension: 65 °C for 5 min.
8. After the PCR amplification, purify the library using AMPure XP beads (1× volume). Then, elute in 20 μL nuclease-free water.

9. Quantify final library using Qubit or qPCR-based library quantification kit. Analyze library size distribution using Bioanalyzer or TapeStation (Expected peak: 300–500 bp (insert + adapters)) (*see* **Note 29**).
10. Normalize each library to 4 nM or 10 nM. Pool libraries equimolarly using unique dual indices (avoid index-hopping).
11. Submit to a core facility for sequencing on NovaSeq SP or S1 flow cell (for moderate sample numbers) in 150 bp paired-end reads or 50 bp single-end (depending on TF footprinting resolution needs).

3.8 Bioinformatic Analyses

1. Use this section as the standard workflow for processing raw ChIP-seq data to identify and characterize protein–DNA binding sites.
2. Start with raw FASTQ files from the sequencing facility and finish with annotated peak lists, motif discovery, and differential binding analysis.
3. Run all steps on an HPC cluster (*see* **Note 30**).
4. Consult the original tool papers and developers' websites for publication-quality examples of expected outputs for each tool.

3.8.1 Quality Control of Raw Data

1. Transfer all FASTQ files delivered by the sequencing center into a secure, backed-up directory.
2. Create a directory for quality-control output.
3. Evaluate each library using FastQC program [9] to inspect per-base quality, adapter content, over-represented sequences, GC distribution, and duplication levels. Per-base Q-score is expected to be ≥30 for ≥90% of read length, and adapter content is expected to be <5%.
4. Aggregate individual FastQC outputs using MultiQC [10] into a single interactive summary and flag libraries that fall below acceptable thresholds (e.g., median Q < 20 in the first 30 bp or >5% adapter contamination) (*see* **Note 31**).
5. Record QC outcomes in the project log for future reference.

3.8.2 Trimming

1. Remove Illumina adapters with "-a your_adapter_sequence" parameter and low-quality bases (≤ Q20) with "-q your_desired_score(commonly 30)" parameter from each read pair and reads shorter than 20 bp with "-m your_desired_lenght" parameter using Cutadapt [11] (*see* **Note 32**). Since NEBNext Ultra II DNA Library Prep Kit was used for library preparation in this protocol, following command can be used for this step: cutadapt -a AGATCGGAAGAGCACACGTCTGAACTC CAGTCA -A AGATCGGAAGAGCGTCGTGTAGGGAAA GAGTGT -q 30 -m 20 -o sample_R1.trimmed.fastq.gz -p sample_R2.trimmed.fastq.gz sample_R1.fastq.gz sample_R2.fastq.gz > cutadapt.log.

2. Repeat the QC assessment with FastQC/MultiQC on trimmed reads to confirm adapter removal and improved quality profiles.
3. Document read loss, average read length, and updated quality metrics for each sample.

3.8.3 Alignment to Reference Genome

1. Obtain the TAIR10 *Arabidopsis thaliana* nuclear genome (from https://www.arabidopsis.org/), verify checksum integrity, and index the reference genome once using bwa index [12].
2. Align each trimmed read pair using bwa mem [12] and retain only primary alignments using samtools view [13].
3. Sort the resulting alignments by coordinate using samtools sort and index the sorted BAM file using samtools index [13].
4. Optionally remove chloroplast and mitochondrial reads to reduce background signal.
5. Record alignment statistics, including total reads, uniquely mapped reads, and mapping rate using samtools flagstats [13]. It is expected that uniquely mapped reads ≥80% and duplication rate ≤30%.

3.8.4 Remove Duplicates and Filter

1. Mark PCR duplicates in each BAM file using Picard MarkDuplicates [14] and note the overall duplication rate.
2. Filter the alignments to retain only reads with mapping quality ≥30 with "-q 30" parameter while removing unmapped, secondary, supplementary, and duplicate reads with "-F 1804" parameter using samtools view [13] (*see* **Note 33**).
3. Index the final filtered BAM file using samtools index [13] and archive both the filtered file and its associated metrics using samtools flagstats [13].

3.8.5 Peak Calling (MACS2)

1. Pair each ChIP BAM (treatment conditions) with its matched input control BAM (control conditions) when available.
2. For transcription factors, call narrow peaks at a false-discovery rate (q-value) of 0.01 using macs2 callpeak [15] with "-g your_genome_size -q 0.01" parameters; for histone modifications, use broad-peak settings with a broad cut-off of 0.05 with "--broad --broad-cutoff 0.05 -g your_genome_size" parameters (*see* **Note 34**). Peak counts, d-value, enrichment will appear in _peaks.xls output file.
3. Inspect peak statistics, which are written by MACS2 into the output file, including total peaks, summit resolution, enrichment scores, and store peak files (*.narrowPeak/*.broadPeak) in the project directory.

4. Verify that the fragment-length estimate reported by the software (sometimes called "d-value" or "shift size") is close to the expected chromatin-fragment size obtained during sonication QC (≈ 200–600 bp for TF ChIP-seq).
5. Rename each output file with a clear, unique identifier.
6. Move all peak files (*.narrowPeak, *.broadPeak, accompanying .xls and _summits.bed) into a dedicated peaks/ sub-directory of the analysis folder.
7. Compile the fragment-length (shift size) values for every biological replicate into a spreadsheet. Confirm that all replicates show similar values (variation should be <50 bp); large discrepancies may indicate uneven chromatin shearing or library preparation problems.

3.8.6 Peak Visualization

1. Convert each filtered BAM into a normalized bigWig coverage track (Reads Per Kilobase per Million mapped reads (RPKM) or Reads Per Genomic Content (RPGC) normalization) with an appropriate bin size using deepTools (specifically bamCoverage with "—normalizeUsing RPKM" —binSize your_desired_window_size —effectiveGenomeSize your_genome_size" parameters) [16]. Smaller bin size gives higher resolution.
2. Load the TAIR10 reference genome, the final BAM and index files, bigWig tracks, peak files, and _summits.bed file into an interactive genome browser (IGV) [17] or the UCSC Genome Browser [18] to visually confirm enrichment over expected loci.
3. Capture representative screenshots to include in the final report.

3.8.7 Peak Annotation

1. Annotate peaks relative to gene features (promoters, exons, introns, intergenic regions) using HOMER (annotatePeaks.pl) [19] with the final peak file for each sample and reference genome or gene-annotation for the same genome build used in alignment. A tab-delimited text file listing each peak with its nearest gene and genomic feature will be created as an output for downstream enrichment analyses (*see* **Note 35**).
2. Generate summary plots showing the distribution of peaks across genomic features and distances to transcription-start sites (TSS) using ChIPseeker package [20, 21] in R.
3. Export the created plots in a high-resolution format for publication or reporting.

3.8.8 Motif Discovery (HOMER)

1. Perform de-novo motif discovery on ±100 bp regions surrounding peak summits, masking repetitive sequences using HOMER (findMotifsGenome.pl with "-size 200 -mask" parameters) [19] with generated peak files and reference genome.
2. Inspect Homer de novo motif results for enriched sequence logos and position-weight matrices.
3. Compare enriched motifs against reference databases of known transcription-factor binding sites to validate ChIP specificity (*see* **Note 36**).
4. Record the top motifs, their match scores, and the percentage of target peaks containing each motif.
5. Save motif logos and position-weight matrices for inclusion in the report.

3.8.9 Differential Binding Analysis

1. Assemble a metadata table listing, sample IDs, BAM files, peak files, experimental conditions, and biological replicates. Also, gather peak files generated for each individual replicate and final, filtered, sorted, and indexed alignment BAM files for each replicate.
2. Quantify read counts within a consensus peak set across all samples, establish contrasts between conditions, and apply an appropriate statistical model to identify sites with significant changes in protein binding (FDR $\leq$0.05) using DiffBind package in R [22] (*see* **Note 37**).
3. Perform the differential binding analysis using DESeq2 [23], which can internally calls by DiffBind [22]. Extract a data frame of all sites, including their genomic coordinates, fold changes, p-values, and FDR.
4. Generate diagnostic plots, such as MA plots, heatmaps, principal-component analysis, to assess replicate concordance and binding changes using DiffBind built-in plotting functions [22].
5. Export a comprehensive table of differentially bound regions for further biological interpretation.

3.8.10 Report

1. Compile all quality-control metrics, alignment statistics, genome-feature distribution, and representative IGV screenshots, peak counts, annotation summaries, motif analyses, and differential binding results into a single, well-structured HTML (R Markdown or Jupyter notebook) or PDF document.
2. Archive the following in a version-controlled repository: raw FASTQ files, trimmed FASTQ files, final BAM files, bigWig tracks, peak BED files, differential binding tables, and the complete analysis workflow.

3. Submit processed data, metadata, and the analysis pipeline to a public repository (e.g., GEO or ENA) in compliance with community guidelines and data-sharing requirements (*see* **Note 38**).

Completion of these steps provides a reproducible pipeline that converts raw sequencing reads into biologically interpretable ChIP-seq results.

3.9 qPCR Enrichment Analysis

1. Perform qPCR enrichment analysis after ChIP-seq to validate binding at selected loci.
2. Design primers to amplify regions with significant ChIP-seq enrichment and include negative control regions with no expected binding.
3. Subject immunoprecipitated (ChIP) DNA and input DNA to qPCR under identical conditions.
4. Calculate enrichment relative to input or a control locus using the ΔCt method.
5. Interpret robust enrichment in ChIP DNA over controls as confirmation of antibody specificity and validation of the ChIP-seq results.
6. Save 1% of the sonicated chromatin before immunoprecipitation as "input DNA." Treat this in parallel: reverse crosslink, RNase, Proteinase K, and purify as described. Input DNA represents total chromatin and is used to calculate % input or fold enrichment.
7. Controls that should be included in the analysis include: ChIP sample with specific antibody, IgG (nonspecific antibody) control, No-antibody mock IP (optional), Input DNA (1%) for normalization.
8. For qPCR reaction, use technical triplicates for each sample (ChIP, IgG, and input). Include no-template controls (NTC) to detect contamination. Prepare a master mix to reduce pipetting error (Table 1).
9. Set up qPCR with the cycling conditions in Table 2 (*see* **Notes 39** and **40**).

3.10 qPCR Data Analysis

3.10.1 Option A: % Input Method

1. Calculate Ct difference (ΔCt) between ChIP and input:

$$\Delta Ct = Ct_{ChIP} - Ct_{Input}$$

2. Adjust for dilution factor (e.g., 1% input = 6.64 in $\log_2$ scale):

$$\%Input = 100 \times 2^{(6.64 - \Delta Ct)}$$

Table 1
Reaction components (per 20 μL reaction) (*see* Note 39)

Component	Volume
SYBR green master mix (2×)	10 μL
Forward primer (10 μM)	0.5 μL
Reverse primer (10 μM)	0.5 μL
Purified ChIP or input DNA	2–5 μL
Nuclease-free water	To 20 μL

Table 2
qPCR cycling conditions

Step	Temperature	Time	Cycles
Initial denaturation	95 °C	3 min	1
Denaturation	95 °C	10 s	
Annealing/extension	60 °C	30 s	40
Melt curve	60–95 °C	0.3 °C/s	

3.10.2 Option B: Fold Enrichment Over IgG

$$\text{Fold Enrichment} = 2 - \left(\text{Ct}_{\text{ChIP}} - \text{Ct}_{\text{IgG}}\right)$$

1. Normalize enrichment of target genes to a negative control region (not bound by the TF).
2. Plot data as % input or fold enrichment ± standard deviation.

4 Notes

1. First, surface-sterilize *Arabidopsis thaliana* (Col-0 or mutant/transgenic lines) seeds using 70% ethanol for 1 min followed by 10% bleach (with 0.1% Tween-20) for 10 min. Then, rinse 5× with sterile distilled water. Next, sow seeds on ½ × Murashige and Skoog (MS) medium supplemented with 1% (w/v) sucrose and solidified with 0.8% agar. Then, stratify them at 4 °C in the dark for 2–3 days to synchronize germination. For the plant growth, transfer plates to a growth chamber set at 22 °C, under long-day conditions (16 h light/8 h dark), light intensity ~100 μmol/m^2 s, and grow for 10–14 days until seedlings develop true leaves. For ABA treatment, prepare a 50 μM ABA solution (or your optimized concentration) in sterile ½ × MS liquid medium. Then, gently transfer seedlings to ABA-containing liquid medium and incubate for 1–3 h,

depending on the target transcription factor's activation kinetics. Include a mock control (solvent only, typically ethanol or DMSO) for untreated samples and perform treatment under the same growth conditions (light, temperature). Immediately after treatment, harvest seedlings (~2 g per sample) and proceed directly to crosslinking with 1% formaldehyde under vacuum.

2. All steps should be performed in a fume hood with appropriate personal protective equipment, including gloves, a lab coat, and safety goggles, due to the toxic and volatile nature of formaldehyde.
3. Keep all buffers and equipment cold (4 °C) throughout the process to prevent protein degradation. Add PMSF and β-mercaptoethanol fresh to every buffer right before use. Avoid vortexing or harsh pipetting since this may rupture nuclei and reduce chromatin integrity.
4. Efficient grinding is critical. Incomplete homogenization reduces nuclei yield.
5. Pellet may be soft or loosely packed. Pour off slowly and keep tube tilted.
6. Tilt the tube and pipette slowly to avoid disturbing the interface.
7. If samples must be stored, do so after the nuclear isolation step, not during intermediate stages.
8. Do not vortex, which may shear chromatin prematurely or rupture nuclei unevenly.
9. Resuspend thoroughly since clumped nuclei result in poor shearing and low ChIP efficiency.
10. Vortex gently before sonication if inconsistent shearing is observed in **step 8**.
11. Key Point: Fragment size is critical. Therefore, optimize sonication conditions with a small test sample and check DNA size on a 1.5% agarose gel.

Example 1 Bioruptor Plus (Diagenode).

30 s ON/30 s OFF; Total: 10–20 cycles (5–10 min total ON time); Medium power (keep temperature low!)

Example 2 Probe Sonicator.

Pulse for 10 s ON/30 s OFF; Repeat for 6–8 pulses; Amplitude: ~30%; Use a water or ice bath to prevent overheating (or work in a cold room).

12. Use 1.5–2% agarose gel for better resolution of small fragments. Run gel in 1× TAE or TBE buffer. Use 100 bp DNA ladder as molecular weight marker. Load 5–10 μL purified DNA (after reverse crosslinking, RNase/Proteinase K treatment, and cleanup). Stain with ethidium bromide or SYBR Safe.
13. Avoid vigorous mixing, as excessive shear can disrupt protein–DNA complexes or damage the beads.
14. After pre-clearing the chromatin with beads (no antibody), separate the beads from the supernatant using one of the following methods, depending on the bead type.
    - If using magnetic beads:

      Place the tube on a magnetic stand for 2–5 min at room temperature until the solution clears and the beads are fully attracted to the magnet wall.

      Carefully transfer the supernatant (containing the pre-cleared chromatin) to a new, pre-chilled tube using a pipette. Avoid touching or disturbing the bead pellet.
    - If using agarose beads (nonmagnetic):

      Centrifuge the tube briefly at 1000–2000 × *g* for 1–2 min at 4 °C to pellet the beads.

      Then, gently transfer the clear supernatant to a clean, cold tube, being careful not to aspirate any beads.

      Important: Bead carryover can increase background noise and interfere with antibody binding. If unsure, leave a small volume behind rather than risk transferring beads.
15. Use ChIP-grade antibody validated for immunoprecipitation.
16. To ensure the specificity and reliability of your ChIP results, it is essential to include appropriate negative controls alongside your experimental samples.
    - No-antibody control (mock IP): This sample contains the chromatin and beads but no antibody is added. It helps detect any nonspecific binding of DNA to the beads or tube surfaces. Any signal observed in this control indicates background noise unrelated to antibody-specific pulldown.
    - IgG control (nonspecific antibody control): Add an equal amount (e.g., 1–5 μg) of nonspecific IgG (from the same host species as your target antibody, e.g., normal rabbit or mouse IgG). This control accounts for any nonspecific interactions between chromatin and antibody Fc regions or beads. It provides a baseline for assessing the specificity of your target antibody enrichment.

Including these controls is critical when interpreting ChIP-qPCR or ChIP-seq data, as they help distinguish true target enrichment from background noise or technical artifacts.

17. Pre-blocking beads helps prevent nonspecific binding of plant-derived nucleic acids, proteins, and secondary metabolites to the beads. This is especially critical in Arabidopsis and other plant systems, where polysaccharides and polyphenols can contribute to high background in ChIP assays. To block the beads:
    - First prepare blocking solution in two alternative ways.

      Option 1: 0.5% BSA in PBS.

      Option 2: 1 mg/mL denatured salmon sperm DNA + 0.1% BSA in PBS
    - Resuspend the desired amount of beads (e.g., 50 μL per sample) in the blocking solution.
    - Incubate for 1 h at 4 °C on a rotator to allow blocking agents to coat the bead surface.
    - After incubation, wash the beads 2–3 times with ChIP dilution buffer or PBS to remove excess blocking material. Keep the beads on ice until use.
    - Once blocked, continue from Subheading 3.4, **step** 7, by adding the beads to the antibody–chromatin mixture and incubate for 2–4 h at 4 °C on a rotator (10–15 rpm) to allow the beads to bind the antibody–chromatin complexes.
18. Incomplete blocking often leads to high background or non-specific qPCR signals. Always use freshly blocked beads and avoid overloading your samples with unblocked material.
19. Avoid bead loss at this stage since beads may be nearly invisible after washes. Leave behind a small amount of buffer rather than risk aspirating the beads.
20. If you need to pause, keep antibody-bead complexes on ice or at 4 °C for short periods before washes. Avoid freezing at this stage since antibody binding may be disrupted.
21. Washing steps follow these general guidelines:
    - Always keep tubes and buffers on ice or at 4 °C to protect protein–DNA interactions.
    - Use pre-chilled buffers and fresh protease inhibitors in each wash if working slowly.
    - Pipette gently or flick tubes to resuspend beads. Avoid vortexing.
    - Use 1 mL of wash buffer per sample.

- Incubate samples with wash buffer on a rotator at 10–15 rpm for the indicated time.
- After incubation, separate beads using a magnetic rack (or centrifuge if agarose beads are used).
- Carefully remove and discard the supernatant without disturbing the beads.
- Proceed immediately to the next wash.

Wash 1: Low Salt Wash Buffer (20 mM Tris–HCl (pH 8.0), 150 mM NaCl, 2 mM EDTA, 0.1% SDS, 1% Triton X-100).

Purpose: Removes loosely bound chromatin or proteins.

Steps:

- Add 1 mL buffer to each tube.
- Rotate at 4 °C for 5 min.
- Collect beads and discard supernatant.

Wash 2: High Salt Wash Buffer (20 mM Tris–HCl (pH 8.0), 500 mM NaCl, 2 mM EDTA, 0.1% SDS, 1% Triton X-100).

Purpose: Disrupts weak protein–DNA and protein–protein interactions.

Steps:

- Add 1 mL buffer.
- Rotate at 4 °C for 5 min.
- Magnet or spin, then discard supernatant (Tip: Beads may clump due to salt; flick tube gently to resuspend).

Wash 3: LiCl Wash Buffer (10 mM Tris–HCl (pH 8.0), 250 mM LiCl, 1 mM EDTA, 1% NP-40 (or Igepal CA-630), 1% sodium deoxycholate).

Purpose: Removes nonspecific protein–protein or nucleic acid–protein interactions (This is the most stringent wash; therefore, ensure gentle but thorough mixing to avoid bead loss while maximizing background removal).

Steps:

- Add 1 mL buffer.
- Rotate at 4 °C for 5 min.
- Separate beads, discard supernatant.

Wash 4: TE Buffer (Final Wash) (10 mM Tris–HCl (pH 8.0), 1 mM EDTA).

Purpose: Removes remaining salts and detergents, prepares beads for elution.

Steps:

- Add 1 mL TE buffer.
- Rotate at 4 °C for 2–3 min.
- Collect beads, discard supernatant.

22. After adding the elution buffer to the beads, you can perform an initial 15-min elution at 65 °C with shaking (before NaCl addition), then collect the supernatant and proceed to reverse crosslinking with NaCl. Alternatively, NaCl can be added directly to the elution buffer upfront.
23. Ensure tubes are tightly sealed. If evaporation is a concern, overlay with mineral oil or use strip caps for PCR tubes.
24. Use RNase A that has been heat-inactivated if needed.
25. Warm elution buffer to 37–50 °C and incubate 2 min on the column before centrifugation to improve yield.
26. DNA should be free of RNA or protein contamination. If impurities are suspected, do a second round of phenol:chloroform extraction.
27. Avoid adapter dimers by using adapter:DNA molar ratio of 10: 1 or optimize according to input.
28. Let beads dry briefly after ethanol wash (do not over-dry or crack the pellet).
29. Over-amplified libraries often show a bump >600 bp. If so, reduce PCR cycles in future preps.
30. Use a reproducible environment (e.g., Conda or Docker) to lock tool versions to prevent version conflicts before you start. This also helps you to repeat the analysis later if necessary (e.g., FastQC 0.12.1, Cutadapt 5.1, BWA 0.7.19, SAMtools 1.22.1, Picard 3.4, MACS2 2.2.9, deepTools 3.5.6, HOMER 4.12, R 4.4 + DiffBind 3.14).
31. Libraries with adapter contamination >5% or per-base quality < Q20 in the first 30 bp require more aggressive trimming or re-sequencing.
32. Adapter sequence should be checked based on library preparation protocol. Different library preparation pipeline can have different adapter sequences.
33. Mark duplicates by default; delete only when duplication rate exceeds 30% and biological signal is strong.
34. Set genome size (-g) to 1.35 × 10^8 bp for the nuclear portion of *A. thaliana*. Adjust FDR (-q) according to sequencing depth and antibody quality.

35. Combine annotated gene lists with clusterProfiler for GO/KEGG enrichment to link peaks with biological pathways.
36. For datasets with <500 peaks, increase motif window size to 500 bp and consider MEME-ChIP as a complementary approach.
37. DiffBind defaults to edgeR for statistical testing; supply replicates' fragment counts to ensure accurate dispersion estimation.
38. Deposit processed data (peaks, bigWigs) and the complete analysis workflow to GEO or ENA under MIAME-compliant guidelines.
39. Primer Design Guidelines for ChIP-qPCR:
    - Amplicon optimal length: 70–150 bp.
    - Shorter amplicons improve amplification efficiency and are better suited for low-input DNA (typical in ChIP).
    - Design primers within promoter regions (especially ~500 bp upstream of TSS) and putative transcription factor binding motifs (based on DAP-seq, literature, or motif prediction tools).
    - Include positive control loci (known binding sites) and negative control loci (nontarget intergenic regions or housekeeping genes not regulated by the TF).
    - Primer characteristics:

Feature	Recommended range
Length	18–24 nucleotides
GC content	40–60%
Melting temperature (tm)	58–62 °C (match forward and reverse within ±1 °C)
Secondary structures	Avoid hairpins, self-dimers, cross-dimers ($\Delta G > -9$ kcal/Mol)
3′-end stability	Prefer G/C clamp at 3′-end to ensure stable binding

    - Use software like Primer3, NCBI Primer-BLAST, or OligoAnalyzer to evaluate primer quality.
    - Avoid placing primers across exon–exon junctions (since ChIP targets DNA, not RNA).
    - Use BLAST to ensure primers are specific to the desired locus.
    - Avoid repeats or regions with high sequence homology.

- Validate primers using standard qPCR on input DNA to ensure single melt curve peak, 90% amplification efficiency (check using serial dilution curve) and no amplification in no-template controls (NTC).

40. Optimize primer annealing temperature and check for single melting peaks (no primer-dimers).

Acknowledgments

This research was funded by The Scientific and Technological Research Council of Türkiye (project number 119Z149).

References

1. Santner A, Estelle M (2009) Recent advances and emerging trends in plant hormone signalling. Nature 459:1071–1078
2. Verma V, Ravindran P, Kumar PP (2016) Plant hormone-mediated regulation of stress responses. BMC Plant Biol 16:86
3. Gendrel A-V, Lippman Z, Martienssen R, Colot V (2005) Profiling histone modification patterns in plants using genomic tiling microarrays. Nat Methods 2:213–218
4. Kaufmann K, Muiño JM, Østerås M, Farinelli L, Krajewski P, Angenent GC (2010) Chromatin immunoprecipitation (ChIP) of plant transcription factors followed by sequencing (ChIP-SEQ) or hybridization to whole genome arrays (ChIP-CHIP). Nat Protoc 5:457–472
5. O'Malley RC, Huang SC, Song L, Lewsey MG, Bartlett A, Nery JR et al (2016) Cistrome and Epicistrome features shape the regulatory DNA landscape. Cell 165:1280–1292
6. Ricardi MM, González RM, Iusem ND (2010) Protocol: fine-tuning of a Chromatin Immunoprecipitation (ChIP) protocol in tomato. Plant Methods 6:11
7. Heyndrickx KS, de Velde JV, Wang C, Weigel D, Vandepoele K (2014) A functional and evolutionary perspective on transcription factor binding in Arabidopsis thaliana. Plant Cell 26:3894–3910
8. Gendrel A-V, Lippman Z, Yordan C, Colot V, Martienssen RA (2002) Dependence of heterochromatic histone H3 methylation patterns on the Arabidopsis gene DDM1. Science 297: 1871–1873
9. Andrews S (2010) Babraham bioinformatics - FastQC a quality control tool for high throughput sequence data [Internet]. [cited 2025 Aug 4]. Available from: https://www.bioinformatics.babraham.ac.uk/projects/fastqc/
10. Ewels P, Magnusson M, Lundin S, Käller M (2016) MultiQC: summarize analysis results for multiple tools and samples in a single report. Bioinformatics 32:3047–3048
11. Martin M (2011) Cutadapt removes adapter sequences from high-throughput sequencing reads. EMBnetjournal 17:10–12
12. Li H (2013) Aligning sequence reads, clone sequences and assembly contigs with BWA-MEM [Internet]. arXiv. [cited 2025 Aug 4]. Available from: http://arxiv.org/abs/1303.3997
13. Li H, Handsaker B, Wysoker A, Fennell T, Ruan J, Homer N et al (2009) The sequence alignment/map format and SAMtools. Bioinformatics 25:2078–2079
14. Picard (2019) "Picard Toolkit." Broad Institute, GitHub Repository. https://broadinstitute.github.io/picard/; Broad Institute. [Internet]. [cited 2025 Aug 4]. Available from: https://broadinstitute.github.io/picard/
15. Zhang Y, Liu T, Meyer CA, Eeckhoute J, Johnson DS, Bernstein BE et al (2008) Model-based analysis of ChIP-Seq (MACS). Genome Biol 9:R137
16. Ramírez F, Ryan DP, Grüning B, Bhardwaj V, Kilpert F, Richter AS et al (2016) deepTools2: a next generation web server for deep-sequencing data analysis. Nucleic Acids Res 44:W160–W165
17. Robinson JT, Thorvaldsdóttir H, Winckler W, Guttman M, Lander ES, Getz G et al (2011)

Integrative genomics viewer. Nat Biotechnol 29:24–26

18. Perez G, Barber GP, Benet-Pages A, Casper J, Clawson H, Diekhans M et al (2025) The UCSC genome browser database: 2025 update. Nucleic Acids Res 53:D1243–D1249
19. Heinz S, Benner C, Spann N, Bertolino E, Lin YC, Laslo P et al (2010) Simple combinations of lineage-determining transcription factors prime cis-regulatory elements required for macrophage and B cell identities. Mol Cell 38:576–589
20. Wang Q, Li M, Wu T, Zhan L, Li L, Chen M et al (2022) Exploring epigenomic datasets by ChIPseeker. Curr Protoc 2:e585
21. Yu G, Wang L-G, He Q-Y (2015) ChIPseeker: an R/Bioconductor package for ChIP peak annotation, comparison and visualization. Bioinformatics 31:2382–2383
22. Stark R, Brown G (2013) DiffBind: differential binding analysis of ChIP-Seq peak data
23. Love MI, Huber W, Anders S (2014) Moderated estimation of fold change and dispersion for RNA-seq data with DESeq2. Genome Biol 15:550

Chapter 15

Detection of tRNA-Derived Small RNAs in *Arabidopsis thaliana* via Northern Blotting: Exploring Potential Biomarkers for Phytohormone Analysis

Dinesh Singh Pujara, Tanzila Kamal Choity, Yogendra Bordiya, Susan M. Magdaleno, and Hong-Gu Kang

Abstract

Noncoding small RNAs (sRNAs) are well-known regulatory molecules that influence a wide range of cellular processes in eukaryotes and often function in concert with phytohormones. Advances in next-generation sequencing (NGS) have further expanded the identification and characterization of sRNAs from diverse sources, highlighting their potential as early biomarkers for hormone and stress signaling. In recent work, we identified transfer RNA-derived fragments (tRFs) as a prominent class of sRNAs, induced as early as 1 h post-infection with *Pseudomonas syringae*, that positively regulates defense gene expression in Arabidopsis. However, verifying these sRNA findings from NGS data presents a technical challenge, as the small size and often low abundance of these sRNAs complicate accurate detection and quantification. Here, we present a procedure to enrich sRNAs and facilitate the detection of sRNAs in Arabidopsis using a sensitive northern blot protocol. We evaluated sRNA enrichment using the conventional Trizol method with and without nucleic acid precipitants, as well as commercial column-based extraction kits. Our northern blot results show that total RNA purified by both methods is enriched with sRNAs. Additionally, the precipitants, such as linear acrylamide and glycogen, improved the detection of tRFs. Column-based kits also yielded comparable enrichment of tRFs and provided a rapid and straightforward protocol essential for preserving RNA integrity. This study thus presents a robust, nonradioactive and PCR-free detection method for characterizing the regulatory functions of sRNAs.

Key words *Arabidopsis thaliana*, Noncoding small RNAs, Transfer RNA-derived fragments, Northern blot, Total RNA isolation method

1 Introduction

Noncoding small RNAs (sRNAs) play pivotal roles in regulating diverse cellular processes in eukaryotes [1]. These sRNAs include microRNAs (miRNAs), small interfering RNAs (siRNAs), piwi-interacting RNAs (piRNAs), and transfer RNA-derived small RNA fragments (tRFs) [1, 2]. Since the discovery of miRNAs in *C. elegans* more than two decades ago, advances in small RNA next-

Shahid Mukhtar (ed.), *Plant Hormones: Methods and Protocols*, Methods in Molecular Biology, vol. 3026,
https://doi.org/10.1007/978-1-0716-5214-5_15,

generation sequencing (NGS) have led to remarkable progress in identifying and exploring several important classes of sRNAs. Recently, tRFs, which are generated from precursor or mature tRNAs, have gained particular attention for their regulatory roles in development, stress responses, and tumorigenesis [1, 3].

Phytohormones are essential signaling molecules involved in nearly all biological processes in plants [4], orchestrating growth, development, and defense. Interestingly, these processes are also regulated by sRNAs, which serve as critical modulators of plant growth and development under both favorable and adverse conditions [5]. Due to this close association with plant biological functions, sRNAs have been proposed to be promising candidates as biomarkers for a wide range of hormone-related responses. While hormone-responsive genes may take hours to be expressed, certain sRNAs can be induced much earlier in response to hormones or environmental cues. For instance, we observed that specific tRFs are induced within 1 h in response to avirulent *Pseudomonas syringae*, whereas defense genes require several hours to activate [6]. This suggests that tRFs, along with phytohormones, act as early regulators in modulating plant biological processes.

When we observed that specific tRFs were rapidly induced under biotic stress using NGS [6], we conducted additional molecular biology experiments, including northern blotting and quantitative reverse transcription PCR (qRT-PCR), to validate these findings. Northern blotting analysis for sRNA is technically challenging due to the small size of the probes, which presents significant obstacles for accurate isolation and quantification, requiring the development of robust protocols to address these challenges. Nevertheless, northern analysis remains a key analytical method for validating sRNAs without PCR [7, 8], thereby avoiding potential PCR biases [9, 10].

The Trizol method for RNA extraction is a widely used approach for isolating sRNAs, including miRNAs, siRNAs, and piRNAs [11]. Low-abundance sRNAs can be enriched using carriers like linear acrylamide and glycogen, making them more detectable by Northern blotting [12]. Linear acrylamide serves as a co-precipitant, binding to RNA and increasing its effective size, which reduces loss during centrifugation and washing steps and improves yield [13]. Similarly, glycogen forms a precipitate with RNA, facilitating pelleting during centrifugation and significantly boosting yield without inhibiting downstream enzymatic reactions [14]. Traditionally, radioisotope-based labeling was used in northern blot analysis. However, to minimize hazards associated with isotope labeling, we chose DIG-labeled probes, which offer comparable sensitivity to radioactive methods [15]. In addition, we used a highly sensitive protocol developed by Kim et al. [7], which utilized the advantages of locked nucleic acid (LNA) oligonucleotides, EDC (1-ethyl-3-(3-dimethylaminopropyl)

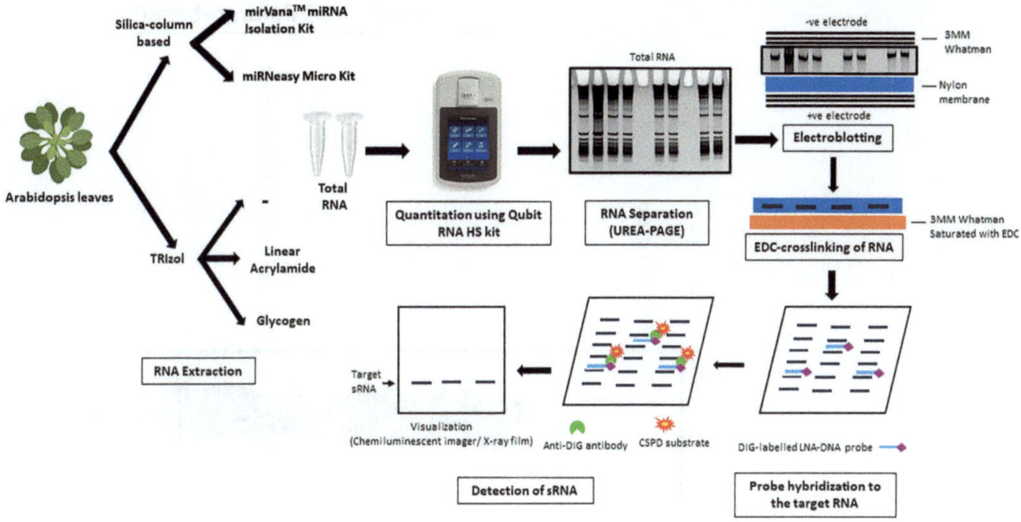

Fig. 1 Schematic diagram summarizing the workflow of small RNA purification and detection using Northern blot. Total RNA, isolated from WT Arabidopsis leaves using the indicated methods, was quantified using Qubit RNA HS assay kit and subjected to Northern analysis. Total RNA was first size-separated using urea-PAGE and transferred to a Nylon membrane via electroblotting. RNA was cross-linked to the membrane using EDC, and a DIG-LNA-DNA probe was hybridized. Target tRF^{31Asp2} was detected by using an anti-DIG antibody, CSPD substrate, and imaging using a chemiluminescent imager or X-ray film

carbodiimide) cross-linking, and DIG-labeling. In particular, LNA oligonucleotides have been shown to display a significantly higher binding affinity for complementary RNA/DNA sequences compared to unmodified counterparts [16].

To evaluate the enrichment and detection efficiency of sRNAs in total RNA purified using the conventional Trizol method and the commercially available silica-column based kits: mirVana™ miRNA Isolation Kit (Thermo Fisher Scientific) and miRNeasy Micro Kit (Qiagen), we performed the northern blot assay as detailed in Fig. 1. Our northern analysis showed that total RNA isolated with both the Trizol and the column-based methods from Arabidopsis leaves detected tRFs, specifically tRF^{31Asp2}, 31nts-long tRF derived from a tRNA^{AspGTC} (Fig. 2). While total RNA precipitated with isopropanol (IPA) alone was adequate, the inclusion of sRNA precipitants, linear acrylamide and glycogen, enhanced the detection of smaller tRF molecules at around 31 nts (Fig. 2). The kit-based sRNA purification methods offer a simpler and faster workflow (mirVana miRNA Isolation Kit: ~30 min, miRNeasy Micro Kit: ~45 min vs Trizol method ~1 h 15 min), allowing for efficient detection of sRNAs and likely avoiding RNA degradation. Note that we have used mirVana miRNA Isolation Kit for our NGS applications [6].

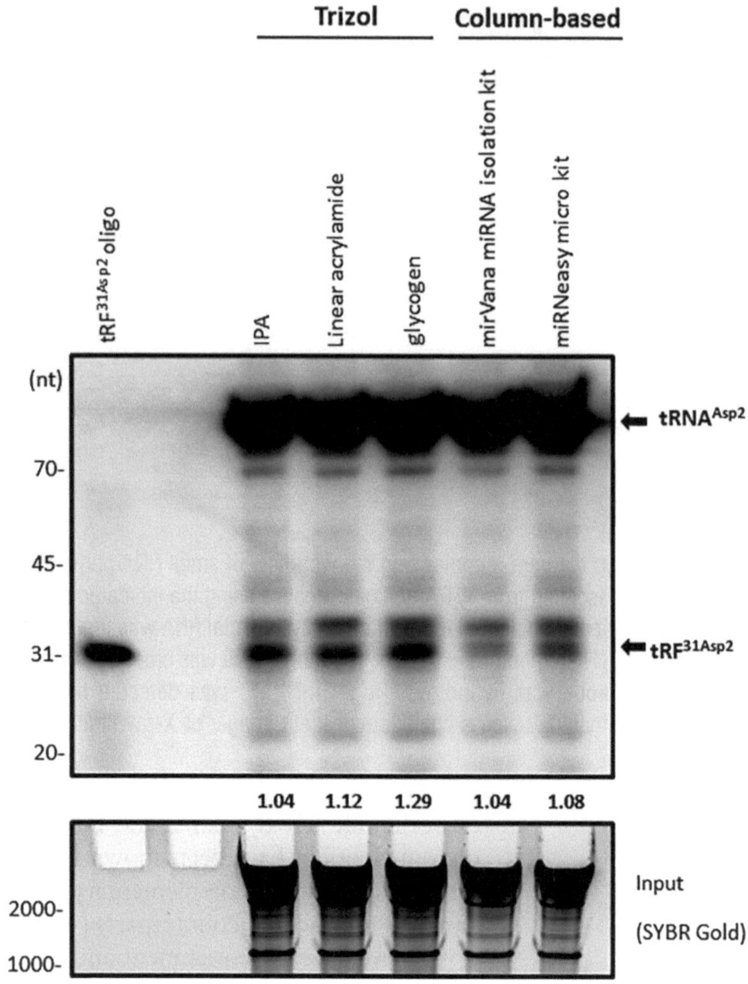

Fig. 2 Northern blot analysis of tRF31^{Asp2} in total RNA isolated using the indicated methods. RNA was separated on a 15% urea-PAGE. The upper panel shows the Northern blot detection of small RNAs using a DIG-labeled LNA probe specific to tRF^{31Asp2}, while the lower panel presents larger RNAs from the same gel stained with SYBR Gold as a loading input. tRNA^{Asp2} and tRF^{31Asp2} are indicated by arrowheads. A tRF^{31Asp2} oligonucleotide served as a positive control. Size markers in nucleotides (nts) are shown on the left of each panel. The values below the Northern blot indicate the ratio of tRF^{Asp2} signal to the loading input

Here, we outline our recent approach for sRNA enrichment and detection using northern blotting, comparing the traditional Trizol method with the commercially available sRNA purification kits (Fig. 2). These findings showed the robustness and sensitivity of northern blotting to detect targeted tRFs, providing a PCR-free molecular tool for studying noncoding sRNAs and elucidating their roles in diverse regulatory processes.

2 Materials

2.1 Plant Materials and Growth Conditions

1. Wild-type *Arabidopsis thaliana* (Col-0) plants, 3.5 weeks old, grown in soil at 22 °C and 60% relative humidity under a 16 hour light period.

2.2 Equipment

1. Paint shaker for a high-throughput homogenization.
2. Tabletop microcentrifuge.
3. Hybridization oven and bottles.
4. Electrophoresis system for PAGE.
5. Power supply.
6. Wet/tank blotting systems.
7. Vortex mixer.
8. Incubator at 37 °C.
9. Saran wrap.
10. 3MM Whatman paper.
11. Rotator.
12. Qubit® 3.0 Fluorometer.
13. Millipore Milli-Q ultrapure water system.
14. Azure imaging station.

2.3 Reagents

1. mirVana™ miRNA Isolation Kit, without phenol (Thermo Fisher, Cat#: AM1561).
2. miRNeasy Micro Kit (Qiagen, Cat#: 217084).
3. Invitrogen™ Acid-Phenol:Chloroform, pH 4.5.
4. Trizol Reagent.
5. Chloroform.
6. Isopropanol.
7. Ethanol, Absolute.
8. Linear Acrylamide.
9. Glycogen.
10. Acrylamide: Bis-Acrylamide 29:1 (40% Solution/ Electrophoresis).
11. Urea.
12. N,N,N′,N′-tetramethylethylene-diamine (TEMED).
13. Ammonium Persulfate—ULTRA PURE.
14. SYBR gold.
15. 1-ethyl-3-(3-dimethylaminopropyl) carbodiimide (EDC) (Thermo Fisher Cat#: 22980).

16. Ultrahyb Ultrasensitive Hybridization Buffer (Thermo Fisher, Cat#: AM8670).
17. Nylon membrane (Roche, Cat#: 209299001).
18. Anti-Digoxigenin-AP, Fab fragments antibody (Roche, Cat#: 11093274910).
19. CSPD (Roche, Cat#: 11655884001).
20. Liquid nitrogen.
21. Gel loading tip.
22. Deionized H_2O.
23. tRF31^{Asp2} Synthetic oligonucleotide (Sigma, [Phos] GUC GUUGUAGUAUAGUGGUAAGUAUUCCCGC).
24. DIG LNA-DNA probe (Sigma, ([DIG]CTTACCACTATAC [+T][+A][+C][+A]A[+C][+G][+A]C).
25. DNA ladders composed of oligonucleotides derived from the 5′ end of the tRNA^{Asp2} sequence with lengths of 20, 31, 45, and 70 nts.
26. Sodium Acetate (3 M, pH 5.2).
27. Elution buffer (10 mM Tris–HCl, pH 8.0).
28. Qubit RNA HS Assay Kit (Thermo Fisher; Cat#: Q32855).
29. 5× Tris-Borate EDTA (TBE) buffer.
30. DIG Wash and Block Buffer Set (Sigma; Cat#: 11585762001).
31. Low Stringent Buffer (2× SSC with 0.1% (wt/vol) SDS), High Stringent Buffer (0.1× SSC with 0.1% (wt/vol) SDS), Washing Buffer (1× SSC).
32. DEPC-treated water.

3 Methods

3.1 RNA Extraction Using Trizol

1. Extract total RNA using Trizol from 50 mg fresh leaf tissue collected from 3.5-week-old WT Arabidopsis plants.
2. Grind leaf tissue by freezing with liquid nitrogen and homogenizing using a paint shaker.
3. Isolate total RNA according to the manufacturer's instructions, with and without the addition of sRNA carriers such as linear acrylamide and glycogen.
4. Add 1 mL of Trizol to the ground tissue, vortex thoroughly, and incubate for 5 min at room temperature to allow complete dissociation of the nucleoprotein complex.
5. Add 0.2 mL of chloroform to the tube, vortex for about 30 s, and incubate for 2–3 min.
6. Centrifuge the lysate at 12,000 × *g* for 15 min at 4°C.

7. Transfer the aqueous phase containing the RNA carefully into a new 1.5 mL tube using a micropipette.
8. Precipitate total RNA by mixing the aqueous phase with 500 μL of isopropanol (IPA) alone or IPA combined with 1:10 volume of 3 M sodium acetate (pH 5.2) and either a 10 μg/mL linear acrylamide or 20 μg/mL glycogen (final concentrations).
9. Incubate the mixture at 4 °C for 10 min, followed by centrifugation for 10 min at 12,000 × *g* at 4 °C to pellet the total RNA.
10. Remove the supernatant carefully using a micropipette. Wash the pellet twice with 70% ethanol by centrifuging at full speed for 1 min per wash.
11. Air dry the pellet for at least 5 min, ensuring that no visible traces of ethanol remain in the tube.
12. Finally, resuspend the dried pellet in 50 μL of Elution Buffer to elute the total RNA.

3.2 RNA Extraction Using mirVana™ miRNA Isolation Kit

1. Isolate total RNA using the mirVana miRNA isolation kit according to the manufacturer's instructions. Vortex the leaf tissue homogenates, prepared in the same manner as described for the Trizol method, with 700 μL of mirVana Lysis/Binding reagent.
2. Add 1/10 volume of miRNA Homogenate additive to the tissue lysate and incubate the mixture on ice for 10 min.
3. Add an equal volume of Acid Phenol-Chloroform, equal to the lysate volume before the addition of miRNA Homogenate Additive, and vortex for 60 s to mix thoroughly.
4. Centrifuge the sample at 10,000 × *g* for 5 min at room temperature. After centrifugation, transfer the aqueous phase carefully into a new collection tube. Mix with 1.25 volumes of room-temperature 100% ethanol and pass through a silica-based membrane filter tube (provided with the kit). Note: Up to 700 μL can be processed through the filter at a time. Repeat until all the mixture has been passed through the membrane.
5. Wash the RNA bound to silica-based membrane filter first with 700 μL of miRNA wash solution 1, followed by two more washes with 500 μL of miRNA wash solution 2/3 by centrifuging for 5–10 s. Use the same collection tube for all washing steps.
6. After discarding the flow-through from the last wash, place the filter cartridge in the same collection tube and centrifuge one more time for 1 min to remove residual washing solution.
7. Place the filter cartridge in a new collection tube and elute the RNA in 50 μL pre-heated Elution buffer (95°C) by centrifuging at maximum speed for 20–30 s.

3.3 RNA Extraction Using miRNeasy™ Micro Kit

1. Isolate total RNA using QIAGEN miRNeasy™ Micro Kit according to the manufacturer's instructions. Mix the leaf tissue homogenates, prepared in the same manner as described for the Trizol method, with 700 μL of QIAzol lysis reagent by vortexing for 1 min, followed by incubation at room temperature for 5 min.
2. Add 140 μL of chloroform to the homogenate, vortex for 15 s, and allow to sit at room temperature for 2 min.
3. Centrifuge the mixture at 12,000 × *g* for 15 min at 4 °C.
4. Transfer the aqueous phase (around 350 μL) carefully and mix with 1.5 volumes (around 525 μL) of room-temperature 100% ethanol. Homogenize the mixture by pipetting several times.
5. Pass the mixture through the RNeasy MiniElute spin column (up to 700 μL at a time) placed in a 2 mL collection tube (provided with the kit) and centrifuge at 8000 × *g* for 15 s. Discard the flow-through. Repeat this step until all the mixture has been passed through the membrane.
6. Wash the column sequentially with 700 μL of Buffer RWT and 500 μL of Buffer RPE (both provided with the kit) by centrifuging at 8000 × *g* for 15 s.
7. Wash the column with 500 μL of freshly prepared 80% ethanol by centrifuging at 8000 × *g* for 2 min.
8. Place the RNeasy MiniElute spin column in a new 2 mL centrifuge tube with caps open, and centrifuge at 8000 × *g* for 5 min to dry.
9. Place the spin columns in a fresh 1.5 mL collection tube and elute the total RNA using 30 μL of Elution Buffer by centrifuging at full speed for 1 min.

3.4 Northern Blotting Analysis

1. Perform Northern blot as described by [7] with the modifications included below. Mix 750 ng of total RNA with Gel Loading Buffer II (provided with the mirVana miRNA Isolation Kit). Denature the mixture at 95°C for 5 min and allow to cool at 4°C for at least 5 min.
2. Separate the denatured RNA sample on a 15% urea-PAGE by running in 1× TBE buffer at 150 V. Note, pre-run the gel at 200 volts for 1 h to clear urea from the wells and wash the wells using a 1 mL micropipette before loading the RNA sample.
3. Use 7.5 fmoles of synthetic tRF^{31Asp2} as a positive control, and 2 μL (10 pmoles) of DNA ladder composed of oligonucleotides derived from the 5′ end of the $tRNA^{Asp2}$ as a size marker.
4. After running, cut the gel into two parts using a sterile razor blade. Stain the upper part, containing larger-sized RNAs, with SYBR Gold as a loading input. Soak the lower part in 1× TBE for 10 min, shaking gently, before transferring.

5. For transfer, cut a nylon membrane and six sheets of 3MM Whatman paper to the gel size and soak in 1× TBE for 5 min with shaking. Prepare a blotting sandwich as shown in Fig. 1, and transfer the RNA to the membrane in the blotting tank containing 1× TBE buffer at 15 V for 90 min at 4 °C. Successful transfer is indicated by the complete transfer of the RNA loading dye (Bromophenol Blue) to the nylon membrane from the gel.
6. Cross-link the membrane using EDC at 60 °C for 2 h. To cross-link, place the membrane on filter paper saturated with EDC (RNA side facing up) and wrap with Saran Wrap.
7. Wash the cross-linked membrane thoroughly with deionized water to remove any residual cross-linking agent. Note that the membrane at this stage can be wrapped in Saran Wrap and stored at −20 °C for later use.
8. Place the membrane in a hybridization bottle, roll with the RNA side facing inward, and pre-incubate in 15 mL of Ultra-Hyb buffer at 55 °C for at least 30 min with constant rotation in a hybridization oven.
9. Add a DIG-labeled LNA-DNA probe for tRF^{31Asp2} to the hybridization buffer to a final concentration of 5 nM. Denature the probe at 95 °C for 1 min before addition and perform hybridization overnight at 55 °C with constant rotation in a hybridization oven.
10. Wash the hybridized membrane sequentially: first, wash twice with Low-Stringency Buffer for 15 min each, followed by two washes with High-Stringency Buffer for 5 min each, and finally wash once with Washing Buffer for 10 min at room temperature.
11. Block the membrane with 15 mL of 1× DIG blocking buffer for 3 h at room temperature.
12. Incubate the blocked membrane with an anti-DIG antibody (1:15,000 dilution) for 30 min at room temperature.
13. Wash the membrane thoroughly with DIG Washing buffer 4 times (15 min each) and then incubate in a detection buffer for 5 min. Note: Do not let the membrane dry.
14. Using clean forceps, place the membrane on a Saran Wrap, and add the chemiluminescent substrate CSPD (1:100 dilution in Detection Buffer) to the surface of the membrane. Incubate for 5 min at room temperature, followed by 15 min at 37 °C in the dark.
15. Detect the presence of tRF^{31Asp2} using the Azure imaging station. Note: Capture multiple exposures to detect both high- and low-abundance small RNAs.

16. Quantify the band intensities using ImageJ [17]. Analyze the ratio of the mean grey value of the target (α-DIG) and the input (SYBR gold).

4 Notes

1. The LNA nucleotides in the DIG-labeled LNA-DNA probe sequence targeting tRF^{31Asp2} are denoted by the [+N] notation.
2. To minimize RNA degradation during extraction, leaf tissue was homogenized under cryogenic conditions using liquid nitrogen and kept frozen until the addition of the lysis buffer.
3. Wells of the PAGE gel were rinsed with 1x TBE buffer from the gel tank using a 1mL pipette to remove excess acrylamide and urea, which is essential for ensuring sharp, undistorted bands.
4. The probe hybridization temperature was systematically optimized by testing sensitivities at 37, 55, and 60 °C. We ultimately used 55 °C as we found reduced background signals without significantly compromising the sensitivity. Nonetheless, we anticipate that testing a range of hybridization temperatures will be necessary to further optimize this nonradioactive detection method for diverse targets.

Acknowledgments

This work is supported by the National Science Foundation (IOS-1553613) and the United States Department of Agriculture (2026-67039-45933) to HGK.

References

1. Li T, Gonzalez N, Inze D, Dubois M (2020) Emerging connections between small RNAs and phytohormones. Trends Plant Sci 25(9): 912–929. https://doi.org/10.1016/j.tplants.2020.04.004
2. Alves CS, Nogueira FTS (2021) Plant small RNA world growing bigger: tRNA-derived fragments, longstanding players in regulatory processes. Front Mol Biosci 8:638911. https://doi.org/10.3389/fmolb.2021.638911
3. Pekarsky Y, Balatti V, Croce CM (2023) tRNA-derived fragments (tRFs) in cancer. J Cell Commun Signal 17(1):47–54. https://doi.org/10.1007/s12079-022-00690-2
4. Shan X, Yan J, Xie D (2012) Comparison of phytohormone signaling mechanisms. Curr Opin Plant Biol 15(1):84–91. https://doi.org/10.1016/j.pbi.2011.09.006
5. Qiao Y et al (2021) Small RNAs in plant immunity and virulence of filamentous pathogens. Annu Rev Phytopathol 59:265–288. https://doi.org/10.1146/annurev-phyto-121520-023514
6. Kim S-I et al (2024) A nuclear tRNA-derived fragment triggers immunity in Arabidopsis. Nat Commun. Third revision under review
7. Kim SW et al (2010) A sensitive non-radioactive northern blot method to detect small RNAs. Nucleic Acids Res 38(7): e98–e98

8. Várallyay É, Burgyán J, Havelda Z (2008) MicroRNA detection by northern blotting using locked nucleic acid probes. Nat Protoc 3(2):190–196. https://doi.org/10.1038/nprot.2007.528
9. Kebschull JM, Zador AM (2015) Sources of PCR-induced distortions in high-throughput sequencing data sets. Nucleic Acids Res 43(21):e143. https://doi.org/10.1093/nar/gkv717
10. Choi YS, Edwards LO, DiBello A, Jose AM (2017) Removing bias against short sequences enables northern blotting to better complement RNA-seq for the study of small RNAs. Nucleic Acids Res 45(10):e87. https://doi.org/10.1093/nar/gkx091
11. Rio DC, Ares M Jr, Hannon GJ, Nilsen TW (2010) Purification of RNA using TRIzol (TRI reagent). Cold Spring Harb Protoc 2010(6): pdb prot5439. https://doi.org/10.1101/pdb.prot5439
12. Song X et al (2012) Rice RNA-dependent RNA polymerase 6 acts in small RNA biogenesis and spikelet development. Plant J 71(3): 378–389. https://doi.org/10.1111/j.1365-313X.2012.05001.x
13. Chomczynski P, Sacchi N (1987) Single-step method of RNA isolation by acid guanidinium thiocyanate-phenol-chloroform extraction. Anal Biochem 162(1):156–159. https://doi.org/10.1006/abio.1987.9999
14. Chomczynski P (1993) A reagent for the single-step simultaneous isolation of RNA, DNA and proteins from cell and tissue samples. BioTechniques 15(3):532–534, 536–537
15. Ramkissoon SH et al (2006) Nonisotopic detection of microRNA using digoxigenin labeled RNA probes. Mol Cell Probes 20(1): 1–4. https://doi.org/10.1016/j.mcp.2005.07.004
16. Grunweller A, Hartmann RK (2007) Locked nucleic acid oligonucleotides: the next generation of antisense agents? BioDrugs 21(4): 235–243. https://doi.org/10.2165/00063030-200721040-00004
17. Schneider CA, Rasband WS, Eliceiri KW (2012) NIH image to ImageJ: 25 years of image analysis. Nat Methods 9(7):671–675. https://doi.org/10.1038/nmeth.2089

Chapter 16

Deciphering Protein Interaction Dynamics Using the Yeast Three-Hybrid System

Qianwei Liu and Jathish Ponnu

Abstract

Proteins often act together with other proteins to regulate crucial biological processes. Beyond binary interactions, they often operate within multi-protein assemblies, where alterations in binding affinity or structural conformation of one or more components can significantly impact the overall function of the complex. The yeast three-hybrid (Y3H) system extends the conventional yeast two-hybrid (Y2H) by adding a bridge protein to test ternary interactions *ex planta*. Here, we present a Y3H protocol utilizing the pBridge and pACT2 AD vectors in the *Saccharomyces cerevisiae* Y2HGold strain and examine the role of SUPPRESSOR OF PHYA-105 1 (SPA1) in mediating the interaction between CONSTITUTIVE PHOTOMORPHOGENIC 1 (COP1) and CRYPTOCHROME 1 (CRY1) in *Arabidopsis thaliana*. COP1 and SPA proteins form a functional E3 ubiquitin ligase complex that prevents photomorphogenesis in darkness, while CRY1 antagonizes this activity under blue light conditions. Using the Y3H system, we show that co-expression of SPA1 enhances the interaction between COP1 and CRY1, as detected via *MEL1* reporter gene activation in a quantitative assay using p-nitrophenyl alpha-D-galactopyranoside (PNPG). This protocol offers a practical framework for dissecting ternary protein interactions using the Y3H assay.

Key words Yeast three-hybrid, Y3H, Y2H, Quantitative assay, PNPG, α-galactosidase, α-gal, COP1, SPA1, CRY1, Ternary, pBridge, pACT, LiAc, PEG

1 Introduction

Interconnected protein interaction networks form the basis of key biological processes that define most cellular functions. The function of a protein is often determined by its interaction with other molecules, particularly other proteins, within its specific cellular environment [1, 2]. While a binary physical interaction between two proteins can serve as a functional unit in certain contexts, proteins commonly operate as components of larger, multiunit assemblies. Within these complexes, even subtle changes in binding affinities can affect the interaction dynamics and thereby influence the overall function [3–7]. Therefore, it is essential to investigate how constituent proteins interact within these complexes.

Shahid Mukhtar (ed.), *Plant Hormones: Methods and Protocols*, Methods in Molecular Biology, vol. 3026,
https://doi.org/10.1007/978-1-0716-5214-5_16,

Several assays are available to investigate protein-protein interactions (PPIs) in plant biology, ranging from methods that detect binary associations to high-throughput approaches that characterize complexes. Each, however, has specific strengths and limitations. Commonly used *in planta* techniques for studying PPIs, such as Bimolecular Fluorescence Complementation (BiFC), Förster Resonance Energy Transfer coupled with Fluorescence Lifetime Imaging Microscopy (FRET-FLIM), Split Luciferase (SplitLUC), and the split ubiquitin systems [8–11], are mostly restricted to pairwise interaction scenarios. In contrast, biochemical approaches such as Co-Immunoprecipitation (Co-IP) and Affinity Purification coupled with Mass Spectrometry (AP-MS) are suitable to study protein complexes in their native cellular contexts [12–14], as they tend to preserve posttranslational modifications that may be critical for interaction specificity. However, these methods primarily aim at identifying the proteins within complexes, which may participate in both direct and indirect interactions. Therefore, further validation is required to determine the nature of these interactions. For example, two proteins found within a complex may either interact physically or their association is facilitated by a third protein (often referred to as a "bridge" protein), which connects them indirectly. The bridge protein may also enhance the interaction between the proteins of interest (POIs) by changing the binding affinity of one or both partners [4, 5, 15]. Common scenarios involving three-protein interactions include one protein acting as a bridge between the other two, facilitating the proper conformation necessary for their interaction (functioning as a chaperone), modifying one or both proteins through posttranslational modifications, and providing an active interaction surface by associating with one of the proteins.

Deciphering the subtleties of protein interaction dynamics *in planta* often requires multistep designs, for example, pairing standard assays with targeted mutagenesis of the POIs or testing candidate bridge proteins. In recent years, researchers have attempted to modify pairwise interaction assays to incorporate bridge proteins. For example, an adaptation of BiFC demonstrated that the interaction between *Arabidopsis thaliana* (Arabidopsis) CONSTITUTIVE PHOTOMORPHOGENIC 1 (COP1) and DELLA proteins is mediated by SUPPRESSOR OF PHYA-105 1 (SPA1) [16]. In this approach, co-expression of mRFP-SPA1 as a bridge protein was necessary to observe YFP fluorescence in the BiFC assay, where COP1 and DELLAs were fused to the N- and C-terminal halves of YFP, respectively. However, this technique has limited applicability for ternary complex formation due to inherent challenges such as the irreversible reconstitution of YFP halves in BiFC [17] and the differences in bridge protein functions. Similarly, three-way protein interactions have been standardized using a modified FRET-FLIM approach [6], which is promising

[5, 18] but requires precise protein labelling and access to advanced, often expensive instrumentation, making it technically demanding.

Although yeast-based interaction assays, such as yeast two-hybrid (Y2H) and three-hybrid (Y3H) systems, are *ex planta* methods, they are relatively easy to implement using basic laboratory equipment and standard molecular biology techniques. These systems offer distinct advantages, particularly the absence of plant-specific proteins and signalling pathways that could otherwise interfere with the proteins of interest (POIs). Additionally, yeast systems are suitable for high-throughput assays such as screening many "prey" proteins for potential interactors using a "bait" protein. A typical Y3H assay involving three proteins can provide preliminary insights into interaction effects and guide subsequent *in planta* validation. Y3H extends Y2H by incorporating a bridge protein together with the POIs. In the GAL4-based Y2H system, POIs are fused to the modular and functionally separable domains of the *Saccharomyces cerevisiae* GAL4 transcription factor [19]. The N-terminal DNA-binding domain (BD) of GAL4 binds to upstream activation sequences (UASs) of reporter genes, while the C-terminal activation domain (AD) initiates transcription. Genes encoding the POIs are cloned and expressed in compatible strains as BD (bait) and AD (prey) fusions. When the bait and prey proteins interact, the BD and AD are brought into proximity, reconstituting the functional GAL4 transcription factor [20]. This leads to the activation of reporter genes (such as *HIS3* or *LacZ*), which can then be detected or quantified, respectively, by yeast growth on selective media (e.g., lacking histidine) and through colorimetric assays using substrates of β-galactosidase (the enzymatic product of *LacZ* expression). In the Y3H system, in addition to BD- and AD-fusions, expression of a bridge protein from a separate expression cassette is employed (for example, using the vector pBridge™, Takara Bio). This design ensures that all three proteins—the bait, prey, and bridge—are simultaneously expressed in a compatible yeast strain. Such a system also utilizes the standard reporter genes available in yeast strains, such as growth on selective media or a colorimetric approach, to detect and quantify the interactions.

In this protocol, we describe Y3H to test the effect of SPA1 in the interaction between COP1 and CRYPTOCHROME 1 (CRY1). COP1 and SPA proteins directly interact to form a functional E3 ubiquitin ligase that degrades many light-responsive transcription factors in dark conditions [21]. CRY1 is a blue light photoreceptor in Arabidopsis [22] and suppresses the activity of the COP1/SPA complex under blue light by directly interacting with both COP1 and SPA1 [6]. Co-expression of SPA1 enhanced the interaction between COP1 and CRY1, as shown in the Y3H assay [4] (Fig. 1). For this assay, the coding sequences (CDSs) of COP1 and SPA1 were cloned into the pBridge vector, serving as the bait

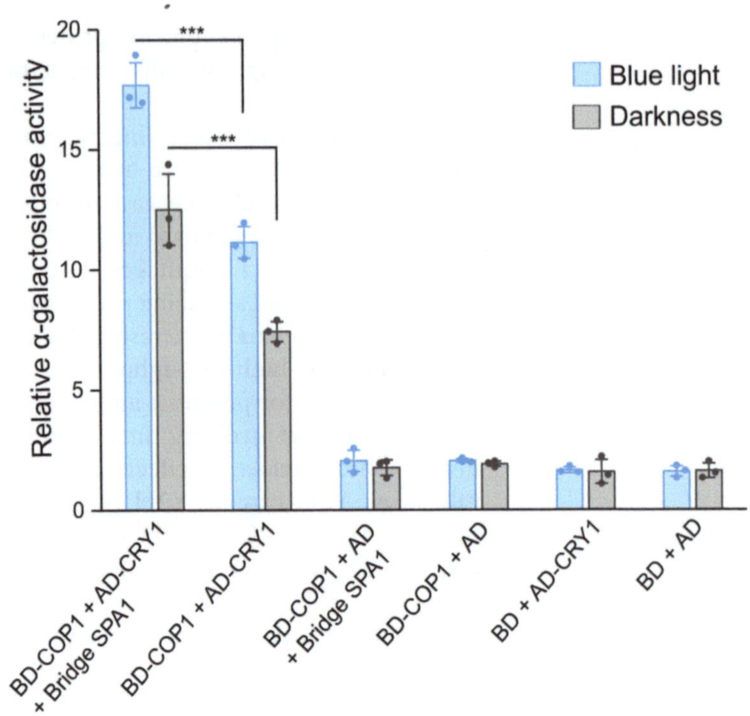

Fig. 1 Co-expression of SPA1 enhances COP1-CRY1 interaction. The yeast three-hybrid assay was employed to evaluate the interaction strength between COP1 and CRY1 in the presence or absence of co-expressed SPA1 in Y2HGold. Co-transformed yeast cells were grown in SD/-Trp/-Leu/-Met liquid media for 48 h in darkness (darkness samples) or 24 h in darkness followed by 24 h in blue light (50 μmol/m^2 s) with shaking at 30 °C. Error bars represent SEM of three biological replicates. Individual data points were displayed superimposed on the bars. Asterisks (***) show significant differences between indicated combinations at $p < 0.001$, as determined by Student's t-test

(BD) and bridge protein, respectively, using the two available multiple cloning sites (MCSs) in pBridge. The CDS of CRY1 was cloned into the pACT2 AD vector as the prey (AD). Since the expression of the bridge protein (SPA1) in pBridge is driven by a methionine-repressible promoter (*MET25*, also known as MET17 [23]), selection was performed on synthetic dropout (synthetically defined; SD) media lacking methionine (-Met) to ensure bridge protein expression. However, we observed that the methionine-repressible promoter in pBridge exhibited leaky expression in the Y190 yeast strain. To account for this, we included an empty vector control in our previous study [4], pBridge expressing BD-COP1 alone, without any bridge protein expression from the second cassette. However, as promoter leakiness could be strain-dependent [24], we found that the *MET25* promoter in pBridge was comparatively tightly regulated in the yeast strain Y2HGold (Takara Bio). Therefore, in this protocol, we demonstrate the Y3H in Y2HGold utilizing the *MEL1* reporter-mediated α-galactosidase assay.

2 Materials

2.1 Common Laboratory Supplies

1. Centrifuge tubes of various sizes (50 mL, 15 mL, and standard microcentrifuge tubes, 1.5 mL, and 2 mL) and the compatible centrifuges.
2. Pipettes that can cover volumes from 2 μL to 1000 μL (e.g., P10, P200, P1000), along with sterile, matching tips.
3. Multidispenser pipettes, useful for parallel processing and synchronizing the start of α-galactosidase assay across combinations.
4. Freezers and refrigerators for short- and long-term storage: − 80 °C for yeast strain, −20 °C for plasmids, 4 °C for plates and reagents.
5. Sterile inoculation loops and sterile toothpicks for colony picking and inoculation; sterile glass beads for spreading the transformed yeast cells.
6. Standard labware for media preparation, such as beakers, measuring cylinders, conical flasks, glass bottles, petri dishes, etc. of various sizes, analytical balances, and pH meter.
7. Incubators with temperature control and shaking capability for yeast culture growth.
8. 96-well half-area flat-bottom plates (e.g., Greiner Bio-One or equivalent) and a 96-well plate reader (e.g., Tecan Spark™) for absorbance measurements.
9. Computer with statistical and plotting software, such as Microsoft Excel® or Origin®Pro, for data analysis and visualization.

2.2 Yeast Strain

1. *Saccharomyces cerevisiae* strain Y2HGold (Takara Bio, a part of Matchmaker® Gold Yeast Two-Hybrid System; cat. no. 630489) has the genotype: *MATα*, *trp1–901*, *leu2–3, 112*, *ura3–52*, *his3–200*, *gal4Δ*, *gal80Δ*, *LYS2::GAL1*$_{UAS}$*-GAL1*$_{TATA}$*-HIS3*, *GAL2*$_{UAS}$*-GAL2*$_{TATA}$*-ADE2*, *URA3::MEL1*$_{UAS}$*-MEL1*$_{TATA}$*-AUR1-C*, *MEL1* (*see* **Note 1**).

2.3 Cloning Vectors

1. pBridge: GAL4 BD fusion vector capable of expressing a bridge protein in a separate cassette (*see* **Note 2**).
2. pACT2 AD: GAL4 AD fusion vector (*see* **Note 3**).

2.4 Solutions and Reagents for Yeast Growth and Selection

1. Glucose, 40% (w/v): Add 400 mL of distilled water to a glass beaker and heat with a magnetic stirrer on a magnetic heating block. Add 200 g of glucose (Sigma Aldrich, cat. no. 49159) and switch on the block to start the magnet stirring. Continue until the glucose is completely dissolved. Make up the final volume to 500 mL with distilled water. Dispense into a glass bottle and filter-sterilize using a 0.22 μm filter (Carl Roth; cat.

no. P666.1). The solution can be stored up to 3 months at room temperature.

2. YAPD broth (1 L): In a glass beaker, mix 20 g of Bacto-Peptone (Thermo Fisher Scientific; cat. no. 211677), 20 g of Bacto yeast extract (Thermo Fisher Scientific; cat. no. 212750), 100 mg of Adenine hemi sulfate (Sigma-Aldrich; cat. no. A9126) and 900 mL of distilled water. Stir to dissolve and adjust the pH to 5.8 using KOH. Make up the volume to 950 mL using distilled water, pour into a 1-L glass bottle and autoclave. Add 50 mL of sterile 40% (w/v) glucose to obtain a final volume of 1 L.

3. SD/-Trp/-Leu/-Met solid medium (1 L): In a glass beaker, combine 6.7 g of Yeast nitrogen base without amino acids (Sigma-Aldrich, cat. no. Y0626), 40 mg of Adenine hemi sulfate (Sigma-Aldrich; cat. no. A9126), 0.7 g of Yeast synthetic drop-out supplement (SD/-Trp/-Leu/-Met/-His; Takara Bio; Cat. no. 6304294), 20 mg of L-histidine (Carl-Roth; cat. no. 1697.1) (*see* **Note 4**) and 900 mL of distilled water. Dissolve the ingredients by stirring. Adjust the pH to 5.8 using KOH. Make up the total volume to 950 mL with distilled water and dispense into a 1-L glass bottle with a magnetic stirrer. Add 18 g of Agar-Agar Kobe I (Carl Roth; cat. no. 5210.2). Autoclave and add 50 mL of sterile 40% (w/v) glucose while the media is still lukewarm and not solidified. Stir well to mix and pour the warm medium into sterile plates (approximately 25 mL per 92-mm-diameter round plate). Let the medium solidify and then store the plates at 4 °C for up to 2 weeks.

4. SD/-Trp/-Leu/-Met liquid medium: Follow the above step for preparing the solid medium, but avoid adding Agar-Agar. Store the liquid medium at 4 °C.

5. SD/-Met medium: The SD/-Met solid and liquid media can be prepared following the above steps for making SD/-Trp/-Leu/-Met media, with the addition of 60 mg leucine (Sigma-Aldrich; cat. no. L8912) and 40 mg tryptophan (Sigma-Aldrich; cat. no. T8941) (*see* **Note 4**).

6. Lithium acetate (LiAc), 1 M: In a glass beaker, combine 10.2 g Lithium acetate dihydrate (Sigma Aldrich, cat. no. 6108-17-4) and 75 mL of distilled water. Stir to dissolve and make up the volume to 100 mL using distilled water. Dispense into a glass bottle and autoclave. The solution can be stored at 4 °C for up to 6 months.

7. Polyethylene glycol (PEG), 40% (w/v): Combine 10 g of PEG (MW 3350; Sigma, cat. no. 202444) and 20 mL of distilled water by continuous stirring and autoclaving. The remaining PEG after stirring will dissolve upon autoclaving.

8. Tris–HCl, 1 M, pH 7.5: Combine 9.1 g of Tris base (VWR, cat. no. 103156X) and 75 mL of distilled water by constant stirring. Adjust the pH to 7.5 using HCl and make up the total volume to 100 mL using distilled water. Autoclave and store at room temperature for up to 1 year.
9. 10× TE buffer, pH 7.5: Combine 10 mL of 1 M Tris–HCl, pH 7.5 (100 mM final concentration), 0.37 g of Disodium EDTA (10 mM final concentration; Carl Roth, cat. no. 8043.2), in 80 mL of distilled water. Dissolve completely by stirring and adjust the pH to 7.5 using HCl. Make up the final volume to 100 mL using distilled water, dispense into a glass bottle and store up to 1 year at room temperature.
10. 1× TE solution (10 mL): Mix 1 mL of 10× TE buffer with 9 mL of distilled water.
11. 1× TE/LiAc solution (10 mL): Mix 1 mL of 10× TE buffer, 1 mL of 1 M LiAc solution, and 8 mL of distilled water
12. TE/LiAc/PEG solution (10 mL): Mix 1 mL of 10× TE buffer, 1 mL of 1 M LiAc and 8 mL of 40% PEG solution.
13. 0.33 M sodium acetate solution, pH 4.5 (100 mL): Combine 2.71 g of sodium acetate (Sigma-Aldrich; cat. no. S8750) and 80 mL of distilled water. Stir well to dissolve. Adjust the pH to 4.5 using glacial acetic acid. Make up the final volume to 100 mL using distilled water and store at room temperature.
14. α-Galactosidase assay buffer (10 mL): In a 15 mL centrifuge tube, add 99.4 mg of p-nitrophenyl-α-D-galactopyranoside (PNPG; Sigma-Aldrich; cat. no. N0877) and 9 mL of 0.33 M sodium acetate solution. Mix by inverting multiple times or by using a rotary shaker. Finally, make up the volume to 10 mL using 0.33 M sodium acetate solution to obtain a concentration of 33 mM PNPG.
15. 2 M Sodium carbonate (Na_2CO_3) solution: In a glass bottle, combine 21.2 g of Na_2CO_3 (Carl Roth, cat. no. P028) and 90 mL of distilled water by stirring. Make up the final volume to 100 mL using distilled water and store at room temperature.

3 Methods

3.1 Cloning the Required Constructs

To generate constructs for Y3H assays, the CloneExpress II One Step Cloning™ kit (Vazyme Biotech, China) is routinely used, following the manufacturer's instructions. It is a seamless cloning method analogous to Gibson Assembly [25]. Below are the general steps used to clone any gene of interest into Y3H vectors.

1. PCR Amplification: Coding sequences (CDSs) of the genes of interest using primers designed according to Gibson Assembly

principles. Primer design can be performed using SnapGene™ software (Dotmatics, USA) (*see* **Note 5**).

2. Cloning as BD (bait) in pBridge: Insert the CDS encoding the bait protein into MCS I of the pBridge vector (*see* **Note 2**).
3. Cloning as Bridge Protein: Clone the CDS of the bridging protein into MCS II of the pBridge vector. This third protein is intended to modulate or facilitate the interaction between bait and prey in the Y3H system (*see* **Note 2**).
4. Cloning as AD (prey) in pACT2 AD: Insert the CDS encoding the prey protein into the pACT2 AD vector, which provides the AD-fusion (*see* **Note 3**).
5. Plasmid Preparation: Prepare midipreps of the resulting plasmids using the NucleoBond Xtra™ Midi Kit (Macherey-Nagel, Germany), following the manufacturer's protocol (*see* **Note 6**).

3.2 Yeast Transformation

Yeast transformation can be performed using the classical lithium acetate (LiAc) and polyethylene glycol (PEG) method, as well as the Frozen-EZ™ Yeast Transformation II Kit (Zymo Research, Germany) (*see* **Note 7**), which allows the preparation of competent yeast cells that can be stored at −80 °C for future use. Below, a step-by-step protocol for the LiAc/PEG method is described. All procedures should be conducted under a sterile hood to prevent contamination.

1. Streak the Y2HGold yeast strain from glycerol stock (stored at −80 °C) (*see* **Note 8**) onto a YAPD agar plate with SD medium lacking methionine (SD/-Met) using a sterile inoculation loop. Incubate at 28 °C for 5–6 days. Note that colony formation is delayed due to slower growth in SD/-Met.
2. Pick 4–5 colonies growing on the SD/-Met plate and inoculate into 30 mL YAPD broth in a sterile conical flask. Incubate overnight (up to 18 h) at 28 °C with shaking at 200 rpm until OD_{600} exceeds 1.5. This will serve as the preculture.
3. For the main culture, inoculate ~20 mL of the preculture into 300 mL YAPD broth in a sterile conical flask to achieve an initial OD_{600} of 0.2–0.3.
4. Incubate the main culture at 30 °C with shaking at 220 rpm for 1–1.5 h or until OD_{600} reaches 0.5–0.6. Avoid exceeding OD_{600} of 0.6 (*see* **Note 12**). During the incubation time, complete **steps 5–7**.
5. Denature salmon sperm DNA (10 mg/mL in sterile distilled water) (*see* **Note 9**) by heating at 95 °C for 5 min, followed by rapid cooling on ice. Repeat this cycle three times to ensure complete denaturation.

6. In a 1.5 mL microcentrifuge tube, mix 1–1.5 μg each of the pBridge construct (containing BD and bridge fusions) and the pACT2 AD construct (containing AD fusion). Include appropriate control combinations in separate tubes (Fig. 1).
7. Prepare 1× TE/LiAc and TE/LiAc/PEG solutions freshly by diluting from the stocks and mixing the additional components.
8. Aliquot the main culture into six 50 mL centrifuge tubes. Centrifuge at 1000 × *g* for 5 min at room temperature. Discard the supernatant.
9. Resuspend each pellet in 3 mL of 1× TE buffer. Gently mix by inverting (avoid vortexing). Pool all fractions into one 50 mL tube, centrifuge again at 1000 × *g* for 5 min, and discard the supernatant.
10. Resuspend the pellet in 20 mL of 1× TE/LiAc buffer by gentle pipetting. Centrifuge at 1000 × *g* for 5 min. Discard the supernatant and resuspend the final pellet in 1.5 mL of 1× TE/LiAc buffer. These are the competent cells, which should be used within 1–1.5 h and are not suitable for long-term storage (*see* **Note 10**).
11. Add 5 μL of denatured salmon sperm DNA to each plasmid DNA tube and mix by vortexing. Add 50 μL of competent yeast cells to each tube and vortex for 10 seconds. Add 300 μL of TE/LiAc/PEG solution to each tube and vortex for 10 seconds until fully homogenized.
12. Incubate the tubes at 28 °C with shaking at 200 rpm for 30 min. While incubating, prepare a water bath at 42 °C.
13. Take out the tubes from the incubator and add 35 μL of DMSO to each tube, and gently invert 2–3 times to mix (do not vortex). Incubate the tubes in the 42 °C water bath for 15 min.
14. Immediately transfer the tubes to ice and incubate for 1–2 min.
15. Centrifuge the tubes at 14,000 × *g* for 10 seconds at room temperature. Carefully remove the supernatant under a sterile hood. Use an additional pipetting step if necessary to remove the remaining supernatant.
16. Resuspend the pellet in 300 μL of 1×TE buffer by gentle pipetting.
17. Spread ~50 μL of the transformed yeast cells onto SD/-Trp/-Leu/-Met selection media using sterile glass beads. Seal the plates with parafilm and incubate at 28 °C for 5–6 days until colonies appear. This step selects for yeast cells transformed with both pBridge and pACT2 AD plasmids.

3.3 Quantitative α-Galactosidase Assay

The yeast strain Y2HGold carries *MEL1* reporter [26], which produces α-galactosidase upon bait-prey interaction, which can be quantified using PNPG (p-nitrophenyl alpha-D-galactopyranoside) (*see* **Note 11**).

1. From SD/-Trp/-Leu/-Met selection plates containing transgenic yeast colonies, pick ten individual colonies per combination using sterile toothpicks or autoclaved 200 μL tips.
2. Resuspend all ten colonies per combination in 1 mL of SD/-Trp/-Leu/-Met liquid medium in 1.5 mL microcentrifuge tubes. Transfer the suspension into an autoclaved 5 mL tube containing an additional 1 mL of SD medium, resulting in a total volume of 2 mL per culture.
3. Ensure that three biological replicates are prepared for each combination. Include all relevant control combinations, such as empty BD, empty bridge, and empty AD, along with their respective pairings (*see* Fig. 1). Incubate the dark samples at 30 °C with shaking at 200 rpm for 48 h in a light-proof shaker. For blue light samples, first incubate in darkness for 24 h and then in blue light (50 μmol m^{-2} s^{-1}) for 24 h at 30 °C with shaking.
4. On the following day, measure and adjust the OD_{600} of each culture to 1.0 in a final volume of 1 mL. Prepare two technical replicates for each biological replicate.
5. Centrifuge the cultures at 14,000 × *g* for 10 seconds at room temperature. Collect 200 μL of the supernatant each and mix with 600 μL of α-galactosidase assay buffer using a multidispenser pipette (*see* **Note 13**). Vortex vigorously for 3 seconds and incubate at 30 °C for 12–24 h, or until the yellow color of p-nitrophenol becomes visible (*see* **Note 14**).
6. Stop the assay by adding 200 μL of 2 M Na_2CO_3 to each tube, using a multidispenser pipette. Mix by vortexing for a few seconds.
7. Aliquot 200 μL of the reaction mixture into each well of a 96-well Greiner half-area flat-bottom plate. Make sure to add three replicates of blank control consisting of SD/-Trp/-Leu/-Met liquid medium without yeast cells, in three of the wells.
8. Measure the absorbance at 410 nm (*see* **Note 15**) using a Tecan Spark® 96-well plate reader. Export the data to Microsoft Excel®. The α-galactosidase activity can be calculated by an arbitrary formula:

   α-gal. Units = $((A_{410,Sample} - A_{410,Blank}) \times V_{Culture})/t) \times K)$, where.

   A_{410} = Absorbance at 410 nm.

$V_{Culture}$ = Volume of the culture (2 mL).

t = Time in min.

K = 10,000 (an arbitrary scaling factor)

9. Subtract the average blank value from all sample readings. Calculate the mean of each biological replicate from the respective technical replicates, followed by the overall average across biological replicates for each combination. Perform statistical analysis using Excel® or other dedicated software such as Origin®Pro, and visualize the results in graphical format.

4 Notes

1. The Y2HGold yeast (*Saccharomyces cerevisiae*) strain is derived from PJ69-2A and incorporates the *MEL1* reporter gene, along with *HIS3* and *ADE2*. For detailed information, refer to the Matchmaker® Gold Yeast Two-Hybrid System user manual (Cat. No. 630489, Takara Bio). In our hands, the Y2HGold strain tends to lose the capacity to survive in SD/-Met if maintained in YAPD without selection. Therefore, Y2HGold colonies should always be verified for growth in SD/-Met before using them for competent cells preparation.
2. The pBridge vector (Takara Bio, Cat. No. 630404) is a shuttle vector featuring two multiple cloning sites (MCS I and MCS II). A downloadable map is available at https://www.takarabio.com/services-and-support/vector-information/vector-document-finder?cat1=21038. The bait protein can be cloned into MCS I using the restriction sites EcoRI, SmaI, SalI, BamHI, and PstI. For bridge protein insertion, MCS II provides NotI and BglII sites. The vector carries an ampicillin resistance marker for selection in *E. coli* and a *TRP1* marker for yeast selection. Expression of the bridge protein is regulated by the *MET25* promoter, which is repressed in the presence of methionine. Therefore, to induce bridge protein expression, SD/-Met medium should be used.
3. The pACT2 AD (Takara Bio; GenBank Accession No. U29899) is a shuttle vector designed for the expression of prey fusion proteins as AD. It carries an ampicillin resistance gene (*bla*) for selection in *E. coli* and a *LEU2* marker for selection in yeast. The MCS of pACT2 includes several restriction enzyme recognition sites to facilitate convenient cloning.
4. To prepare SD/-Trp/-Leu/-Met medium, we routinely use SD/-Trp/-Leu/-Met/-His (Takara Bio; Cat. No. 6304294) and supplement it with L-histidine (Carl Roth; Cat. No. 1697.1). Alternatively, ready-to-use SD/-Trp/-Leu/-Met medium is available from other vendors. Similarly, for

SD/-Met medium, we add leucine (Sigma-Aldrich; Cat. No. L8912) and tryptophan (Sigma-Aldrich; Cat. No. T8941) in addition to L-histidine. However, pre-formulated SD/-Met media can also be sourced from suppliers other than Takara Bio.

5. When designing primers for seamless cloning, we use the default parameters for Gibson Assembly as provided in the SnapGene® software. These include a primer overlap of 15–20 base pairs, a total primer length of 18–25 base pairs, and a melting temperature of approximately 60 °C at the target-specific region. The GC content is maintained between 40% and 60%, and care is taken to avoid secondary structures.
6. Although minipreps typically yield sufficient plasmid DNA for yeast transformation, we routinely perform midipreps in the context of Y3H assays. This is due to the comparatively large size of the pBridge vector following the insertion of both BD and bridge sequences. In our experience, midipreps consistently provide higher DNA yields and improved reliability in generating transformed yeast colonies compared to minipreps in Y3H.
7. Yeast transformation can be performed using −80 °C frozen competent cells prepared with the Frozen-EZ™ Yeast Transformation II Kit (Zymo Research, Germany). The manufacturer's protocol for preparing competent cells is available on the Zymo Research website. In addition to the standard procedure, we also employ a slightly modified protocol that has been previously published in detail [27].
8. For long-term storage of yeast strains as glycerol stocks, we follow a protocol adapted from the Yeast Protocol Handbook (Clontech). After selecting the desired strain on appropriate selection media, a single colony is picked and resuspended in 500 μL of YAPD liquid medium in a 1.5 mL screw-cap microcentrifuge tube. The suspension is vortexed to disperse the cells thoroughly. Subsequently, 500 μL of 50% (v/v) glycerol is added to the tube. The cap is tightly closed, the contents are mixed by repeated inversion, and the tube is properly labelled before storage at −80 °C.
9. Salmon sperm DNA serves as carrier DNA and is believed to enhance the efficiency of plasmid uptake during yeast transformation. In routine Y3H transformations, we typically achieve sufficient transformation efficiency without the use of carrier DNA. However, if salmon sperm DNA is desired, it is essential to store the stock solution at −20 °C and to dispense aliquots into the working solution under a sterile hood. This precaution minimizes the risk of bacterial contamination of yeast colonies on selection plates.

10. All solutions required for yeast transformation should be freshly prepared from stock solutions.
11. For long-term storage and consistent activity, PNPG should be stored at −20 °C. When required, the necessary quantity should be weighed promptly and the stock bottle returned to −20 °C to minimize degradation over time. It is crucial to prepare the PNPG solution freshly just before performing the quantitative assay. Given that PNPG is an organic compound with limited solubility in water, repeated inversion or vigorous shaking is recommended for complete dissolution.
12. When preparing the main culture for yeast competent cell generation, it is critical to ensure that the OD_{600} does not exceed 0.6, as transformation efficiency declines significantly beyond this threshold.
13. Using multi-dispenser pipettes not only accelerates workflow but also ensures synchronized sample processing. For example, when initiating an analysis, reagents can be dispensed across all combinations within a short span, thereby enhancing accuracy and minimizing errors.
14. The intensity of yellow color formation in the quantitative assay depends on several factors, including the incubation time, the concentration of PNPG added to the reaction mixture, and the extent of molecular interactions being measured. Therefore, the incubation period must be optimized for each specific assay. In the protocol described here, an incubation time ranging from 12 to 24 h was found to be optimal for consistent color development.
15. PNPG is a chromogenic substrate for α-galactosidase, the enzyme encoded by the *MEL1* gene [26], which is typically secreted into the growth medium by yeast cells. α-Galactosidase cleaves PNPG to release p-nitrophenol, a yellow-colored compound, and D-Galactose. The amount of p-nitrophenol produced is directly proportional to the enzymatic activity and thus reflects the extent of molecular interactions in the assay. Under alkaline conditions (when 2 M Na_2CO_3 is added), p-nitrophenol is deprotonated and exhibits strong absorbance at 410 nm [28]. This allows for reliable spectrophotometric quantification of α-galactosidase activity.

Acknowledgments

J.P. gratefully acknowledges Peter Nick for hosting his laboratory at the Department of Molecular Cell Biology, JKIP, Karlsruhe Institute of Technology (KIT).

References

1. Schreiber G (2020) Protein–protein interaction interfaces and their functional implications. https://doi.org/10.1039/9781788016544-00001
2. Cuadrado AF, Van Damme D (2024) Unlocking protein–protein interactions in plants: a comprehensive review of established and emerging techniques. J Exp Bot 75:5220–5236
3. Kastritis PL, Bonvin AMJJ (2013) On the binding affinity of macromolecular interactions: daring to ask why proteins interact. J R Soc Interface 10:20120835
4. Holtkotte X, Ponnu J, Ahmad M et al (2017) The blue light-induced interaction of cryptochrome 1 with COP1 requires SPA proteins during Arabidopsis light signaling. PLoS Genet 13:e1007044
5. Trimborn L, Kuttig F, Ponnu J et al (2025) Cryptochrome 1 promotes photomorphogenesis in Arabidopsis by displacing substrates from the COP1 ubiquitin ligase. Plant J 121: e70071
6. Ponnu J, Riedel T, Penner E et al (2019) Cryptochrome 2 competes with COP1 substrates to repress COP1 ubiquitin ligase activity during *Arabidopsis* photomorphogenesis. PNAS 116: 27133–27141
7. Ponnu J (2020) Molecular mechanisms suppressing COP1/SPA E3 ubiquitin ligase activity in blue light. Physiol Plant 169:418–429
8. Ren H, Ou Q, Pu Q et al (2024) Comprehensive review on bimolecular fluorescence complementation and its application in deciphering protein–protein interactions in cell signaling pathways. Biomolecules 14:859
9. De Los Santos C, Chang C-W, Mycek M-A et al (2015) FRAP, FLIM, and FRET: detection and analysis of cellular dynamics on a molecular scale using fluorescence microscopy. Mol Reprod Dev 82:587–604
10. Wang L, Yu G, Macho AP et al (2021) Split-luciferase complementation imaging assay to study protein-protein interactions in Nicotiana benthamiana. Bio Protoc 11:e4237
11. Thaminy S, Miller J, Stagljar I (2004) The split-ubiquitin membrane-based yeast two-hybrid system. In: Fu H (ed) Protein-protein interactions: methods and applications. Humana Press, Totowa, NJ, pp 297–312
12. Mair A, Xu S-L, Branon TC et al (2019) Proximity labeling of protein complexes and cell-type-specific organellar proteomes in Arabidopsis enabled by TurboID. eLife 8:e47864
13. Lin J-S, Lai E-M (2017) Protein-protein interactions: co-immunoprecipitation. Methods Mol Biol 1615:211–219
14. Yugandhar K, Gupta S, Yu H (2019) Inferring protein-protein interaction networks from mass spectrometry-based proteomic approaches: a mini-review. Comput Struct Biotechnol J 17:805–811
15. Lee B-D, Kim MR, Kang M-Y et al (2017) The F-box protein FKF1 inhibits dimerization of COP1 in the control of photoperiodic flowering. Nat Commun 8:2259
16. Blanco-Touriñán N, Legris M, Minguet EG et al (2020) COP1 destabilizes DELLA proteins in Arabidopsis. Proc Natl Acad Sci USA 117:13792–13799
17. Kudla J, Bock R (2016) Lighting the way to protein-protein interactions: recommendations on best practices for bimolecular fluorescence complementation analyses. Plant Cell 28: 1002–1008
18. Cañibano E, Bourbousse C, García-León M et al (2021) DET1-mediated COP1 regulation avoids HY5 activity over second-site gene targets to tune plant photomorphogenesis. Mol Plant 14:963–982
19. R B', Ptashne M (1985) A eukaryotic transcriptional activator bearing the DNA specificity of a prokaryotic repressor. Cell 43:729–736
20. Fields S, Song O (1989) A novel genetic system to detect protein–protein interactions. Nature 340:245–246
21. Hoecker U (2017) The activities of the E3 ubiquitin ligase COP1/SPA, a key repressor in light signaling. Curr Opin Plant Biol 37: 63–69
22. Ponnu J, Hoecker U (2022) Signaling mechanisms by Arabidopsis cryptochromes 13
23. D'Andrea R, Surdin-Kerjan Y, Pure G et al (1987) Molecular genetics of met 17 and met 25 mutants of Saccharomyces cerevisiae: intragenic complementation between mutations of a single structural gene. Mol Gen Genet 207: 165–170
24. Tominaga M, Shima Y, Nozaki K et al (2024) Designing strong inducible synthetic promoters in yeasts. Nat Commun 15:10653
25. Gibson DG, Young L, Chuang R-Y et al (2009) Enzymatic assembly of DNA molecules up to several hundred kilobases. Nat Methods 6: 343–345
26. Aho S, Arffman A, Pummi T et al (1997) A novel reporter Gene*MEL1*for the yeast two-hybrid system. Anal Biochem 253:270–272

27. Trimborn L, Hoecker U, Ponnu J (2022) A simple quantitative assay for measuring β-galactosidase activity using X-gal in yeast-based interaction analyses. Curr Protoc 2:e421

28. Syedd-León R, Sandoval-Barrantes M, Trimiño-Vásquez H et al (2020) Revisiting the fundamentals of p-nitrophenol analysis for its application in the quantification of lipases activity. A graphical update. Uniciencia 34:31–43

Chapter 17

Simplified In Vitro Assay for Testing Chemicals with Auxin Activity Using Direct Protein-Protein Interaction

Heshini N. Weerakkody, Ikechukwu E. Opara, Prianka Acharjya, Sunethra Dharmasiri, and Nihal Dharmasiri

Abstract

Auxin, a major phytohormone, plays a pivotal role throughout the plant life cycle. While indole 3-acetic acid (IAA) is the major natural auxin found in plants, there are other natural and many synthetic chemicals with auxin activity. Ability to test the auxin activity in a simple cell-free assay will facilitate the identification of different chemicals with auxin activity. Moreover, the same assay can be used to identify the chemicals with antiauxin activity. Here we describe an in vitro pull-down assay that can be easily used to test auxin activity of different chemicals. Further, the concept of this assay can be used to test the activity of other hormones with similar mode of action or small ligand-based protein-protein interactions.

Key words GST-Aux/IAA, TIR1/AFBs, Pull-down, Cell-free assay, Auxin, Jasmonic acid

1 Introduction

Auxin, a critical plant hormone, facilitates diverse developmental processes by modulating the expression of many genes. At the molecular level this is achieved through the degradation of a group of transcriptional repressor proteins known as Aux/IAAs (Auxin/Indole-3-Acetic Acid) with the help of 26S proteasome and a family of F-box proteins, which comprises TIR1/AFB (Transport Inhibitor Response 1/Auxin Signaling F-Box) in Arabidopsis [1, 2]. The TIR1/AFBs are components of the SCF (SKP1/CULLIN/F-BOX) complex, which is involved in ubiquitin mediated protein degradation [3]. In response to high level of auxin, Aux/IAAs are degraded through the ubiquitin proteasome pathway involving the $SCF^{TIR1/AFBs}$. In this process, Aux/IAA repressor proteins along with.TIR1/AFBs, form a co-receptor complex, which is central to auxin signaling in plants [4, 5]. Auxin directly interacts with Aux/IAAs and TIR1/AFBs promoting the direct interaction between these two groups of proteins [6].

Shahid Mukhtar (ed.), *Plant Hormones: Methods and Protocols*, Methods in Molecular Biology, vol. 3026,
https://doi.org/10.1007/978-1-0716-5214-5_17,

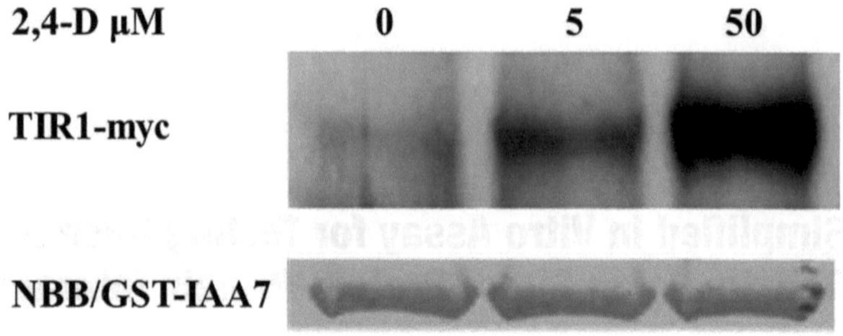

Fig. 1 Interaction of GST-IAA7 with TIR1-myc in response to different 2,4-D (synthetic auxin) concentrations. Bacterially expressed GST-IAA7 was affinity purified with glutathione-agarose beads and incubated with crude protein extracts from *35S::TIR1-myc* Arabidopsis seedlings in the presence of different concentrations of 2,4-D. GST-IAA7 interacting proteins were recovered by centrifugation, separated by SDS-PAGE and subsequently analyzed by Western blotting. TIR1-myc protein was detected by anti-myc primary antibody and HRP conjugated anti-mouse secondary antibody. Signal was detected using BioRad (Clarity™)—ECL kit. The upper panel shows the Western blot results while the lower panel shows the naphthalene blue-black (NBB) stained PVDF membrane to indicate the GST-IAA7 protein level used in each pull-down reaction. Results indicate that the interaction between GST-IAA7 and TIR1-myc is dependent on 2,4-D concentration

In vitro pull-down assays, whose working principle is that one binding partner is immobilized on to resins or beads, which are recoverable through centrifugation allows the co-purification of the other protein partner to be assayed [7]. This technique has emerged as an important tool for investigating protein-protein interactions under controlled in vitro conditions.

Studies have demonstrated that auxin stabilizes direct interactions between TIR1 and Aux/IAA proteins without requiring stable posttranslational modifications. This has been validated through pull-down assays involving purified components and affinity-tagged proteins [8] (also *see* Fig. 1). This interaction has also been shown between AFB1, AFB2, & AFB3 and different AUX/IAA proteins with varying affinity [6] (also *see* Fig. 2).

There are several natural auxins as well as many synthetic chemicals with auxin-like activities. Some of these synthetic chemicals are being widely used as selective herbicides [9]. Availability of a simple in vitro pull-down assay coupled with Western blotting enables to test auxin agonist and antagonist chemicals in a relatively short time in any laboratory (Figs. 1, 2, and 3).

Moreover, another plant hormone, jasmonic acid, which plays an important role in plant response to environmental stress, has a similar mode of action as auxins [10]. Thus, similar to auxin, pull-down assays can be used to disclose the hormone dose-dependent interaction between JASMONATE ZIM-DOMAIN (JAZ)

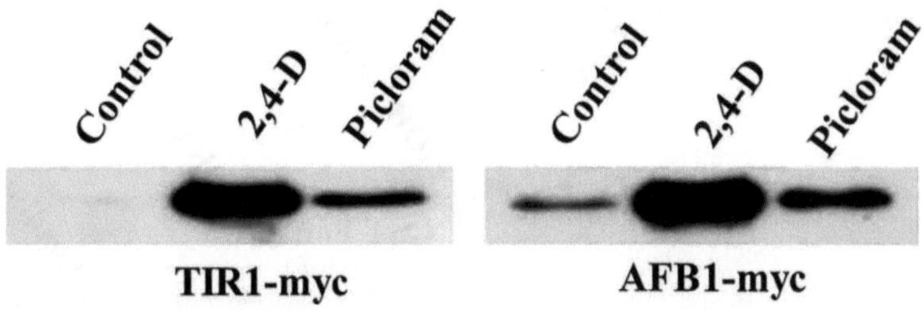

Fig. 2 Interaction of GST-IAA7 with TIR1-myc and AFB1-myc in response to 50 μM 2,4-D or 50 μM picloram. Bacterially expressed GST-IAA7 was affinity purified with glutathione-agarose and incubated with crude protein extracts from either *35S::TIR1-myc* or *35S::AFB1-myc* Arabidopsis seedlings in the presence of different synthetic auxins. GST-IAA7 interacting proteins were recovered by centrifugation, separated by SDS-PAGE and analyzed by Western blotting. TIR1-myc and AFB1-myc proteins were detected by anti-myc primary antibody and HRP conjugated anti-mouse secondary antibody. Signal was detected using BioRad (Clarity™)—ECL kit. Results indicate that 2,4-D promotes interaction between GST-IAA7 and TIR1-myc/AFB1-myc more efficiently than picloram

repressors and the F-box protein CORONATINE INSENSITIVE 1 (COI1), indicating JAZ-COI1 as a JA-Ile (Jasmonate-isoleucine) co-receptors [11]. Therefore, the concept of this technique can be used in many other studies with minor modifications.

2 Materials

2.1 Protein Expression in E. coli *and Purification*

1. Protein expression construct pGEX-4 T-3 plasmid vector containing in-frame coding sequence of Arabidopsis Aux/IAA7 cDNA (*see* **Note 1**).
2. *Escherichia coli* protein expression strain BL21(DE3) competent cells (*see* **Note 2**).
3. Glutathione-agarose resin.
4. Isopropyl β-D-1-thiogalactopyranoside (IPTG) (500 mM stock solution).
5. Lauria Broth (LB) liquid media.
6. Ampicillin or Carbenicillin (1 mg/mL stock solution).
7. Phosphate buffered saline (PBS) (pH 7.4): 137 mM NaCl, 2.7 mM NaH_2PO_4, 2 mM KH_2PO_4.
8. Phenyl methylsulfonyl fluoride (PMSF) at a final concentration of 1 mM.
9. Tween 20 at a final concentration of 0.1% (v/v).

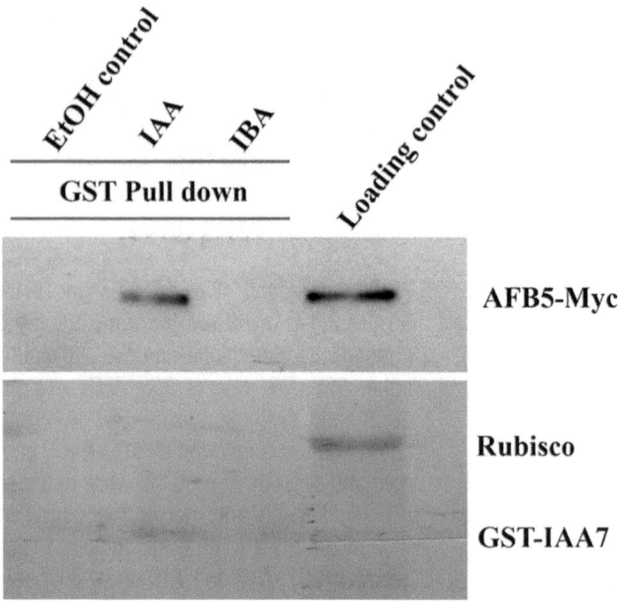

Fig. 3 Interaction of GST-IAA7 with AFB5-myc in response to natural auxins, IAA and IBA. Bacterially expressed GST-IAA7 was affinity purified using glutathione-agarose and incubated with crude protein extracts from *35S::AFB5-myc* Arabidopsis seedlings in the presence of 50 μM IAA or 50 μM IBA. GST-IAA7 interacting proteins were recovered by centrifugation, resolved by SDS-PAGE and analayzed by Western blotting. AFB5-myc protein was detected using anti-myc primary antibody and HRP conjugated anti-mouse secondary antibody. Signal was detected using BioRad (Clarity™)—ECL kit. The upper panel shows the Western blot results while the lower panel shows the naphthalene blue-black stained PVDF membrane to indicate the GST-IAA7 protein level used in each pull-down reaction. Results indicate that while IAA promotes the interaction between GST-IAA7 and AFB5-myc, IBA does not behave as a regular auxin

2.2 Extraction of Total Plant Protein

1. 10- to 12-day-old seedlings of transgenic Arabidopsis line expressing *35S::TIR1-myc* (*see* **Note 3**).
2. Liquid nitrogen.
3. Tris–HCl extraction buffer: 50 mM Tris–HCl buffer (pH 7.5), 150 mM KCl, 10% glycerol (v/v), 0.1% Tween-20 (v/v), 1 mM PMSF, 10 μM MG132, and protease inhibitor cocktail (Roche) (Alternatively, 50 mM Tris–HCl can be replaced with 50 mM HEPES buffer (pH 7.5) (*see* **Note 4**).

2.3 In Vitro GST Pull-Down

1. Affinity purified GST-IAA7 using glutathione-agarose beads.
2. Tris–HCl wash buffer: [50 mM Tris–HCl buffer (pH 7.5), 100 mM KCl, 10% glycerol (v/v), 0.1% Tween-20 (v/v)].

3. Indole-3-acetic acid (IAA) at a final concentration of 50 μM or any other auxin at a desired concentration.

2.4 Detection of Interacting Proteins

1. 2× Laemmli Sample Buffer (2× LSB): 100 mM Tris–HCl (pH 6.8), 4% SDS (w/v), 20% Glycerol (v/v), 10% β-mercaptoethanol (v/v), 0.1% Bromophenol blue (w/v).
2. 10–12% Sodium dodecyl sulfate–polyacrylamide gel electrophoresis (SDS-PAGE) gel.
3. SDS running buffer: 25 mM Tris–HCl 192 mM glycine, 0.1% SDS (w/v).
4. Vertical protein mini gel electrophoresis apparatus.
5. Wet blotting protein transfer apparatus.
6. Polyvinylidene fluoride (PVDF) protein transfer membrane.
7. Transfer buffer (pH 8.3): 25 mM Tris–HCl 192 mM glycine, 20% methanol (v/v).
8. TBST: [Tris buffered saline with Tween-20: 50 mM Tris–HCl (pH 8.0), 150 mM NaCl, 0.1% Tween-20 (v/v)].
9. Anti-Myc antibody and corresponding HRP conjugated secondary antibody.

3 Methods

3.1 Protein Expression in E. coli and Purification of GST-IAA7

1. Clone the IAA7 (Aux/IAA) cDNA in frame with pGEX-4 T-3 plasmid vector, thus GST-IAA7 recombinant protein can be expressed in *E. coli*. Transform the protein expression vector (GST-IAA7) into *E. coli* BL21(DE3) strain. Select a single colony of transformed bacteria by spreading the transformed competent cells on LB + carbenicillin (100 μg/mL) solid media (*see* **Note 5**).
2. Inoculate the selected GST-IAA7 (Aux/IAA) clone into 3–4 mL liquid LB medium with 100 μg/mL carbenicillin (or ampicillin) and grow overnight in a shaking incubator at 37 °C.
3. Add the overnight GST-IAA7 culture into 100 mL liquid LB containing 100 μg/mL Carbenicillin (or ampicillin) and grow at 37 °C in a incubator shaker until the OD_{600} reaches approximately 0.6.
4. Add IPTG to the culture medium to a final concentration of 1 mM and incubate the bacterial culture for further 4 h at 30 °C in an incubator shaker to induce expression of GST-IAA7 recombinant protein.
5. Pellet the bacterial cells by centrifugation at 10,000 × *g* for 10 min at 4 °C and remove the supernatant (*see* **Note 6**).

6. Resuspend the bacterial pellet in 7 mL of 1× PBS in a 15 ml Falcon tube on ice and lyse the cells by sonication (*see* **Note 7**).
7. Add PMSF and Tween-20 to a final concentration of 1 mM and 0.1% (v/v) respectively soon after sonication and incubate the lysate on a rocker at 4 °C for 10 min.
8. Remove the bacterial cell debris by centrifugation at 10,000 × *g* at 4 °C for 10 min. Transfer the supernatant to a new tube and incubate with 30 μL of pre-hydrated glutathione-agarose beads overnight on a rocker at 4 °C to affinity purify GST-IAA7 protein.
9. Centrifuge the supernatant at 1000 × *g* for 1 min at 4 °C. Wash glutathione-agarose beads with 10 ml of 1× PBS + 0.1% Tween-20 three times, 5 min each at 4 °C. Collect glutathione beads after each wash by centrifugation at 1000 × *g* at 4 °C.
10. Remove all the liquid and resuspend washed beads in 100–150 μL of 1× PBS. Affinity purified GST-IAA7 can be directly used for the pull-down assay or store at 4 °C for a few days if necessary (*see* **Note 8**).

3.2 Plant Protein Extraction

1. Germinate surface sterilized Arabidopsis seeds of *35S::TIR1-myc* line on nutrient agar medium for 10–12 days in vertically oriented Petri plates in a growth chamber at 21 °C with continuous light. Harvest whole seedlings and briefly rinse with water and directly hormogenize in Tris–HCl extraction buffer. A ratio of 1 g plant tissue to 5 ml of extraction buffer works best in our hands.
2. Keep the homogenates at 4 °C for 10 min on a rocker and centrifuge at 6000 × *g* for 10 min at 4 °C to remove cell debris. Recover the supernatant (crude protein extract) and estimate the total protein content using Bradford assay. Aliquot (about 350 μL each) the crude protein extract into 1.5 mL microcentrifuge tubes and directly use in the pull-down assay or store at −80°C until further use.

3.3 In Vitro GST Pull-Down Assay

1. Incubate 350–1000 μL (depending on protein concentration) of *35S::TIR1-myc* crude protein extract with 10–25 μL (depending on protein concentration) of glutathione-agarose beads previously bound with GST-IAA7 recombinant protein, by gently rocking at 4 °C for 1 h. Add IAA (or any other auxin-like chemical need to be tested) directly to the pull-down mixture. Carry out a parallel appropriate control pull-down reaction without auxin (*see* **Note 9**).
2. Spin down the glutathione-agarose beads in a microcentrifuge for 10–15 s and remove the supernatant carefully without disturbing the beads.
3. Wash the beads three times with 1000 μL of Tris–HCl extraction buffer containing IAA (or any other auxin to be tested) of the same concentration as in pull-down mixture, for 5–10 min

each. Control pull-down mixture should be washed with the extraction buffer without auxin but the same volume of the solvent (e.g. ethanol or DMSO) used to dissolve IAA or any other auxin.

4. After final spin down, completely remove the washing buffer using a flat-head tip. Resuspend the beads in 15–20 μL of 2× LSB and boil the tube for 5 min in a boiling water bath to release proteins in to the 2× LSB buffer.

3.4 Detection of Interacting Proteins

1. Load the boiled pull-down samples onto a 10–12% SDS-PAGE gel using a flat-head tip to avoid glutathione beads. Also, load a lane with protein molecular weight markers (pre-stained markers will be a better choice).
2. Resolve proteins in the SDS-PAGE gel using SDS running buffer at 90 V for 10 min, and then at 160 V for approximately 1 h or until the loading dye front reaches the bottom of the mini gel.
3. Assemble the protein transfer cassette according to the manufacturer's instructions and transfer the separated proteins from the SDS-PAGE gel on to the pre-wetted PVDF membrane using the wet transfer blotting apparatus and ice cold transfer buffer at 90 V for about 1.5 h.
4. Disassemble the transfer cassette and carefully remove the PVDF membrane and rinse it with milli-Q water. Incubate the PVDF membrane in 5% nonfat dry milk in TBST on a rocker for 1 h and then wash the membrane 3 times with TBST, 5 min each.
5. Incubate the membranes with primary antibody (Anti-Myc antibody: 1:5000 in TBST) at room temperature for 1 h by gently rocking the membrane in the antibody solution. Wash the membrane three times with TBST, 5 min each.
6. Incubate the membrane in appropriate secondary antibody that is conjugated to HRP (Anti-Mouse IgG: 1:10,000 in TBST) for 1 h at room temperature. Then wash the membrane with TBST for 15–20 min, followed by three more washes for 5 min each.
7. Flood the protein side of the PVDF membrane with Enhanced Chemiluminescence (ECL) reagent (BioRad) for 5 min according to the manufacturer's instructions. Remove the excess ECL solution from membrane and sandwich the membrane between two thin plastic sheets.
8. Expose the ECL treated membrane to sensitive X-ray film for about 3–4 min and develop the X-ray film either using an automated X-ray film developer or manually according to manufacturer's instructions (*see* **Note 10**).

4 Notes

1. While the cDNA of any canonical Aux/IAA protein with conserved domain II can be used to prepare the GST-Aux/IAA construct, we prefer to use GST-IAA7 as this Aux/IAA protein has been shown to interact with TIR1/AFBs efficiently (Figs. 1, 2, and 3). Other canonical Aux/IAAs would also interact with TIR1/AFBs but the efficiency of interaction might be different in each case.
2. We have used other *E. coli* host strains with variable success including Top10 cells. Most of the *E. coli* strains work fine for Aux/IAA expression.
3. While we frequently use *35S::TIR1-myc* transgenic Arabidopsis line, other transgenic lines of *35S::AFB1-myc*, *35S::AFB2-myc*, and *35S::AFB3-myc* or *35S::AFB4-myc/35S::AFB5-myc* transgenic lines can also be used for the assay depending on the type of auxin used.
4. 50 mM HEPES buffer can also be used in place of 50 mM Tris–HCl buffer. We use both buffers with similar results.
5. It is better to confirm the GST-IAA7 (Aux/IAA) clone using PCR before using it for protein expression.
6. Generally, recombinant Aux/IAA proteins express very well in *E. coli*. A portion of the pellet may be sufficient for several pull-down assays. Extra bacterial pellet can be stored at −80 °C for a couple of months without affecting the function.
7. Conditions (frequency and duration) for sonication should be worked out carefully. Most of the Aux/IAA proteins are cleaved easily due to mechanical damage during the sonication but generally this cleavage does not affect the auxin induced interaction with TIR1/AFBs. Nevertheless, careful sonication conditions will help to recover intact GST-Aux/IAA proteins.
8. If GST-Aux/IAA proteins are purified carefully, they can be stored at 4 °C for about a week and used in pull-down assays. However, it would be better to check the quality of the purified protein by running on a SDS-PAGE gel and staining it just before the pull-down assay. This will also help to approximately quantify the GST-Aux/IAA protein by comparison with known protein standard.
9. The Aux/IAA-TIR1-myc interaction occurs within 5–10 min, but to detect better interaction, incubation of the pull-down reaction for 1 h would be sufficient. Most of the auxins can be dissolved in ethanol or DMSO. Make sure to use concentrated stock solutions of auxins to avoid solvent effect on the pull-down reaction, and to use the same volume of the solvent in the control pull-down reaction.

10. How long the PVDF membrane should be exposed to the X-ray film will depend on the efficiency of the interaction between Aux/IAA protein and TIR1/AFBs. The efficiency of the interaction will depend on the concentration and the type of auxin, as well as the particular combination of Aux/IAA and the F-box protein used. GST-IAA7 interacts very efficiently with TIR1-myc and the interaction is very efficiently promoted by IAA compared to 2,4-D or picloram.

Acknowledgments

This research was supported by National Science Foundation Grant IOS 0845305 (to ND), Research Enhancement Grant from Texas State University (to ND and SD), One-time Grant 90000525 from Texas State University (to ND).

References

1. Gray WM et al (2001) Auxin regulates SCF^{TIR1}-dependent degradation of AUX/IAA proteins. Nature 414:271–276
2. Dharmasiri N, Estelle M (2004) Auxin signaling and regulated protein degradation. Trends Plant Sci 9(6):302–308. https://doi.org/10.1016/j.tplants.2004.04.003
3. de Roij M, Borst JW, Weijers D (2024) Protein degradation in auxin response. Plant Cell 36(9):3025–3035
4. Dharmasiri N, Dharmasiri S, Estelle M (2005) The F-box protein TIR1 is an auxin receptor. Nature 435:441–445
5. Kepinski S, Leyser O (2005) The Arabidopsis F-box protein is an auxin receptor. Nature 435: 446–451
6. Dharmasiri N, Dharmasiri S, Weijers D, Lechner E, Yamada M, Hobbie L, Eismann JS, Jurgens G, Estelle M (2005) Plant development is regulated by a family of auxin receptor F box proteins. Dev Cell 9:109–119
7. Lapetina S, Gil-Henn H (2017) A guide to simple, direct, and quantitative in vitro binding assays. J Biol Methods 4(1):e62
8. Dharmasiri N, Dharmasiri S, Jones AM, Estelle M (2003) Auxin action in a cell-free system. Curr Biol 13(16):1418–1422
9. Kathare PK, Dharmasiri S, Vincil ED, Routray P, Ahmad I, Roberts DM, Dharmasiri N (2020) Arabidopsis PIC30 encodes a Major Facilitator Superfamily (MFS) transporter responsible for the uptake of picolinate herbicides. Plant J 102:18–33
10. Wang J, Li Song L, Gong X, Xu J, Li M (2020) Functions of jasmonic acid in plant regulation and response to abiotic stress. Int J Mol Sci 21: 1446
11. Monte I, Caballero J, Zamarreño AM, Fernández-Barbero G, García-Mina JM, Solano R (2022) JAZ is essential for ligand specificity of the COI1/JAZ co-receptor. Proc Natl Acad Sci U S A 119(49):e2212155119

Chapter 18

Root Growth and Leaf Senescence Assays as Physiological Readouts of Cytokinin Response in *Arabidopsis thaliana*

Grace A. Johnston, Omar Hasannin, Risheek R. Khanna, Aaron M. Rashotte, and Cristiana T. Argueso

Abstract

Cytokinin (CK) is a plant hormone that regulates multiple aspects of growth and development throughout the plant. Here we provide protocols to examine these effects in roots and leaves. In roots, basal levels of CK, in conjunction with other hormones like auxin, are key in normative structural organization by maintaining the boundary between the meristematic zone and the elongation zone and post-transition cell wall stiffening. To quantify CK effects on root growth, we present a simple plate-based assay in which *Arabidopsis thaliana* seeds are surface-sterilized, plated on phytagel media, and grown vertically with or without CK. In this protocol, 6-benzylaminopurine (BA) is used as a representative CK (though other CKs can be substituted), and primary root length is measured after a defined growth period. Exogenous CK treatment suppresses root elongation, therefore, differences in root length between treated and nontreated roots directly reflect hormone sensitivity and signaling. In leaves, CKs act to preserve photosynthetic capacity by delaying senescence. To assess this function, this chapter presents a simple detached-leaf bioassay wherein excised Arabidopsis leaves are incubated in CK-containing solutions. CK efficacy is evaluated by measuring chlorophyll fluorescence and pigment retention, which reflect photosystem II activity and chlorophyll stability, respectively. Further analyses of cell damage and oxidative stress allow for comprehensive assessment of the state of senescence. Thus, these quantitative assays provide robust methods to determine how Arabidopsis plants respond to CK in a dose-dependent manner, and can be used as a proxy of CK signaling. Furthermore, these assays can be used to determine how CK perception and signaling responds to environmental or genetic perturbations, and provide reproducible measurements that can help elucidate hormone physiology and signaling in either root and shoot tissues.

Key words Cytokinin, Root growth, Leaf senescence, Protocols, *Arabidopsis thaliana*

1 Introduction

Plant root architecture and development are determined by integrated phytohormone signals to maintain environmental plasticity [1, 2]. The hormone signaling network during root growth is

Grace A. Johnston, Omar Hasannin and Risheek R. Khanna contributed equally with all other contributors.

Shahid Mukhtar (ed.), *Plant Hormones: Methods and Protocols*, Methods in Molecular Biology, vol. 3026,
https://doi.org/10.1007/978-1-0716-5214-5_18,

complex due to spatiotemporal crosstalk between different hormonal classes, with much of canonical growth governed by the mutual antagonism of auxin and cytokinin (CK) [3–7]. In the *Arabidopsis thaliana* (hereafter Arabidopsis) primary root, CK maintains a coordinated balance between cell proliferation and differentiation, as it defines the transition zone between the meristematic and elongation zones [3, 4] and regulates cellular expansion [8]. As a result, elevated CK levels inhibit primary root elongation. Exogenous CK treatment strongly reduces cell division in the root meristem of wild-type seedlings, leading to attenuated primary root growth [3, 9]. Conversely, mutants impaired in CK perception or signaling (such as loss-of-function alleles of type-B *ARABIDOPSIS RESPONSE REGULATOR* genes) exhibit CK insensitivity, often maintaining greater root lengths under CK treatment compared to wild type [9].

To study exogenous treatment effects quantitatively, researchers frequently use a vertical-plate root assay. This chapter describes a protocol in which Arabidopsis seeds are surface sterilized and grown on phytagel plates supplemented with defined concentrations of CK [e.g., 6-benzylaminopurine (BA), a synthetic CK analog] or a mock control. The plates are oriented vertically, so that root systems grow on the surface of the media, yielding relatively straight roots that can be easily imaged and measured. The experiment is conducted in a plant growth chamber to control environmental conditions. The primary roots of each seedling are marked once post-germination and again after a specified period of time (commonly between 5 and 10 days). The data is collected by digitally tracing the root between both markings. Root lengths can be compared between treatments and genotypes, given the correct controls are present, and can be expressed as a percentage of the control value, enabling the construction of dose-response curves or statistical comparisons across genotypes.

During the final stage of plant leaf development, senescence, there is a systematic degradation of cellular machinery, including the photosynthetic framework. Endogenous CKs have significant control over the onset and progression of leaf senescence [10]. Following perception of CK by ER/PM-localized receptors, the two-component signaling components regulate photosynthetic components, metabolism-related gene expression, and senescence [11, 12]. Specifically, CK aids in chloroplast development and functional responses to environmental stimuli, as well as chlorophyll production [13]. CK further acts as an antisenescence signal by maintaining photosynthetic performance, delaying programmed cell death, and alleviating oxidative stress-induced damage [14]. Thus, leaves with exogenous CK treatment exhibit retention of chlorophyll pigment and prolonged leaf functionality [15].

This chapter also describes a second protocol that monitors CK influence on photosynthesis-related physiological parameters, such as amount of chlorophyll to evaluate leaf senescence progression. Briefly, detached leaves of Arabidopsis are floated on a buffer solution containing CK. Samples are placed in dark to halt photosynthesis and assessed for photosystem performance at regular intervals using nondestructive chlorophyll fluorescence imaging (photosystem II efficiency quantification). After the functional status of the photosynthetic framework is analyzed, leaves can be destructively sampled to assess chlorophyll degradation (chlorophyll content analysis), cell membrane integrity (leaf tissue ion leakage assay), and oxidative damage (DAB staining for H_2O_2 detection in leaves) [16–19].

Taken together, both protocols in this chapter can be applied to address a variety of scientific questions, including how CK impacts root growth by quantifying the extent of inhibition (or stimulation) and leaf senescence status by quantitative and qualitative assessment of photosynthetic activity at different doses or how genetic perturbations alter CK responses. Although these protocols are designed to assess CK sensitivity, they can be readily adapted to test other hormones or chemical treatments, provided their exogenous concentrations are researched. Additionally, environmental conditions or temporal constraints can be modified to investigate other hypotheses.

2 Materials

2.1 Root Growth Assay

2.1.1 Reagents and Supply Items

1. 1.7 mL Microcentrifuge tubes.
2. Tube rack.
3. Sharpie Industrial Permanent marker (*see* **Note 1**).
4. Chlorine bleach (*see* **Note 2**).
5. 12 M Hydrochloric acid (HCl).
6. 250 mL Glass beaker.
7. 10 mL Serological pipette.
8. Closed plastic container (*see* **Note 3**).
9. Deionized water (DI).
10. 1 L Pyrex bottles with gaskets and caps (*see* **Note 4**).
11. Murashige and Skoog powder (MS).
12. 2-Morpholinoethanesulfonic acid (MES).
13. Sucrose.
14. Potassium hydroxide (KOH).
15. Phytagel.
16. Sterile DI.
17. 6-Benzylaminopurine (BA) or other cytokinin (CK).

18. Dimethyl sulfoxide (DMSO).
19. 100 mm x 100 mm Square petri plates (*see* **Note 5**).
20. Sterile 200 μL pipette tips.
21. Micropore tape.
22. Vertical plate rack(s).

2.1.2 Specialist Equipment

1. Chemical fume hood.
2. Magnetic stirrer and stir bar.
3. pH meter.
4. Scale.
5. Autoclave.
6. Water bath.
7. Vortex machine.
8. Laminar flow hood.
9. 4 °C Refrigerator.
10. Plant growth chamber.
11. Dissecting stereomicroscope.
12. Flatbed scanner.

2.1.3 The Composition of All Buffers, Media, Solutions

1. 1 M KOH (50 mL): Dissolve 2.805 g of KOH pellets in approximately 40 mL of DI. Bring up to volume.
2. 1 × MS media: 1 × MS powder, 0.05% MES, and 1% sucrose, pH 5.8. Added 0.6% phytagel to each bottle.

2.2 Leaf Senescence Assay

2.2.1 Reagents and Supply Items

1. 6- or 12-Well plates or petri plates.
2. 2-Morpholinoethanesulfonic acid (MES).
3. Glass media bottle.
4. Weighing scale.
5. Weigh boats.
6. Sodium hydroxide (NaOH).
7. Transfer pipette.
8. Distilled water (DW).
9. Pipettes and tips (1000 μL and 2–20 μL).
10. Dimethyl sulfoxide (DMSO).
11. Cytokinin (CK) (synthetic or natural hormone form).
12. Scissors.
13. Tweezers.
14. Cork borer.
15. Dry wipes/filter paper.

16. Aluminum foil.
17. 1.7 mL Mircocentrifuge tubes.
18. Methanol (100%).
19. Cuvettes.
20. 15 mL Centrifuge tubes.
21. 3,3′-Diaminobenzidine (DAB).
22. Plastic/glass beaker.
23. Tween 20.
24. Sodium phosphate, dibasic (Na_2HPO_4).
25. Ethanol.
26. Acetic acid.
27. Glycerol.
28. Potting mix.

2.2.2 Specialist Equipment

1. Magnetic stirrer.
2. Dark adaptation box.
3. Chlorophyll fluorometer (e.g., FluorCam).
4. Centrifuge.
5. Spectrophotometer.
6. Conductivity meter.
7. Water bath.
8. Vacuum infiltration system.
9. Plant growth chamber.

2.2.3 Optional Reagents and Equipment

1. Orbital shaker or rocker.
2. Sodium chloride (NaCl).
3. 12 M Hydrochloric acid (HCl).
4. Hand-held chlorophyll meter.

2.2.4 The Composition of All Buffers, Media, Solutions

1. 3 mM MES: Weigh MES powder on a weighing scale and transfer to a glass beaker/bottle. Dissolve in DW while continuously stirring to generate a 300 mM stock solution. Dilute with DW for 3 mM final concentration. Using NaOH/HCl solution, adjust the MES solution to have a final pH of 5.7.
2. 1 mg/mL DAB: Dissolve 1 mg/mL DAB (e.g., 50 mg DAB in 45 mL sterile DW). Add a stir bar and adjust pH to 3.0 with HCl.
3. Bleaching solution: ethanol:acetic acid:glycerol, 3:1:1.

3 Methods

3.1 Root Growth Assay

3.1.1 Gas Sterilization of Seeds

1. Aliquot seeds to be sterilized into microcentrifuge tubes. Write the genotype and other pertinent information on the cap and side of the tube, using an industrial-strength marker (*see* **Note 1**).
2. Place tubes in a tube rack onto an open plastic container in the chemical hood. Open the caps of all the tubes with seeds.
3. Add 100 mL of bleach to a 250 mL beaker. Move beaker to the plastic container, next to the tube rack.
4. Add 3 mL of 12 M HCl to the bleach. It is imperative to close the lid of the plastic container as soon as the HCl is added because the production of chlorine generated from the chemical reaction between the bleach and HCl will start immediately (*see* **Note 3**).
5. Sterilize seeds for 5–6 h. Do not leave it for longer, the seeds may die.
6. After the sterilization time, carefully open the plastic container and close the caps of all Eppendorf tubes immediately to maintain sterility of the seeds.
7. Leave beaker and plastic container in the chemical fume hood for a couple of days to air out. The remaining liquid in the beaker will crystalize and, at which point, can be rinsed down the sink.

3.1.2 Growth Media Preparation

1. Fill Pyrex bottles with approximately 500 mL of DI. Sterilize bottles in an autoclave using a liquid cycle at 121 °C for 1 h (*see* **Note 6**). Dump water out and let bottles completely dry.
2. For 1 L of media, add 4.3 g MS, 0.5 g MES, and 10 g sucrose to approximately 800 mL DI. Stir to dissolve.
3. Adjust pH to 5.8 using 1 M KOH.
4. Bring up to final volume of 1 L with DI.
5. Add 6 g/L of phytagel to the bottles and MS solution.
6. Weigh and record each bottle.
7. Autoclave media using a liquid cycle at 121 °C for 1 h.
8. After autoclaving, move bottles into the 65 °C water bath. Wait for them to cool down to the point you're able to hold them for 5 s (*see* **Note 7**).
9. Prepare cytokinin stock solutions. BA is dissolved in DMSO. It is important to have the same volume of DMSO or BA solution in each bottle. As such, prepare the 100× solutions of BA appropriate for each concentration.

10. Once cool enough, weigh the first bottle of media to be poured and calculate the amount of water that evaporated during autoclaving (1 g = 1 mL). In the sterile hood, add sterile DI to the bottle to restore it to its original volume using a sterile serological pipette.
11. Add BA or DMSO to bottle. Swirl very well for good mixing. Pour media into square plate (approximately 40 mL per plate), avoiding bubbles (*see* **Note 8**). Proceed to the next bottle/BA concentration. Remove one bottle at a time from the water bath, as needed, to avoid phytagel solidification.
12. Keep plates closed after pouring and mark the sides to identify BA concentration. Allow for plates to dry overnight, covered. Use the plates the next day (or within 2–3 days).

3.1.3 Seed Plating

1. Plate sterilized seeds over the second horizontal gridline from the top of the plate, using a sterile pipette tip (*see* **Note 9**). Evenly space 12 seeds across this horizontal gridline. Mark the genotype used on the plate with a marker. Prepare at least three plates/genotype/BA concentration.
2. Tape plates closed with micropore tape. Store them in vertical plate racks in the dark in a 4 °C refrigerator for 2 days for stratification.

3.1.4 Seedling Growth

1. After stratification, move plates into growth chamber with either a long-day (16 h light/8 h dark) or 24-h photoperiod. Ensure seeds are oriented at the tops of the plates in the vertical plate rack.
2. Four days after light exposure, remove plates from the growth chamber and mark the tips of each root with a marker on the back (media side) of the plate. It helps if this is done under a dissecting stereomicroscope. Move plates back to the growth chamber.
3. Nine days after light exposure, remove plates from the chamber and mark the tips of each root with a marker again. The two markings per root indicate the amount of root growth in 5 days.

3.1.5 Data Collection and Analysis

1. Scan plates, media side down, on a flatbed scanner. Multiple plates can be scanned at once depending on scanner size. Ensure BA concentration and genotype are visible on each plate. Images should be taken in color, with a resolution of 300+ dpi, and saved as TIFF files. After scanning, plates can be discarded.
2. Using ImageJ/Fiji software [20, 21], open scanned files. Use the Segmented Line or Freehand Line tool to trace each root from the 4-day marking to the 9-day marking.

3. Statistical analysis and data visualization can be performed using preferred software. Plotting the root length either as total root length or percentage of root length in relation to control (DMSO) plants is common.

3.2 Leaf Senescence Assay

3.2.1 Plate and Buffer Preparation

1. Determine samples (leaves) to be used and set out number of multi-well plates (standardly 6-well cell culture plates) to allow multiple leaves to be examined across treatments. Larger petri dishes are fine to use for large leaves, as long as sufficient buffer is present to float samples (*see* **Note 10**).
2. Freshly prepare 3 mM MES buffer, adjust pH to 5.7 with NaOH.
3. Pipette 3 mL of MES buffer into each plate well or enough to allow tissue in a plate to freely float.
4. Prepare CK working solution (e.g., tZ, BA) by diluting from stock to 1 mM and add 3 μL of working solution to get final concentration of 1 μM or desired concentration.

3.2.2 Sample Collection and Senescence Assay

1. Select and cut with scissors, healthy, fully expanded leaves from 4-week-old, soil-grown Arabidopsis plants (*see* **Note 11**). Select specific developmental stage for the experiment (*see* **Note 12**). Handle leaves by the petiole to avoid wounding the lamina.
2. If using leaf discs, punch with a 5-mm cork borer or other standardized cutting edge avoiding major veins (*see* **Note 13**).
3. Rinse leaves or discs in DW to remove debris; blot dry on filter paper.
4. For whole leaves or disks, record fresh weight (FW) for later normalization. It can also be useful to take pictures of all collected samples with a scale for any potential future analysis or surface area calculation.
5. Place one leaf or disc (adaxial side up) into each well containing MES buffer. For chlorophyll measurements, only use one leaf per well to easily keep track of individual leaf FW. Label each well according to genotype, treatment, replicate, and/or FW. While we recommend using FW for greater accuracy, chlorophyll levels can be presented as per leaf or by leaf disk surface area calculation in which case up to three leaves per well or multiple leaf disks can be used.
6. Add 3 μL of 1 mM CK to each well already containing buffer to reach a final concentration 1 μM (*see* **Note 14**).
7. Once lid is placed on plate, ensure samples are kept dark by wrapping plates in aluminum foil, then placing them back in the previous plant growing environment. Ensure that all plates

remain in the dark before taking measurements. Prior light exposure prior can reduce true F_v/F_m values, leading to false readings below 0.8 even if the leaves are healthy.

3.2.3 Photosystem II Efficiency Quantification

1. Place samples in a dark box for at least 20 min prior to quantification. If plates are already wrapped in aluminum foil, then samples are already dark adapted and this step can be skipped. Although, a dark box is helpful to examine one plate at a time without unnecessary, ambient light exposure. Perform measurements in a dark room if possible.
2. Power on the fluorometer (FluorCam); select the F_v/F_m or similar protocol.
3. Gently swirl buffer to center leaf samples away from plate edges. Position each well under the sensor; ensure full coverage of the leaf area.
4. Record initial F_v/F_m values (Time 0) for all wells. Exclude any sample <0.80 relative units, indicating preexisting stress.
5. Return plate to dark box immediately after measurement. Repeat with remaining plates.
6. After initial measurements, incubate plates at room temperature on a shaker (50–75 rpm) to ensure uniform solution contact. Maintain dark conditions.
7. At defined intervals (e.g., 24, 48, 72 h), repeat fluorometer measurements. Maintain consistent order and timing to minimize variability.
8. At end of the desired experimental length, typically 6 days (*see* **Note 15**), take final measurements and complete protocol below or proceed with destructive sample analyses (*see* **Note 16**).

3.2.4 Chlorophyll Content Analysis

1. Collect leaves/discs in labeled 1.7 mL microcentrifuge tubes and record FW from each. This can be conducted after the final nondestructive analysis of F_v/F_m for tissues from Subheading 3.2.3.
2. Add 100 μL of 100% methanol (or other extraction solvent) per 10 mg FW (e.g., 50 mg FW → 500 μL methanol) to each sample.
3. Incubate in the dark at 4 °C overnight (max 72 h to avoid chlorophyll degradation).
4. Remove leaf material with tweezers, then centrifuge at 10,000 × *g* for 5 min to pellet any plant debris.
5. Collect supernatant for spectrophotometry into a new tube, avoiding pelleted debris.

6. Blank the spectrophotometer with 400 μL of methanol (or other extraction solvent) in a cuvette.
7. Pipette 400 μL of the supernatant into a quartz cuvette and measure A_{652} and A_{665} (*see* **Note 17**). Repeat for all samples.
8. Calculate chlorophyll concentrations for each sample using the following equations:

$$\text{Chl a } (\mu g/mL) = 16.29 \times A_{665} - 8.54 \times A_{652}$$

$$\text{Chl b } (\mu g/mL) = 30.66 \times A_{652} - 13.58 \times A_{665}$$

$$\text{Total Chl } (\mu g/mL) = \text{Chl a} + \text{Chl b}$$

9. Chlorophyll concentrations can be normalized to FW by:

$$\text{Chl } (\mu g/g \text{ FW}) = \frac{\text{Chl Conc } (\mu g/mL) \times \text{Extract Volume } (mL)}{\text{FW } (g)}$$

3.2.5 Leaf Tissue Ion Leakage Assay

1. After desired hormone or other treatment and dark-adaptation, remove whole leaves from the plate and gently blot dry on filter paper.
2. Wash leaves twice with DW to remove surface electrolytes; blot dry.
3. Using a cork borer (e.g., 5 mm diameter), punch uniform discs. Consistency in diameter and number of discs per sample is critical. We recommend 3–5 discs per replicate; keep the same number across treatments.
4. Place discs into 15 mL tubes containing 5 mL DW and incubate on a shaker at 100–150 rpm, in the dark, at room temperature for 2 h.
5. Immerse probe of conductivity meter in water and record conductivity as C_1 (initial).
6. Using the same solution, place the samples in a 95–100 °C water bath and boil for 15–20 min to release all electrolytes.
7. Cool to room temperature, then mix gently. Measure conductivity as C_2.
8. Calculate percent of ion leakage using the formula below:

$$\text{Ion Leakage } (\%) = \frac{C_1}{C_2} \times 100$$

3.2.6 DAB Staining for H_2O_2 Detection in Leaves

1. Dissolve 1 mg/mL DAB (e.g., 50 mg DAB in 45 mL sterile DW). Add a stir bar and adjust pH to 3.0 with HCl (DAB•HCl may be pre-acidified).

2. Add 25 μL Tween 20 and 2.5 mL of 200 mM Na_2HPO_4 to buffer to improve wetting. Wrap DAB solution tube in foil to protect from light.
3. Place up to three leaves in 15 mL tubes containing 3 mL DAB solution. If leaves are large in size (e.g., you're using leaves 5 or 6 from ~4-week-old plants), then we highly recommend using one leaf per tube to prevent leaf damage (*see* **Note 18**).
4. Vacuum infiltrate for 5 min to ensure dye uptake into tissues.
5. Incubate on a shaker at 100–150 rpm, in the dark, for 12 h at room temperature (*see* **Note 19**).
6. Carefully remove DAB solution and replace with bleaching solution (ethanol:acetic acid:glycerol, 3:1:1).
7. Incubate in a 95–100 °C water bath for 15 min to clear chlorophyll, leaving brown DAB precipitate.
8. Replace with fresh bleaching solution and store at 4 °C in the dark for up to 72 h.
9. Gently transfer leaves to filter paper for imaging.

4 Notes

1. The marker will fade due to the sterilization process, so it is important to use an industrial-strength permanent marker.
2. Standard Clorox Disinfecting bleach (with 7.5% sodium hypochlorite) can be used. Use bleach opened within a couple of months, otherwise, the bleach will lose potency and sterilization is less effective.
3. A Sterilite cake saver with locking handles is ideal. Although the seed sterilization is performed in a chemical fume hood, the plastic container is an added level of protection from the chlorine gas.
4. The number of bottles required will depend on the experimental design (i.e., number of genotypes and CK concentrations). Use one bottle per concentration with a maximum of 600 mL of media per bottle.
5. Square petri plates with grid lines assist in even seed spacing during plating.
6. Bottles are sterilized before media preparation to rid them of any chemical residues.
7. It is helpful to heat the sterile DI water in the hot water bath at this time. Phytagel solidifies very easily, thus adding preheated water to the media prevents nonhomogeneous media solidification.

8. Bubbles in media can be popped using a sterile 200 mL pipette tip.
9. Seeds will generally stick to the pipette tips by static. If not, gently dab the pipette tip onto the corner of the media on the plate to moisten the end, which will assist adhesion of seeds to the tip.
10. Non-sterile petri dishes/plates can be used for this assay as long as they are washed with mild soap, rinsed thoroughly with DW, and wiped with 70% ethanol.
11. Sow desired *Arabidopsis thaliana* genotype(s) seeds on potting mix (Promix-BX or lab standard soil). Transfer the pots to a standard lab growth conditions (growth chamber set at 22 °C/16 °C with a 16 h/8 h day/night cycle and a light intensity of 100 μmol/m^2/s). Cover pots with transparent lids for 3–4 days to retain humidity for better germination. Grow plants under standard conditions for ~4 weeks with regular watering. Different ages of plants or environmental conditions can be used as needed.
12. Harvest leaves from similar developmental stages, for example, leaves that emerge and develop around the same time (e.g., leaves 3 and 4, 5 and 6, 7 and 8). Using leaves from widely different developmental stages (e.g., leaf 3 and leaf 7) can introduce heightened variability in results. Also note that cotyledons can be used.
13. We recommend using whole leaf to avoid stressing fragile Arabidopsis leaves. If using leaf discs is necessary, handle them with extra care to prevent mechanical damage.
14. This bioassay is sensitive to low CK concentrations (as low as 10 nM) and thus varying amounts and types of cytokinin can be added at this step. Additionally, different stressors, such as NaCl, can be added into this assay, in order to test for CK effects on delaying specific stress senescence [22]. Strong stresses often accelerate timing of senescence, so this should be taken into account.
15. This is a modification of an original CK senescence bioassay standardly conducted over 6 days with cut leaves in the dark [17]. This modified version can still be used for endpoint analysis at 6 days, but also at earlier timepoints in a nondestructive nature. In addition, this assay can be conducted under light growth conditions with similarly observed senescence effects; however, this extends the time it takes for senescence to occur by at least several more days.
16. Note that leaves can be analyzed for various nondestructive and destructive physiological parameters throughout this time. Measurements made before the completion of this assay should

be done carefully to avoid light exposure. Additionally, leaves can be collected for downstream analysis, such as gene expression or metabolic examinations.

17. A_{652} and A_{665} are absorbances specific to 100% methanol extractions. Absorbance wavelengths vary based on extraction solvent (methanol, acetone, ethanol, DMSO) used [22].
18. Ensure that leaves are fully submerged in the DAB solution and that all leaf surfaces are evenly covered. Keep tubes out of the light to prevent DAB degradation—even if wrapped in aluminum foil, we recommend placing the entire rack in a dark box for better protection.
19. We recommend incubating samples in 1 mg/mL DAB solution for at least 12 h. Incubation for less than 6 h often results in weak staining. If an additional stressor (e.g., salt or pathogen) is applied, 4–6 h of incubation may be sufficient.

References

1. De Vos D, Vissenberg K, Broeckhove J, Beemster GT (2014) Putting theory to the test: which regulatory mechanisms can drive realistic growth of a root? PLoS Comput Biol 10(10): e1003910
2. Jung JK, McCouch S (2013) Getting to the roots of it: genetic and hormonal control of root architecture. Front Plant Sci 4:186
3. Dello Ioio R, Linhares FS, Scacchi E, Casamitjana-Martinez E, Heidstra R, Costantino P et al (2007) Cytokinins determine Arabidopsis root-meristem size by controlling cell differentiation. Curr Biol 17(8):678–682
4. Dello Ioio R, Nakamura K, Moubayidin L, Perilli S, Taniguchi M, Morita MT et al (2008) A genetic framework for the control of cell division and differentiation in the root meristem. Science 322(5906):1380–1384
5. Bishopp A, Help H, El-Showk S, Weijers D, Scheres B, Friml J et al (2011) A mutually inhibitory interaction between auxin and cytokinin specifies vascular pattern in roots. Curr Biol 21(11):917–926
6. Muller B, Sheen J (2008) Cytokinin and auxin interaction in root stem-cell specification during early embryogenesis. Nature 453(7198): 1094–1097
7. Ruzicka K, Simaskova M, Duclercq J, Petrasek J, Zazimalova E, Simon S et al (2009) Cytokinin regulates root meristem activity via modulation of the polar auxin transport. Proc Natl Acad Sci USA 106(11): 4284–4289
8. Liu S, Strauss S, Adibi M, Mosca G, Yoshida S, Dello Ioio R et al (2022) Cytokinin promotes growth cessation in the Arabidopsis root. Curr Biol 32(9):1974–85 e3
9. Argyros RD, Mathews DE, Chiang YH, Palmer CM, Thibault DM, Etheridge N et al (2008) Type B response regulators of Arabidopsis play key roles in cytokinin signaling and plant development. Plant Cell 20(8):2102–2116
10. Woo HR, Kim HJ, Lim PO, Nam HG (2019) Leaf senescence: systems and dynamics aspects. Annu Rev Plant Biol 70:347–376
11. Argueso CT, Kieber JJ (2024) Cytokinin: from autoclaved DNA to two-component signaling. Plant Cell 36(5):1429–1450
12. Zwack PJ, Rashotte AM (2013) Cytokinin inhibition of leaf senescence. Plant Signal Behav 8(7):e24737
13. Cortleven A, Schmulling T (2015) Regulation of chloroplast development and function by cytokinin. J Exp Bot 66(16):4999–5013
14. Janeckova H, Husickova A, Ferretti U, Prcina M, Pilarova E, Plackova L et al (2018) The interplay between cytokinins and light during senescence in detached Arabidopsis leaves. Plant Cell Environ 41(8):1870–1885
15. Hallmark HT, Rashotte AM (2020) Cytokinin isopentenyladenine and its glucoside isopentenyladenine-9G delay leaf senescence through activation of cytokinin-associated genes. Plant Direct 4(12):e00292
16. Keshishian EA, Hallmark HT, Ramaraj T, Plackova L, Sundararajan A, Schilkey F et al

(2018) Salt and oxidative stresses uniquely regulate tomato cytokinin levels and transcriptomic response. Plant Direct 2(7):e00071

17. Letham DS (1967) Regulators of cell division in plant tissues: V. A comparison of the activities of zeatin and other cytokinins in five bioassays. Planta 74(3):228–242
18. Daudi A, O'brien JA (2012) Detection of hydrogen peroxide by DAB staining in Arabidopsis leaves. Bio-protocol. 2(18):e263
19. Zhang Z, Li W, Gao X, Xu M, Guo Y. (2020) DEAR4, a member of DREB/CBF family, positively regulates leaf senescence and response to multiple stressors in Arabidopsis thaliana. Frontiers in Plant Science. 11:367
20. Schindelin J, Arganda-Carreras I, Frise E, Kaynig V, Longair M, Pietzsch T et al (2012) Fiji: an open-source platform for biological-image analysis. Nat Methods 9(7): 676–682
21. Schneider CA, Rasband WS, Eliceiri KW (2012) NIH image to ImageJ: 25 years of image analysis. Nat Methods 9(7):671–675
22. Porra RJ (2002) The chequered history of the development and use of simultaneous equations for the accurate determination of chlorophylls a and b. Photosynth Res 73(1–3): 149–156

INDEX

Shahid Mukhtar (ed.), *Plant Hormones: Methods and Protocols*, Methods in Molecular Biology, vol. 3026,
https://doi.org/10.1007/978-1-0716-5214-5,

H

I

J

L

M

N

P

Q

R

S

MIX
Papier aus verantwortungsvollen Quellen
Paper from responsible sources
FSC® C105338

If you have any concerns about our products,
you can contact us on
ProductSafety@springernature.com

In case Publisher is established outside the EU,
the EU authorized representative is:
Springer Nature Customer Service Center GmbH
Europaplatz 3, 69115 Heidelberg, Germany

Printed by Libri Plureos GmbH
in Hamburg, Germany